Carotenoids

Renan Campos Chisté
Eloisa Helena de Aguiar Andrade
Mozaniel Santana de Oliveira
Editors

Carotenoids

Trends and Advances

Springer

Editors
Renan Campos Chisté
Faculty of Pharmacy
Universidade Federal de Minas
Gerais (UFMG)
Belo Horizonte, Minas Gerais, Brazil

Mozaniel Santana de Oliveira
Programa de Pós-Graduação em Ciências
Biológicas - Botânica Tropical Universidade
Federal Rural da Amazônia/Museu Paraense
Emílio Goeldi
Belém, Brazil

Eloisa Helena de Aguiar Andrade
Museu Paraense Emílio Goeldi
Belém, Brazil

ISBN 978-3-031-75321-3 ISBN 978-3-031-75322-0 (eBook)
https://doi.org/10.1007/978-3-031-75322-0

This Springer imprint is published by the registered company Springer Nature Switzerland AG
The registered company address is: Gewerbestrasse 11, 6330 Cham, Switzerland

This book is dedicated to all the scientists worldwide who devoted their lives to making available all the existing broad information related to carotenoid science. All the contributions made by all these crucial professionals are certainly bringing light to the scientific community, and much work has been carried out in different areas of knowledge and in various industrial sectors, focused on improving the overall life quality of people around the world. There is still a lot of work to be done!

The Editor Renan Campos Chisté. *This book is especially dedicated to my lovely family: my mom, Nilza; my father, Alvanir (in memoriam), my sisters Fabiane and Cristiane; my nieces Mariana, Geovana, and Helena; my nephew João Victor; my partner Éder Junior, and my all dearest friends. My family and friends always support me in achieving my goals and give me understanding during my eventual absence every time my job takes up my family's hours. This book is also dedicated to all my former scientific advisors who taught and inspired me so much during my academic* *journey: Adriana Mercadante, Eduarda Fernandes, Lênio de Faria, Alessandra Lopes, and Kelly Cohen.*

The Editor Dr. Mozaniel Santana de Oliveira. *This academic and scientific book is dedicated, to my family, brothers, friends, and my parents Maria de Oliveira and Manoel de Oliveira, and my lovely wife Joyce Fontes and beautiful daughter Isabela de Oliveira.*

Foreword

People worldwide are improving their health standards, and scientific research on new sources of bioactive compounds with relevant physiological properties is always a hot topic, as well as the synthesis of these bioactive compounds. In this scope, carotenoids can be described as a group of compounds structurally varied in terms of chemical structures and biological properties. This class of secondary metabolites is well known for being one of the largest groups of bioactive compounds ubiquitous in nature, with applications in different human activities, including food science and technology, nutrition, pharmacology, and cosmetics, among others. The frequent intake of carotenoids from foods, associated with healthy habits, has shown high potential to reduce damages caused by oxidative stress, increase immunity, and decrease the risk of the development of noncommunicable diseases, such as cardiovascular diseases, neurodegenerative disorders, inflammations, and cancer.

Currently, most carotenoids are produced by chemical synthesis in many industries. However, extraction methods and the search for natural sources with high contents of these compounds are still needed, especially when biotechnology and engineering strategies are approached to extract carotenoids more sustainably. With this in mind, the book "Carotenoids: Trends and Advances" presents a detailed description of new aspects related to the biosynthesis of carotenoids in different organisms and their uses in different fields of knowledge. Furthermore, the book provides information on the distribution of carotenoids in various products and more recent and environmentally friendly carotenoid syntheses.

In this sense, this book aims to present carotenoids and their applications. Consisting of 12 chapters, "Carotenoids: Trends and Advances" can be considered a very informative and valuable source of recent information for professionals and theorists. It is useful for researchers and professionals in chemistry, biochemistry, biotechnology, pharmacy, food, and agriculture sciences because it presents broad information about the biosynthesis, application, and health benefits of carotenoids, from the basic level of knowledge to the most current developments. At the same time, we expect it will be a helpful tool for specialists from various branches of science and nonprofessionals who want to receive accessible information about carotenoids.

Belo Horizonte, Minas Gerais, Brazil — Renan Campos Chisté
Belém, Brazil — Eloisa Helena de Aguiar Andrade
Belém, Brazil — Mozaniel Santana de Oliveira

Preface

Carotenoids, as one of the most widespread groups of natural pigments, are not only essential to plant physiology but also contribute significantly to the biological functions of animal organisms. These pigments, which collectively contribute to the color of fruits and vegetables, play a crucial role in photosynthesis and protection against the harmful effects of light. In recent decades, studies have revealed that carotenoids are not just secondary metabolites for plant health. They have also demonstrated a growing array of effects on human health, mainly due to their numerous structural analogs and/or carotenoid derivatives with various biological properties. These properties include antioxidant, antiproliferative, antimicrobial, and neuroprotective effects, among others, either individually or in combination.

In the same respect, carotenoids can be used in food science and technology as natural food additives either in the isolated form (high purity degree), as extracts with mixed carotenoid composition, or can be encapsulated to be delivered to a molecular target (drug delivery) or for stability purposes. In this sense, the book "Carotenoids: Trends and Advances" collects recent contributions and important findings in the science of carotenoids. Counting on contributions from renowned experts, it provides an in-depth look at these bioactive pigments' biosynthesis, technological applications, and health benefits. The book is a unique collection of 12 chapters: Introduction to Carotenoids and Overview of Their Biosynthesis, Biosynthesis of Carotenoids in Plants and Microorganisms, Biosynthesis of Carotenoids in Insects: What We Currently Know?, C30 Carotenoids: Biosynthesis, Chemical and Biological Aspects, Overview of Recent Trends on Non-conventional Approaches to Extract Carotenoids, Encapsulation Techniques for the Preservation of Carotenoids' Bioactivity, Recent Approaches for the Bioaccessibility and Bioavailability of Carotenoids, Carotenoids from Marine Sources: General Aspects, Biological Properties, and Industrial Applications, Uses of Microorganisms for Carotenoid Production: Contribution of Modern Technologies in Food, Sustainability, Composition and Health Benefits of Carotenoids in Foods and Feeds, Therapeutic Effects of Carotenoids: Trends and Advances, and Neuroprotective

Potential of Carotenoids. Therefore, "Carotenoids: Trends and Advances" brings recent insights into the applications of these natural pigments and can serve as a solid reference for both beginning and experienced researchers, as well as a key text for the development of future research.

Belo Horizonte, Minas Gerais, Brazil — Renan Campos Chisté
Belém, Brazil — Eloisa Helena de Aguiar Andrade
Belém, Brazil — Mozaniel Santana de Oliveira

Acknowledgments

We are thankful to all the authors that accepted to contribute to our book project, and also to all the support provided by the Springer Nature team, the section heads and their colleagues on the editorial board. Special thanks to Daniel Falatko, Editor for Food Science, who believed in the potential of this project.

The Editor Renan Chisté thanks *Conselho Nacional de Desenvolvimento Científico e Tecnológico* (CNPq, Brazil) for the Research Productivity Scholarship (Project 314929/2021–5), and both the Universidade Federal do Pará (UFPA) and Universidade Federal de Minas Gerais (UFMG) for providing academic support and stimulating the development of scientific research in the Institution.

The Editor Mozaniel Santana de Oliveira thanks PDPG-POSDOC—Programa de Desenvolvimento da Pós-Graduação (PDPG) Pós-Doutorado Estratégico, as well as CAPES for the scholarship (Process number: 88887.852405/2023-00).

Contents

Chapter 1
Introduction to Carotenoids and Overview of Their Biosynthesis

George Britton

1 Introduction

The carotenoids are based on or derived from a C_{40} tetraterpene skeleton, with a long central polyene chromophore that is responsible for their typical yellow-orange colours. Carotenoid hydrocarbons are known as '**carotenes**', ones containing hydroxy groups or other oxygen substituents, as '**xanthophylls**'. Hundreds of individual compounds have been isolated and identified (Britton et al. 2004), many of which are present in a normal balanced human diet. Extensive tables are available listing the HPLC analysis of the main carotenoids in a wide range of fruits, vegetables and other food sources. The precise numerical figures given are for a particular sample analysed, and the carotenoid content may be reported for wet weight or dry weight of plant material, or in relation to content of protein or some other component. Carotenoid content and composition are affected by environmental factors, cultivation practice and storage. Water content can also differ widely between samples, even ones from the same plant at the same time. Analysis of another sample of the same plant material grown in different locations or at different times should not be expected to give the same numerical values. The figures are indicative, however, and can show if that material is a good source of the desired carotenoid.

Some carotenoids, particularly the main leaf pigments β-carotene and lutein, have been in the natural diet of humans and their ancestors throughout evolution, so it is not surprising that special functions have evolved for them, *e.g.* for β-carotene as provitamin A, and for lutein in eye and brain health. Other carotenoids, however, even lycopene, which is now a major dietary carotenoid, have only been part of the human diet in comparatively recent times, and are thus 'foreign' substances for the human body, so no general functions can be expected to have evolved for them, and

G. Britton (✉) (retired)
The University of Liverpool, School of Biological Sciences, Liverpool, UK

R. Campos Chisté et al. (eds.), *Carotenoids*,
https://doi.org/10.1007/978-3-031-75322-0_1

any biological action is likely to be pharmacological rather than a true natural function.

Over the years, the carotenoid field has diversified into many aspects of biology. Many natural roles have been suggested for carotenoids and their breakdown products and metabolites. Such biological actions must be consistent with their chemical and physical properties, which follow from their chemical structures. The long conjugated-double-bond system is crucial to many of these properties, notably light-absorption, colour, and photoprotection, but also to reactivity. The different end groups determine the subtle differences in properties between individual carotenoids, especially in terms of their solubility, molecular interactions, ability to fit correctly into biological structures, and detailed chemical reactions.

The main topic of this chapter is to present an overview of the current state of knowledge and understanding of carotenoid biosynthesis but, to put this topic into perspective, the article begins with a brief survey of the development of carotenoid research over the past 200 years. The history of knowledge and research on carotenoids has been described in detail in a 236-page book (Britton et al. 2017). Only the highlights are summarised in this chapter.

Reports of the more recent work, with references, can be found in volumes of the *Carotenoids* book series (Britton et al. 1995, 1996a, b, 1998, 2008, 2009) and in other compilations (von Lintig and Quadro 2020; Wurtzel 2022a, b, c). Many of the references given are to these and other books, chapters in books, and review articles, which provide details of the various topics.

Biological effects that are claimed for carotenoids must be consistent with the properties of the carotenoid but, in the aqueous environment of cells and tissues, these properties may be modified by the medium and the immediate molecular surroundings of the carotenoid, so they may not be the same as the known properties in true solution in organic solvent. Some aspects of this are discussed in Sect. 2.5 in order to give a deeper insight into how carotenoids may perform the many functions and actions that are now attributed to or suggested for them. These factors are also relevant to the biosynthesis of these fat-soluble substances *via* water-insoluble intermediates. The structures of carotenoids and related compounds mentioned in Sects. 1 and 2 are shown in Fig. 1.1.

2 How Carotenoid Research Has Progressed

2.1 First Recognition of Carotenoids

Many natural materials with yellow, orange or red colours due to carotenoids have been known and used for thousands of years. Some of the most familiar examples of food coloration, however, *e.g.* annatto, tomato, *Capsicum* peppers, are native to the Americas, and were only introduced into Europe at the beginning of the sixteenth century after Columbus 'discovered' the New World in 1492.

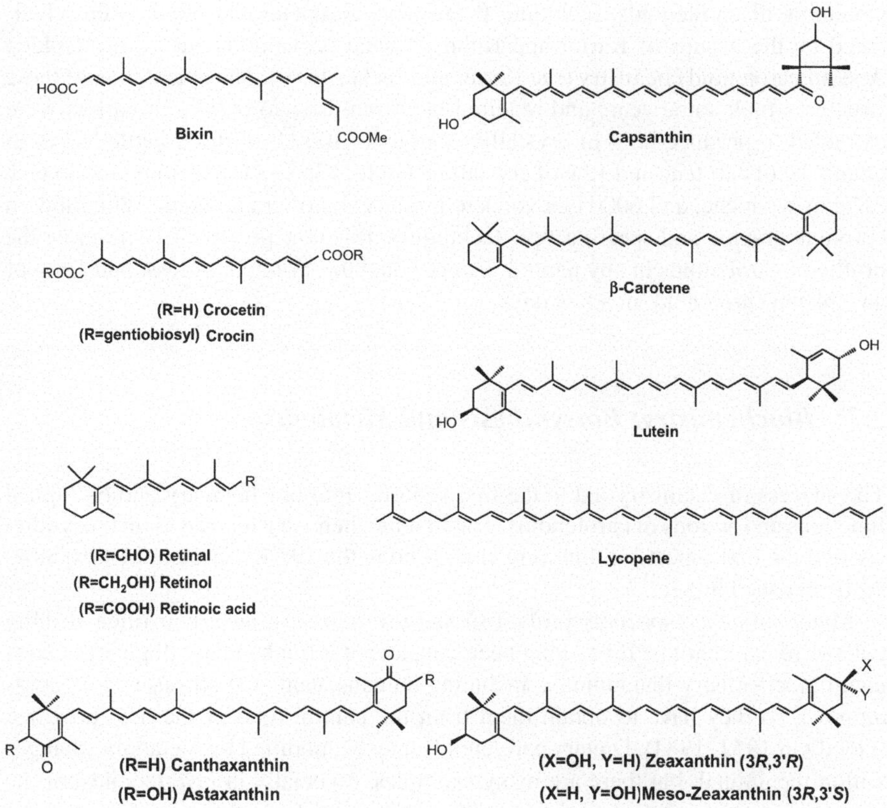

Fig. 1.1 Structures of carotenoids and related compounds discussed in Sects. 1 and 2

Some 200 years ago, it was recognised that the colours are due to the presence of yellow, fat-soluble substances (lipochromes or chromolipoids) that could be extracted with organic solvents. The first reported examples were extracts of annatto (source of bixin), paprika (source of capsanthin *etc*), and the 'Old-World' colorant saffron (source of crocin, carbohydrate esters of the apocarotenoic acid crocetin). The first isolations of carotene from carrots, xanthophyll (lutein) from autumn leaves and, later, from egg yolk, and lycopene ('lycopin') from *Tamus communis,* tomato and water melon, followed, also in the nineteenth century.

2.2 *The 'Golden Age' of Carotenoid Structure Elucidation*

In spite of the considerable progress made by early chemists, notably by Willstätter who, in the early 1900s, showed that one example, carotin (carotene), is a hydrocarbon with molecular formula $C_{40}H_{56}$, it was not until about 100 years ago that the first structures of carotenoids, including β-carotene, lycopene and lutein, were elucidated by the groups of Karrer and Kuhn, in what became known as the 'Golden Age' of carotenoid chemistry (see Karrer and Jucker 1950). The elucidation of these structures took some years and required large samples, so 600 kg of carrots were extracted to produce 20 g of crystalline carotene, 100 kg of dried nettle leaves to obtain 3 g of carotene and 12 g of crystalline lutein, 135 kg of fresh tomatoes to give 2.7 g of lycopene, and 6000 egg yolks to produce 4 g of crude lutein. With modern chromatographic and spectroscopic techniques, it is now possible to determine the profile of carotenoids in any natural material, and the structure of less than 1 mg of an isolated carotenoid, in a few days.

2.3 *Biochemistry: Biosynthesis and Metabolism*

The success of chemistry led to the first ventures into biochemistry and biological functions and actions of carotenoids. The first biochemistry focus was on biosynthesis, and the first aspects of this were elucidated in the 1950s. Carotenoid biosynthesis is reviewed in Sect. 3.

Many animals, especially birds, fish and invertebrates, owe their often striking colours to carotenoids. It has long been considered a fundamental dogma of carotenoid biochemistry that animals, including humans, cannot synthesise carotenoids *de novo*, so they have to obtain them from the natural food or the feed provided (Goodwin 1952, 1984). Dietary carotenoids may be modified by metabolic changes within the animal, but there are many examples, especially in invertebrates, of the presence of carotenoids that would not be expected to be in the natural food of the animal, or to be easily formed from ones that are. In recent years, the lateral transfer of microbial genes into the genome of some insects has been demonstrated (Moran and Jarvik 2010), so the presence of unexpected carotenoids in animals can be rationalised. The carotenoids can indeed be biosynthesised within the animal.

2.4 *Carotenoids and Human Health*

The study of roles of carotenoids in human health began in the 1930s when the structural relationship between β-carotene and vitamin A (retinol) was established and the conversion demonstrated. For several decades, this, including the role of retinal in vision, was the only aspect of carotenoids and human health known and discussed. Later, one of the first successful applications of carotenoids was the demonstration that β-carotene could relieve symptoms of the light-sensitivity condition erythropoietic protoporphyria, caused by the accumulation in the skin of free porphyrins that could photosensitise production of singlet oxygen $^1O_2$, the agent causing skin damage (Mathews-Roth et al. 1970).

In the 1970s, the link between carotenoids—not just β-carotene but also non-provitamin A carotenoids—and protection against cancer and other serious degenerative diseases was first suggested.

2.4.1 Carotenoids in Human Food and Nutrition

Today we know a lot about the presence of carotenoids in fruit and vegetables, and much publicity is given to the idea that these carotenoids are among the many substances present in our food that have an important influence on our health.

The presence of β-carotene in blood and tissues of humans and other mammals was first reported more than 100 years ago. The carotenoids are derived from food, and HPLC reveals the presence of about 20 carotenoids in human blood and tissues, from the 100 or so different carotenoids eaten in a 'normal' varied diet. Being hydrophobic substances, carotenoids from food have to be dissolved in the lipid of mixed micelles in order to be absorbed in the intestine, taken up by blood lipoproteins and transported to various tissues (Canene-Adams and Erdman 2009).

The structure of the food matrix and the physical state of the carotenoid in the food determine how easily the carotenoid may be released and solubilised by lipid. Cooking or processing causes some disruption of food structure so solubilisation and absorption are much improved. There is also wide variation in the ability of individuals to absorb dietary carotenoids and deposit them in tissues.

2.4.2 Vitamin A

In the recorded history of prominent early civilisations, there are references to eye disorders that can now be attributed to vitamin A deficiency, and to the toxicity of vitamin A with consequences such as defective bone structure, liver damage and death, though this toxicity is now known to be due to retinoic acid and not to vitamin A itself. The provitamin carotene is not toxic, and the conversion is controlled so that the retinoids do not reach toxic levels. Part of this control is due to specific proteins such as plasma retinol-binding protein (RBP), for the transport of vitamin A in the body. Related binding proteins for retinoic acid were discovered in the 1970s.

A relationship between vitamin A and growth, general health and disease resistance was recognised in the 19th and early 20th centuries (for details of this and later developments see McCollum and Davis 1915; Moore 1930; Sommer and West 1996; McLaren and Frigg 1997; McLaren 2004; Packer et al. 2004). In the 1960s, details of involvement in growth and development and on reproduction, mediated by retinoic acid, began to be revealed. The hormone-like action of retinoic acid, and the role of nuclear receptors, RAR and RXR for (all-*trans*)- and (9-*cis*)-retinoic acid, respectively, in the regulation of many fundamental molecular and physiological processes, discovered in the 1980s, has now become a major research topic.

2.4.3 Macular Pigments

The distinctly yellow spot in the centre of the retina in humans and other primates, the macula lutea, was described at the end of the eighteenth century (see Schalch et al. 2009, 2010), and in 1945, the yellow colour was shown to be due to a carotenoid of the 'leaf xanthophyll' type, later identified as a mixture of lutein and zeaxanthin, including the (3*R*,3′*S*)-isomer *meso*-zeaxanthin. A link between low macular carotenoid concentration and risk of Age-related Macular Degeneration (AMD), a major cause of impaired vision and blindness in the elderly, was first substantiated in the 1980s. Recent research has revealed the presence of the macular xanthophylls in the brain, and this is correlated with stimulating and maintaining cognitive function in the elderly (Johnson 2011), and with brain development in infants (Vishwanathan et al. 2011).

2.4.4 Protection Against Cancer

From the 1950s, reports began to appear of anti-tumour effects of vitamin A in animal models. Attention then turned to the provitamin A, β-carotene, and, in 1973, skin tumours that were induced in rats and mice were shown to develop more slowly when the animals were fed a diet containing 'unlimited amounts of red carrots'. Further experimental reports of delaying or protective effects against tumours induced in animals followed in the late 1970s and early 1980s (Mathews-Roth 1985). The first tentative suggestions of effects on cellular processes, protection against lipid peroxidation, or stimulation of immunological parameters, were made at that time.

A big impetus for studies of carotenoids and human health came in 1981 when a paper was published that evaluated the evidence from major epidemiological studies, and asked whether dietary β-carotene can materially reduce human cancer rates (Peto et al. 1981). This paper stimulated major research efforts, eventually leading to a second 'Golden Age' of carotenoid research, in which the powerful techniques of molecular biology have led to demonstrations that carotenoids can affect many cellular and molecular processes in cells in culture (Krinsky et al. 2004; Mayne et al. 2009).

In the 1990s, two extensive epidemiological studies to look for effects on cancer incidence in subsequent years after administration of carotene supplements, the Alpha-Tocopherol Beta-Carotene (ATBC) trial and the Beta-Carotene and Retinol Efficacy Trial (CARET), were cut short when they showed that supplementation with β-carotene was associated with a small increase in risk of lung cancer or coronary heart disease, in smokers (Alpha-Tocopherol Beta-Carotene Cancer Prevention Study Group 1994). The supplements used, however, provided up to ten times the normal daily dietary carotene intake, leading to abnormally high concentrations of carotene in the lungs and other tissues. The lungs of smokers are a particularly hostile environment, in which high levels of oxidising free radicals are generated, so the high level of carotene given could be degraded to products that may have harmful effects. No adverse effects were found with non-smokers.

2.4.5 Other Protective Effects

Work on protection against coronary heart disease (Johnson and Krinsky 2009) and skin damage (Goralczyk and Wertz 2009), and immune stimulation (Bendich and Shapiro 1986; Chew and Park 2009) by carotenoids and their breakdown products has expanded rapidly since the mid-1990s. It has been recognised for more than 100 years that a serious consequence of vitamin A deficiency is decreased resistance to disease, especially in young children, due to severe impairment of the immune response. Since the 1970s, β-carotene and some non-provitamin A carotenoids, particularly lutein, canthaxanthin and astaxanthin, have been reported to have similar stimulatory actions (Bendich and Shapiro 1986). It is now thought, however, that most effects on the immune system are mediated by retinoic acid and retinoic acid receptors and not by intact carotenoids.

Lycopene has come under scrutiny for possible protective effects against atherogenesis and coronary heart disease (Johnson and Krinsky 2009), and cancer of the prostate (Obermüller-Jevic and Packer 2004; Bowen 2010).

2.4.6 Metabolic Syndrome

Recent years have also seen a rise in interest about the location of carotenoids in mitochondria, and possible effects on regulation of metabolism, especially of lipids, and this may have consequences for control of obesity and diabetes, and for enhancement of athletic performance.

2.4.7 Antioxidant/Pro-Oxidant Properties

In 1984, it was reported that β-carotene could be an unusual kind of lipid antioxidant, with peroxidation-chain-breaking properties at low oxygen concentrations (Burton and Ingold 1984). This led to major research activity because of the possibility that carotenoids might, therefore, be important dietary antioxidants, with obvious benefits for human health.

Carotenoids are very susceptible to oxidation and can also have pro-oxidant properties. Under appropriate conditions, they can undergo peroxidation and breakdown into products that may have biological actions. They can form transient neutral radicals, radical cations or radical anions, stabilised by delocalisation of the unpaired electron over the conjugated double-bond system (El-Agamey and McGarvey 2008; Edge and Truscott 2010). The possible relevance of oxidation by free-radical chemistry or by reactions in excited singlet or triplet states is not well understood but its relevance to antioxidant or pro-oxidant properties of carotenoids should not be overlooked.

2.5 *Properties of Carotenoids* In Vivo *and* In Vitro

Biological effects that are claimed must be consistent with the properties of the carotenoids but, in the mainly aqueous environment of cells and tissues, these properties may be modified by the medium and the immediate molecular surroundings of the carotenoid, so they may not be the same as the known properties in true solution in organic solvent.

2.5.1 Carotenoids in Aqueous Media

Aggregation and Isomers In organic solvent, unless they are at high concentration, carotenoids are present in solution as single molecules. If water is added to such a solution in a water-miscible solvent, *e.g.* acetone, however, the molecules form aggregates, and a change in colour is seen, showing different light-absorption properties (Britton 1996). In a typical aggregate, the π-electrons of the conjugated double-bond system are shared between molecules so the electron distribution is altered. This determines the chemical properties, so the chemistry of carotenoid aggregates may be different from the known chemistry of the same compound in solution (Köhn et al. 2008).

The formation of aggregates is fundamentally important for functioning in living organisms and in cells and other systems isolated from them. Two well-documented examples of aggregation in food materials are β-carotene in carrots and lycopene in tomatoes, which microscopy has revealed are present largely as crystals (Britton et al. 2002). The carotenoid crystals are difficult to solubilise in lipid so the absorption of β-carotene from raw carrots and of lycopene from unprocessed tomatoes or tomato juice is very low. Cooking or processing causes some disruption of the crystalline aggregates so solubilisation and absorption are much improved.

When carotenoids are incubated with cell cultures, or added as substrates to cell-free systems in biosynthesis studies, they are often added as an organic solution to the aqueous medium, where they will rapidly aggregate, with serious consequences for their delivery.

Aggregation also has a bearing on isomerisation. Carotenoids undergo isomerisation to give an equilibrium between the all-*trans* form and a mixture of *cis*-isomers (Liaaen-Jensen and Lutnæs 2008). The isomerisation equilibrium is usually strongly in favour of the all-*trans* isomer, but the unaggregated form may be enriched in *cis* isomers which are more readily dissolved and absorbed. The *cis*-isomers may, therefore, be present in tissues, including the human body, cells and blood, at higher concentrations than expected. Isomer compositions are determined by two equilibria, that between the aggregated and the dissociated unimolecular forms, and the thermodynamic equilibrium between *trans* and *cis* geometrical isomers.

2.5.2 Interactions with Other Molecules

Lipids and Membranes In living organisms, the only places carotenoids may be present as single molecules is likely to be in association with protein, *e.g.* in the functional pigment-protein complexes in photosynthetic systems, or with other lipids, *e.g.* in fat, membranes or lipoproteins such as chylomicrons, low-density lipoprotein (LDL), high-density lipoprotein (HDL), on which carotenoids are 'dissolved' in the lipid.

In natural and model membranes, *e.g.* liposomes, the carotenoid molecules are associated with the lipid core. β-Carotene and lycopene lie entirely within this core, but hydroxy groups *etc.* in the xanthophylls associate with the polar surfaces of the membrane, so that the diol zeaxanthin, for example, spans across the bilayer (Gruszecki 2010). Lutein, with its different shape and orientation of its hydroxy groups, is believed to occupy two different orientations, one across the membrane, like zeaxanthin, but also one with both hydroxy groups interacting with the same polar surface of the membrane, perpendicular to the membrane lipid chains.

Membrane structures are dynamic, however; at sufficiently high concentrations, xanthophyll molecules will tend to form aggregates or rafts within the membrane, similar to the well-known cholesterol rafts. Temperature and carotenoid structure are major factors in determining this aggregation in membranes, which may also modify the properties and dynamics of the membrane and membrane-associated processes.

Many of the key steps in carotenoid biosynthesis take place within or associated with membranes, allowing easier access and movement of intermediates.

Proteins Direct interactions of carotenoids with proteins can have a large effect on their properties, the most striking examples being the stable blue carotenoproteins such as crustacyanin, found in invertebrate animals (Britton and Helliwell 2008). In these, specific interactions between the carotenoid, usually astaxanthin, and a specialised protein cause a large change in colour and light absorption maxima, indicating a fundamentally different electronic structure in the ground and/or excited state.

Although not so spectacular, interactions between carotenoids and any proteins *etc.*, in their immediate molecular environment may cause changes in the electronic structure of the carotenoid *in vivo* and hence of its properties and reactivity and of the functional properties of the protein. In photosynthetic pigment-protein complexes, for example, carotenoid molecules are held by the proteins in positions that allow efficient energy transfer between carotenoids and chlorophylls and give protection against photodamage.

2.5.3 Consequences of Peroxidation

The electron-rich carotenoids readily react with oxygen or other oxidising agents. This may lead to the addition of oxygen to double bonds and formation of carotenoid peroxides which may cause oxidising chain-reactions, even in carotenoid samples stored in a freezer, leading to bleaching and destruction. Traces of such peroxides in carotenoid samples can seriously influence biological actions of carotenoids and give them strong pro-oxidant rather than antioxidant properties, thereby producing conflicting results. They may also lead to breakdown of the carotenoid molecule into products that may have biological activity, and must be excluded for any biological action to be attributed to the carotenoid itself. Carotenoid samples used in clinical trials and in studies of biological effects must, therefore, be pure and free from pro-oxidant peroxides and stabilising antioxidants and preferably freshly purified immediately before use. Incubation of carotenoids with cell cultures commonly takes place in air, and this exposure to oxygen can also lead to formation of peroxides.

3 Carotenoid Biosynthesis

3.1 Introduction and Overview

The natural production of carotenoids on Earth is enormous—over 100 million tonnes per year. There is great structural variety; the natural occurrence of about 700 different carotenoid structures has been reported. The aim of this section is to summarise overall knowledge about carotenoid biosynthesis in the plant kingdom—in higher plants, algae, fungi and bacteria. This is not an exhaustive survey of all published work but it integrates results from all kinds of investigations to describe the reactions, sequences and regulation of carotenoid biosynthesis. Details can be found in several reviews (Britton 1998; Hirschberg 1998; Cunningham and Gantt 1998; Fraser and Bramley 2004; DellaPenna and Pogson 2006; Moise et al. 2013; Shumskaya and Wurtzel 2013; Giuliano 2014; Nisar et al. 2015). More recent advances are covered in three volumes of *Methods in Enzymology* (Wurtzel 2022a, b, c) and in a review by Stra et al. (2023). Other references given in the text are generally to the first significant reports of specific findings.

In higher plants and algae, carotenoids are commonly present in well-defined subcellular organelles, the plastids. In photosynthetic tissues, they are located in pigment-protein complexes in inner membranes (thylakoids) of the chloroplast, where they function in photosynthesis. Carotenoid biosynthesis is an integral part of the construction of these pigment-protein complexes, and its regulation is closely linked to that of other components, namely chlorophylls, proteins and lipids. The carotenoids are biosynthesised within the chloroplast, though the whole process of chloroplast development, including the expression of the genes for carotenogenesis, is under direct nuclear control.

The most spectacular examples of carotenoids in higher plants are in non-photosynthetic tissues, especially in yellow, orange or red flowers and fruits. They are synthesised rapidly and on a large scale in chromoplasts which develop as the flowers open or the fruits ripen. The synthesis of carotenoids in ripening fruit is controlled by fruit-ripening-specific genes that are independent of those that control the synthesis of chloroplast carotenoids during the green stage (Fray and Grierson 1993; Gillaspy et al. 1993; Giuliano et al. 1993; Fraser et al. 1994; Pecker et al. 1996). Activation of flowering-specific genes is also involved in the massive carotenoid synthesis that occurs in chromoplasts during the development of many yellow-orange flowers (Fraser et al. 1994; Pecker et al. 1996) and levels of mRNA for the main carotenogenic enzymes increase, as they do during fruit ripening. In fruit and flowers, it is common for the xanthophylls to be present as fatty acid esters. Also C(3) acetates, and esterified 19-hydroxy carotenoids are abundant in some algae.

3.1.1 Regulation

The accumulation of carotenoids is influenced by many factors. All of these influences are often considered together as 'Regulation'. There have been many studies of the effects of environmental and nutritional factors on the qualitative or quantitative composition of green plants, especially in relation to crop productivity. For a broad survey, see the book by Goodwin (1980).

In all classes of algae, carotenoids are located in chloroplasts, in photosynthetic pigment-protein complexes, and are biosynthesised as part of chloroplast development, as in higher plants. Many microalgae, especially green algae, however, may accumulate sometimes high concentrations of carotenoids when they are subjected to environmental stress. These 'secondary carotenoids' accumulate outside the chloroplast, *e.g.* in oil droplets, and their synthesis depends on environmental factors, such as nitrogen depletion or high salinity, light intensity or temperature. The regulation of their synthesis is different from that of the chloroplast carotenoids. The concentration of these carotenoids can be very high, and commercial processes have been developed for the production of β-carotene from *Dunaliella salina* and astaxanthin from *Haematococcus pluvialis.*

3.1.2 Stages of Carotenoid Biosynthesis

The carotenoids are C_{40} tetraterpenoid compounds and are biosynthesised by the great isoprenoid or terpenoid biosynthetic pathway from the basic C_5 isoprene unit, as isopentenyl diphosphate (IDP) and its isomer dimethylallyl diphosphate (DMADP). All the 700 or so natural carotenoids are biosynthesised by this basic pathway, with later modifications to elaborate the variety of structural features. Formation of rings in the carotenoid skeleton is limited to the ends of the molecule. The most common structural feature of the xanthophylls of plants is the presence of a hydroxy group at C(3) of one ring or both rings. Introduction of further oxygen

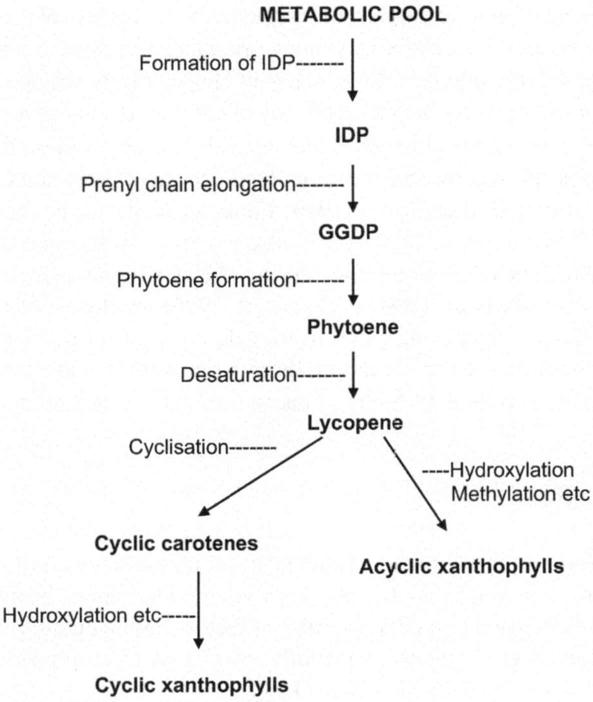

Fig. 1.2 Stages of carotenoid biosynthesis

functions, and other modifications, give rise to the many different xanthophylls that have been identified. The biosynthesis is conveniently dissected into a series of stages (Fig. 1.2). The early stages involve water-soluble precursors and take place in the aqueous medium of the cell, whereas later reactions involving less soluble or insoluble intermediates occur in association with or inside membranes (see Sect. 3.10).

3.2 *Origin of the Isoprene Unit*

The intermediates from which the isoprene unit is made come from the general carbon pool of basic intermediary metabolism. This can vary from one organism to another and even from one kind of tissue, cell or organelle to another. It may also change with time or at different stages of development.

Following the identification of mevalonic acid (MVA) it was widely assumed that all isoprenoids are formed from acetate by the classical route *via* MVA. It has now become clear, however, that there are alternative routes for the formation of the isoprene unit that do not involve mevalonate, and that this is particularly important in relation to carotenoid biosynthesis in plastids. Early work on the formation of the isoprene unit and isoprenoid compounds *via* MVA is summarised in a book edited by Porter and Spurgeon (1981).

3.2.1 The Acetate—Mevalonate Pathway

Much information on details of the formation of mevalonate from acetate was obtained not with carotenogenic organisms but with much more amenable systems, namely yeast and mammalian liver, in relation to sterol biosynthesis. Similar work with carotenogenic systems confirmed the view that the same pathway operates in carotenoid biosynthesis. Three molecules of acetate, as acetyl-CoA, combine to form 3-hydroxy-3-methylglutaryl-CoA (HMGCoA) which then undergoes a two-step reduction to generate mevalonate (Fig. 1.3). The specific inhibition of the enzyme, HMG-CoA reductase, blocks the formation of mevalonate, and has frequently been used in biosynthesis studies.

The formation of IDP from mevalonate is a three-step, three-enzyme process, each step of which requires one molecule of ATP and a divalent cation, Mg^{2+} or Mn^{2+} (Fig. 1.4).

3.2.2 The Mevalonate-Independent Pathway

Many reports that mevalonate is not incorporated into carotenoids and other isoprenoids in some bacteria and in plastids of higher plants and algae suggested that there may be alternative pathways in which various aldehyde or ketone intermediates in general metabolism could provide routes to IDP (Pandian et al. 1981). Labelling patterns in isoprenoid compounds biosynthesised in the presence of ^{13}C-labelled general carbon-source substrates such as glucose were consistent with the formation of sterols *etc*, from acetate and mevalonate and this was also clearly demonstrated for carotenoids biosynthesised in some systems. In other examples, however, *e.g.* in triterpenes of the hopane series, the isoprenoid sidechain of ubiquinone in several bacteria, and the carotenoids of the green alga *Scenedesmus obliquus* (Rohmer et al. 1993, 1996; Schwender et al. 1996), a different labelling pattern was observed that was consistent with the condensation of pyruvate with glyceraldehyde

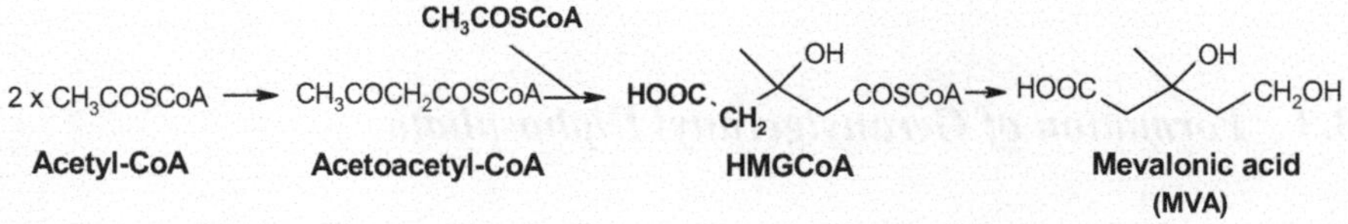

Fig. 1.3 Formation of mevalonate from acetate

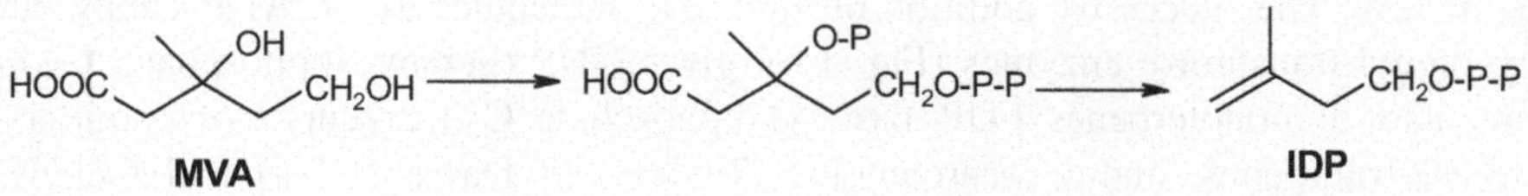

Fig. 1.4 Conversion of mevalonate (MVA) into isopentenyl diphosphate (IDP)

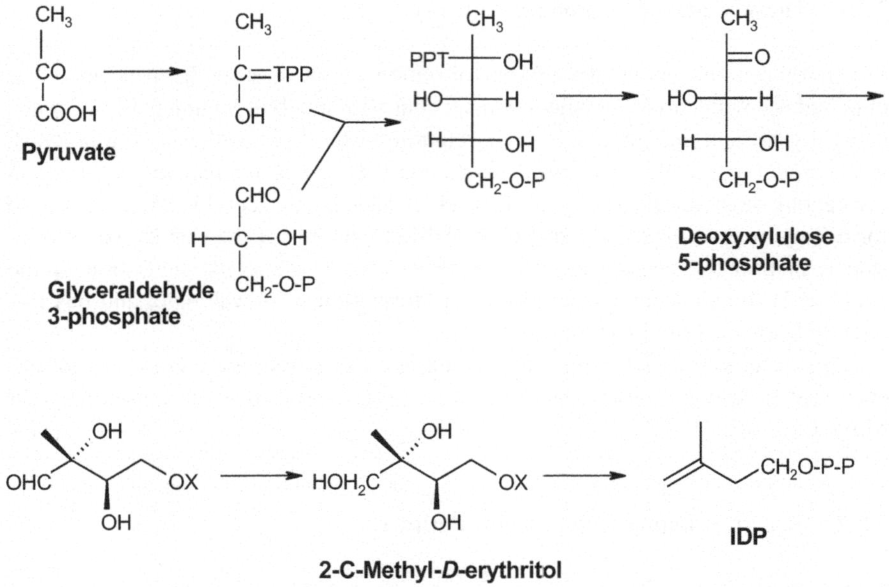

Fig. 1.5 The mevalonate-independent pathway to IDP

3-phosphate to give 1-deoxyxylulose 5-phosphate, *en route* to 2-C-methyl-D-erythritol 4-phosphate and thence to IDP (Rohmer et al. 1996; Rohmer 1999; Eisenreich et al. 2004; Hunter 2007). This methylerythritol phosphate (MEP) or deoxyxylulose phosphate (DXP or DOXP) pathway (Fig. 1.5), has now been shown to be the major route to isoprenoid compounds, including carotenoids, in many carotenogenic systems, especially in many bacteria and in chloroplasts and chromoplasts of plants. In plants, both pathways may operate, the mevalonate pathway outside the plastids and the mevalonate-independent pathway inside (Joyard et al. 2009). Intermediates such as prenyl diphosphates made outside may be taken into the plastid and used in later steps.

3.3 Formation of Geranylgeranyl Diphosphate

The C_{40} skeleton of carotenoids is formed from two molecules of the C_{20} intermediate geranylgeranyl diphosphate (GGDP). The isoprene unit, IDP, is isomerised to the allylic dimethylallyl diphosphate (DMADP) through the action of IDP:DMADP isomerase. The successive addition of three IDP molecules to DMADP, catalysed by prenyl transferase enzymes (Fig. 1.6), gives GDP (geranyl diphosphate, C_{10}), precursor of monoterpenes, FDP (farnesyl diphosphate, C_{15}), precursor of sesquiterpenes, triterpenes, and C_{30}-carotenoids (Taylor and Davies 1974) and GGDP,

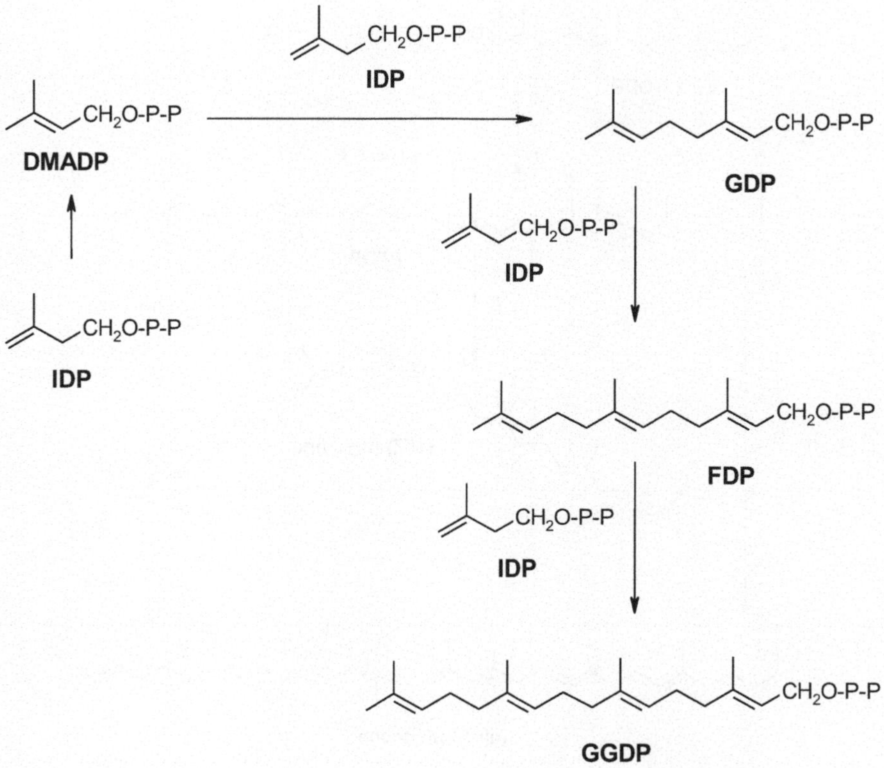

Fig. 1.6 Formation of geranylgeranyl diphosphate (GGDP)

precursor of diterpenes, two molecules of which undergo condensation to give the symmetrical C_{40} carotenoid skeleton as phytoene.

The GGDP synthase from the stromal fraction from chromoplasts of *Capsicum annuum* fruit has been purified to homogeneity and a cDNA encoding the enzyme has been cloned in *E. coli*. The enzyme contained a transit peptide and could use DMADP, GDP or FDP as co-substrate with IDP to give GGDP as the final product (Dogbo and Camara 1987; Kuntz et al. 1992).

3.4 Formation of Phytoene

The C_{40} carotenoid skeleton is formed from two molecules of the C_{20} GGDP to give phytoene, which has a central double bond, and not lycopersene, as once expected by analogy with the formation of squalene, the C_{30} precursor of triterpenes and sterols, from two C_{15} FDP molecules. The apparently simple formation of phytoene,

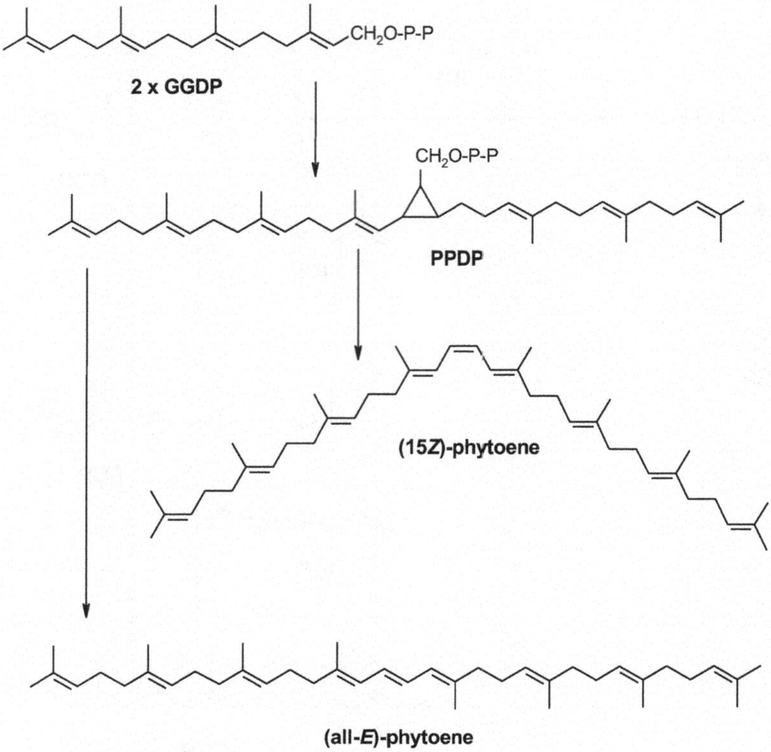

Fig. 1.7 Formation of (all-*E*)- or (15*Z*)-phytoene from GGDP *via* prephytoene diphosphate (PPDP)

however, is complex and occurs *via* a cyclopropyl intermediate, prephytoene diphosphate (PPDP) (Altman et al. 1972) (Fig. 1.7). The C_{30} carotenoids that occur in some bacteria (*e.g. Staphylococcus aureus* and *Streptococcus faecium* (Taylor and Davies 1974) are biosynthesised in an analogous way from two molecules of the C_{15} FDP via the C_{30} analogue of phytoene.

In plants, fungi and some bacteria, the phytoene product is the central-*cis* (15Z) isomer whereas in many bacteria the product is all-*trans* (all-*E*) phytoene (Spurgeon and Porter 1983). Different genes code for phytoene synthase in plants (*Psy*) and in bacteria (*crtB*), but these share conserved amino acid sequence regions. Fruit-producing plants contain two phytoene synthase genes; *Psy1* and *Psy2* are both transcribed in leaves and seedlings, but *Psy2* transcripts are more abundant in fruits and account for the increased level of mRNA and hence increased carotenoid synthesis during fruit ripening (Bartley and Scolnik 1993; Giorio et al. 2008). The situation in flowers as chromoplasts develop is similar. The enzyme products of the two *Psy* genes are highly conserved and may be 95% identical in their amino acid sequence.

3.5 Desaturation

Formation of the coloured carotenoids from colourless phytoene requires extension of the conjugated double-bond system by a series of desaturation reactions, in each of which two hydrogen atoms are removed and the chromophore is extended by two conjugated double bonds. Most commonly, four successive desaturations, occurring alternately in the two halves of the molecule, give the sequence phytoene—phytofluene—ζ-carotene—neurosporene—lycopene, in agreement with the principle first suggested in the classical Porter-Lincoln scheme (Porter and Lincoln 1950) (Fig. 1.8).

In bacteria, the desaturation steps, which may stop at the neurosporene stage and may proceed first in one half of the molecule and then in the other, *i.e. via* unsymmetrical ζ-carotene, are catalysed by a single enzyme, CRTI, the product of the gene

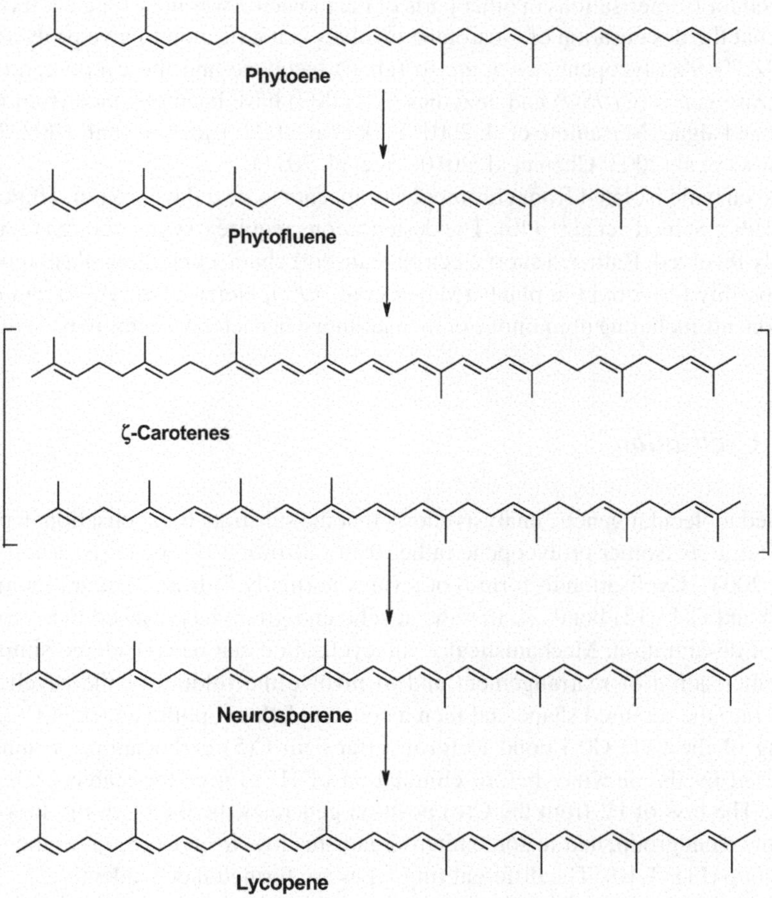

Fig. 1.8 The desaturation sequence from phytoene to lycopene

crtI (Armstrong et al. 1989; Linden et al. 1991). In plants and algae, however, the desaturation requires two genes and two enzymes. The first enzyme, phytoene desaturase PDS, catalyses the conversion of phytoene into ζ-carotene, *via* phytofluene (Pecker et al. 1996), but a second enzyme, ζ-carotene desaturase, ZDS, is needed for the two steps from ζ-carotene to lycopene (Albrecht et al. 1995). Sequence conservation between CRTI and PDS enzymes is low, but there is significant similarity between PDS and ZDS.

The conversion of phytoene into lycopene appears to be a simple series of *trans*-eliminations, in which two hydrogen atoms are removed in each step, consistent with the involvement of FAD, as indicated by the presence of an FAD/NAD(P) binding motif (Pecker et al. 1992). The process in plants and algae is more complicated, however, and involves several double-bond isomerisations (Fig. 1.9). The discovery of a gene coding for an isomerase enzyme to convert the C(15,15′)-double bond from the *Z* configuration in phytoene into the *E* configuration in the coloured carotenoids is not surprising, but the discovery that other genes and enzymes are required for isomerisations in other parts of the molecule was not expected. It is now clear that the desaturation of ζ-carotene involves isomerisation to give prolycopene [(7Z,9Z,7′Z,9′Z)-lycopene] *en route* to (all-*E*)-lycopene and the cyclic carotenes. Isomerase genes (*CrtISO*) and enzymes (CRTISO) have been obtained from many plants and algae (Masamoto et al. 2001; Park et al. 2002; Isaacson et al. 2002, 2004; Matthews et al. 2003; Chen et al. 2010; Yu et al. 2011).

The enzyme isolated from chromoplasts appears to be a flavoprotein (Hugueney et al. 1992; Schledz et al. 1996). The desaturations require oxygen, though O_2 is not directly involved. Rather, a short electron-transport chain, containing plastoquinone and possibly ferredoxin in plants (Mayer et al. 1990; Norris et al. 1995), but other components including ubiquinone or menaquinone in bacteria, seems to be involved.

3.6 *Cyclisation*

Detailed molecular genetic analysis shows that the substrate for cyclisation in plants is the tetra-*cis* isomer prolycopene rather than (all-*trans*)-lycopene (Isaacson et al. 2002, 2004). Cyclisation to form β or ε rings normally follows desaturation of the C(7,8) and C(11,12) bonds so that the acyclic end group has reached the lycopene level of desaturation. Mechanistically, the cyclisation can be considered simply as an isomerisation or rearrangement and to involve the folding of the acyclic end group into the required shape and then a concerted electrophilic attack at C(2) and closing of the C(1)-C(6) bond to form a transient C(5) carbocation, presumably stabilised by the enzyme, before elimination of H^+ to give the stable cyclic end group. The loss of H^+ from the C(6) position generates the β end group, loss from C(4) the ε endgroup, and in some microorganisms loss from C(18) gives the rare γ end group (Fig. 1.10). The different ring types are formed independently.

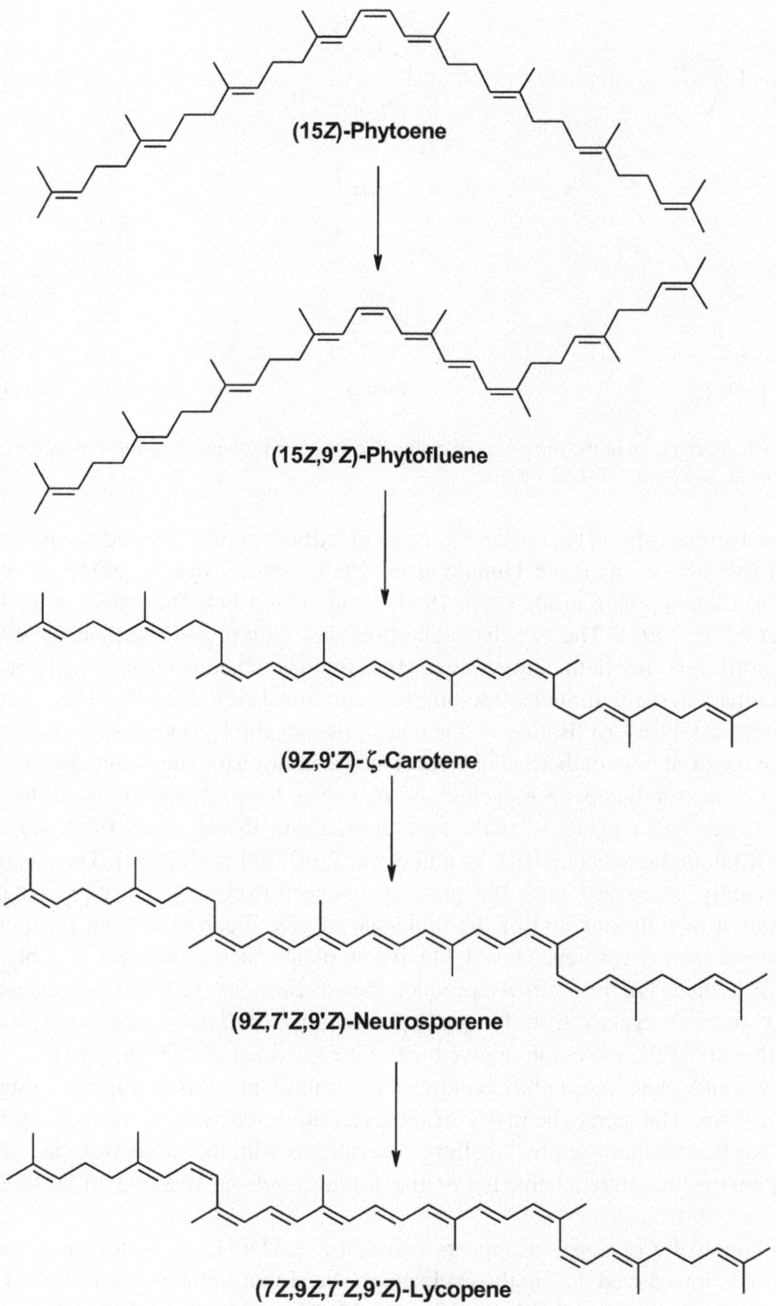

Fig. 1.9 The involvement of *Z*-isomers in the desaturation of phytoene in plants: desaturation sequence to form prolycopene [(7Z,9Z,7′Z,9′Z)-lycopene]

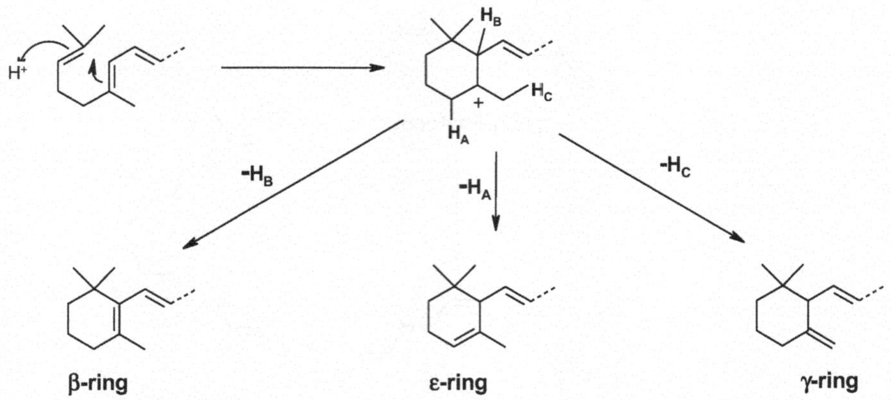

Fig. 1.10 Mechanism of the formation of cyclic β, ε and γ end groups from the acyclic end group of lycopene [drawn as (all-*E*)-lycopene]

The lycopene-β-cyclase gene has been identified in and isolated from bacteria (*crtY*) (Misawa et al. 1990; Hundle et al. 1993, 1994; To et al. 1994) and plants (*crtL-b*) (Cunningham et al. 1993, 1994; Sandmann 1994; Hugueney et al. 1995; Pecker et al. 1996). The two lycopene-β-cyclase polypeptide sequences are distinctly different, but both contain three short regions of sequence similarity, including a characteristic dinucleotide-binding site for FAD/NAD(P). The lycopene β-cyclase catalyses cyclisation at one end of the acyclic lycopene molecule to give γ-carotene or at both ends to give β-carotene (Cunningham and Gantt 2001).

The gene for lycopene ε-cyclase (*CrtL-e*) has been cloned from *Arabidopsis thaliana* and other plants (Cunningham et al. 1996; Ronen et al. 1999; Bai et al. 2009; Mialoundame et al. 2010; Yu and Beyer 2012; Shi et al. 2014). The enzyme is significantly conserved with the plant lycopene β-cyclase but usually catalyses cyclisation of only one end of the molecule to give δ-carotene from lycopene or α-carotene from γ-carotene (Fig. 1.11). A few plant species, however, notably varieties of lettuce (*Lactuca sativa*), produce the ε,ε-carotene-3,3′-diol lactucaxanthin. The lycopene ε-cyclase from lettuce expressed in *E. coli* does catalyse cyclisation of both ends of the molecule to give ε-carotene *via* δ-carotene (Fig. 1.11).

The ε end group has a chiral centre at C(6); in plants, this is normally found as the 6*R* form. The stereochemistry of the cyclisation reactions to form the β and ε rings has been established by labelling experiments with the stable isotopes ^{13}C and ^{2}H, allowing the stereochemistry of the folding and the reaction to be deduced (Britton 1998).

Although the cyclisation appears to require NADPH, no hydrogen from this cofactor is introduced during the cyclisation; the H that initiates cyclisation at C(2) comes from water in the medium (Hornero-Mendez and Britton 2002). It has been suggested that the NADPH is involved in the isomerisation of prolycopene to (all-*trans*)-lycopene prior to or concomitant with cyclisation.

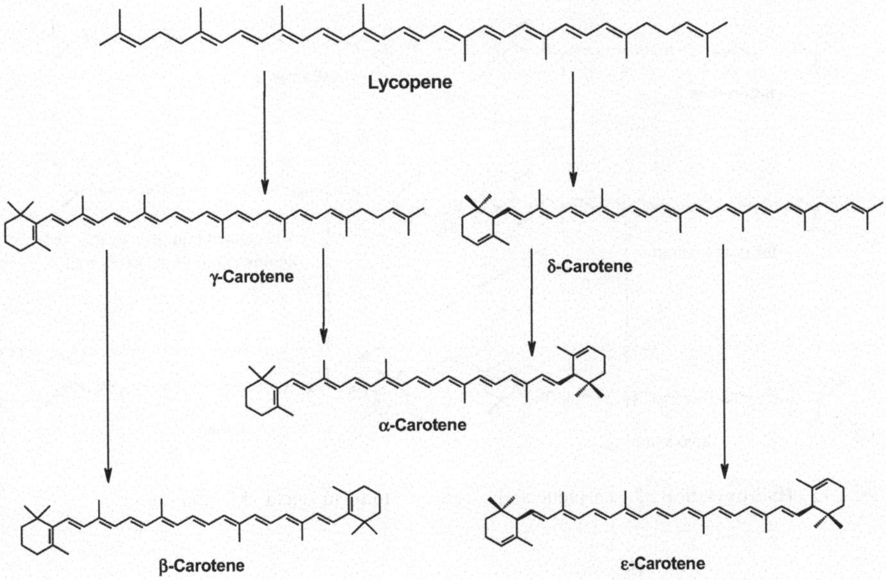

Fig. 1.11 Formation of β-carotene, α-carotene and ε-carotene from lycopene

3.7 Introduction of O Functions

3.7.1 Hydroxylation at C(3)

Hydroxy groups are introduced at C(3) by direct replacement of a hydrogen atom of the corresponding cyclic carotene. The O introduced comes from molecular oxygen not from water (Yamamoto et al. 1962). Hydroxylation of β-carotene thus gives the dihydroxy carotenoid zeaxanthin *via* the monohydroxy intermediate β-cryptoxanthin, whereas lutein is formed by two successive hydroxylations of α-carotene (Fig. 1.12). Separate genes encode hydroxylation of the β and ε rings in plants (*CrtR-b* and *CrtR-e*, respectively) (Pogson et al. 1996; Sun et al. 1996; Kim et al. 2009). Bacteria have only one hydroxylase gene, *crtZ*, which encodes a β-hydroxylase enzyme (Armstrong 1994). The plant and bacterial β-hydroxylases share some sequence similarity and are specific for the β-ring (Fraser et al. 1997). The deduced amino acid sequences contain several regions predicted to form trans-membrane helices, as well as conserved histidine motifs characteristic of non-haem iron enzymes (Hirschberg 1998). A gene for a member of the cytochrome P450 family is required for ε-hydroxylase activity (Tian et al. 2004).

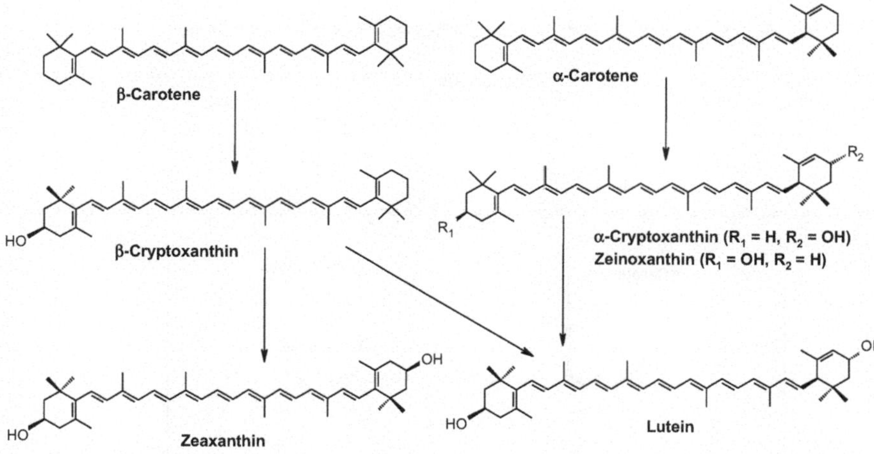

Fig. 1.12 Hydroxylation of β-carotene and α-carotene to form zeaxanthin and lutein

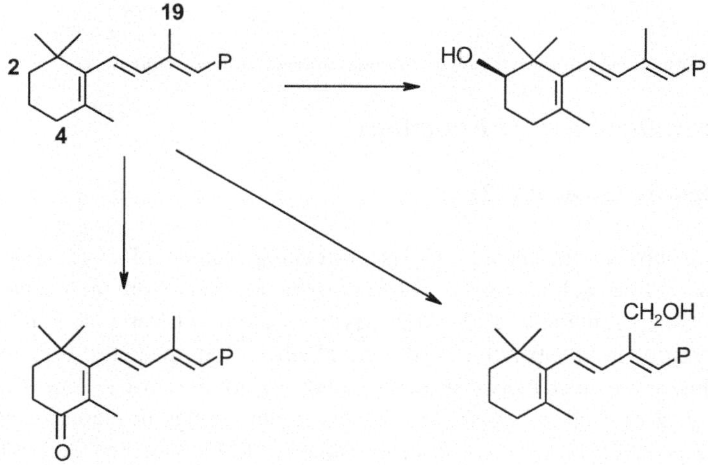

Fig. 1.13 Introduction of oxygen functions at C(2), C(4), or C(19)

3.7.2 Introduction of Hydroxy and Keto Groups at Other Positions

Introduction of oxygen functions at C(4) and C(2) of the β ring and oxidation of the C(19) or C(20) sidechain methyl substituents are features of the characteristic structures of several carotenoids, but, with the exception of the introduction of the keto group at C(4), their biosynthesis has not been studied (Fig. 1.13). The gene coding for the C(4)-oxygenase (sometimes termed 'ketolase') from the astaxanthin-accumulating green alga *Haematococcus pluvialis* (as *Crt O*) (Lotan and Hirschberg 1995; Kajiwara et al. 1995) and that from some bacteria (*crtW*) (Misawa et al. 1995a, b; Harker et al. 1998) has been cloned.

3.8 *Later Reactions*

3.8.1 Epoxidation

Zeaxanthin epoxidase genes, *Zep*, have been identified in several plants (Bouvier et al. 1996; Marin et al. 1996); some may be important in the biosynthesis of the growth regulator abscisic acid. The *Zep* genes share considerable sequence similarity with other mono-oxygenases and oxidases and indicate that the enzyme has an ATP-binding fold and an FAD-binding domain. The 5,6-epoxidation of zeaxanthin to violaxanthin is illustrated in Fig. 1.14.

3.8.2 Formation of Other End Groups

Plausible schemes have been proposed for the conversion of the 5,6-epoxy-β or 3-hydroxy-5,6-epoxy-β end group (or the related 5,6-peroxides) into the characteristic end groups of other xanthophylls, including the cyclopentyl κ end group characteristic of capsanthin and capsorubin (Fig. 1.15). Little biochemical or molecular evidence has been obtained to support most of these proposals, except for the enzyme capsanthin-capsorubin synthase (CCS) that has been characterised as a 50 kDa protein (Bouvier et al. 1994; Hugueney et al. 1995). The corresponding

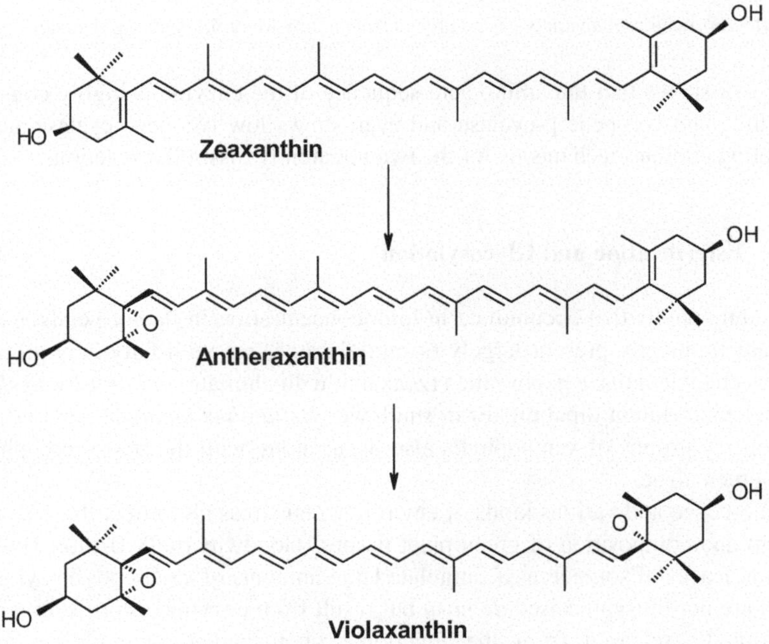

Fig. 1.14 5,6-Epoxidation of the β-ring, illustrated by the conversion of zeaxanthin into violaxanthin *via* antheraxanthin

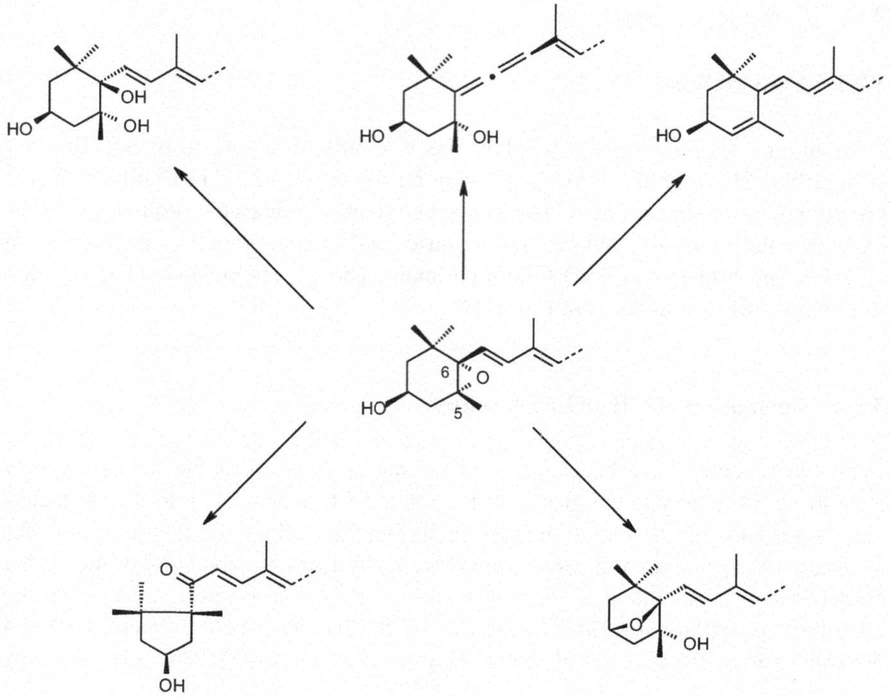

Fig. 1.15 Formation of a variety of end groups from a 3-hydroxy-5,6-epoxy-β structure

gene, *Ccs* shows that the amino acid sequence of the enzyme is highly conserved with the plant lycopene β-cyclase and even shows low lycopene cyclase activity, indicating similar mechanisms for the two apparently dissimilar reactions.

3.8.3 Esterification and Glycosylation

The xanthophylls that accumulate in high concentration in flower petals and ripe fruit are frequently present largely or entirely as esters with fatty acyl groups of various chain lengths, *e.g.* physalien (zeaxanthin dipalmitate) in *Physalis alkekengii* and helenien (lutein dipalmitate) in sunflower *Helianthus annuum.* Acyl esters of 19-hydroxy groups of xanthophylls also occur along with the free xanthophyll in some green algae.

Senescence and various kinds of environmental stress also affect the carotenoid content and composition of green plant tissues (Goodwin 1980; Britton 1998). In autumn, leaves of some trees accumulate large amounts of xanthophyll acyl esters, which are not biosynthesised *de novo* but result from enzymic esterification of the xanthophylls released from the functional photosynthetic complexes, as the chloroplasts degrade. Xanthophyll esters can also accumulate in some plants, *e.g.* barley, treated with diphenylether herbicides, exposed to atmospheric pollutants such as NO_2 and ozone, or under drought stress. It is assumed that the esters are formed by esterification of the hydroxy group of the parent xanthophyll by acyl transferase enzymes but, in most cases, no biochemical or molecular studies have been reported.

Xanthophylls present in some bacteria and Cyanobacteria may also be present as glycosides, whereas crocin, the main componemt of the pigment of saffron, is a glycosyl ester of the apocarotenoid acid crocetin. The gene *crtX* of *Erwinia herbicola* contains a sequence corresponding to a binding domain for UDP-glucose and has been shown to code for a glucosyl transferase that converts zeaxanthin into its 3,3′-diglucoside (Hundle et al. 1994; To et al. 1994) The conversion of crocetin into its glysosyl esters by a glucosyl transferase enzyme isolated from stigmata of the saffron crocus *Crocus sativus* has been demonstrated (Moraga et al. 2004).

3.9 Carotenoid Cleavage Enzymes and Apocarotenoids

Fragments of carotenoid molecules have important natural functions and actions. The importance of the C_{20} product vitamin A (retinol) in animals is well known. In plants, breakdown of the carotenoid skeleton gives apocarotenoids, and smaller fragments that are powerful aroma compounds or may have other functions, *e.g.* abscisic acid and others as growth regulators. Further modifications, especially rearrangements, may then give various other aroma compounds *etc*.

The polyene chain of a carotenoid molecule may undergo oxidative cleavage of any double bond (Fig. 1.16). Genes and enzymes have been identified for cleavage of the C(11,12), C(9,10), and C(7,8) double bonds, and of C(5,6) of the acyclic end group (Auldridge et al. 2006a, b; Fleischmann and Zorn 2008; Walter and Strack 2011; Ilg et al. 2014). Thus, cleavage of β-carotene and other carotenoids at C(9,10) affords β-ionone and other C_{13} aroma compounds (Fleischmann and Zorn 2008).

Similar cleavage of the C(11,12) double bond of (9*Z*)-violaxanthin or the C(11′,12′) double bond of (9′Z)-neoxanthin gives rise to the sesquiterpene plant growth regulators xanthoxin and abscisic acid together with a C_{25} apocarotenoid (Ilg et al. 2014) (Fig. 1.17). In the saffron crocus, *Crocus sativus*, oxidative cleavage of both the C(7,8) and C(7′,8′) double bonds of zeaxanthin gives the C_{20} diapocarotenoid crocetindial which is then further oxidised to the dicarboxylic acid crocetin and its glycosyl esters such as crocin, *etc*, that give saffron its orange colour (Bouvier et al. 2003a) (Fig. 1.18). The C_{10} fragment 3-hydroxy-β-cyclocitral and thence saffranal, give the saffron its characteristic spicy flavour. Similar cleavage of the C(5,6) and C(5′,6′) double bonds of lycopene gives the C_{24} bixin dialdehyde which is

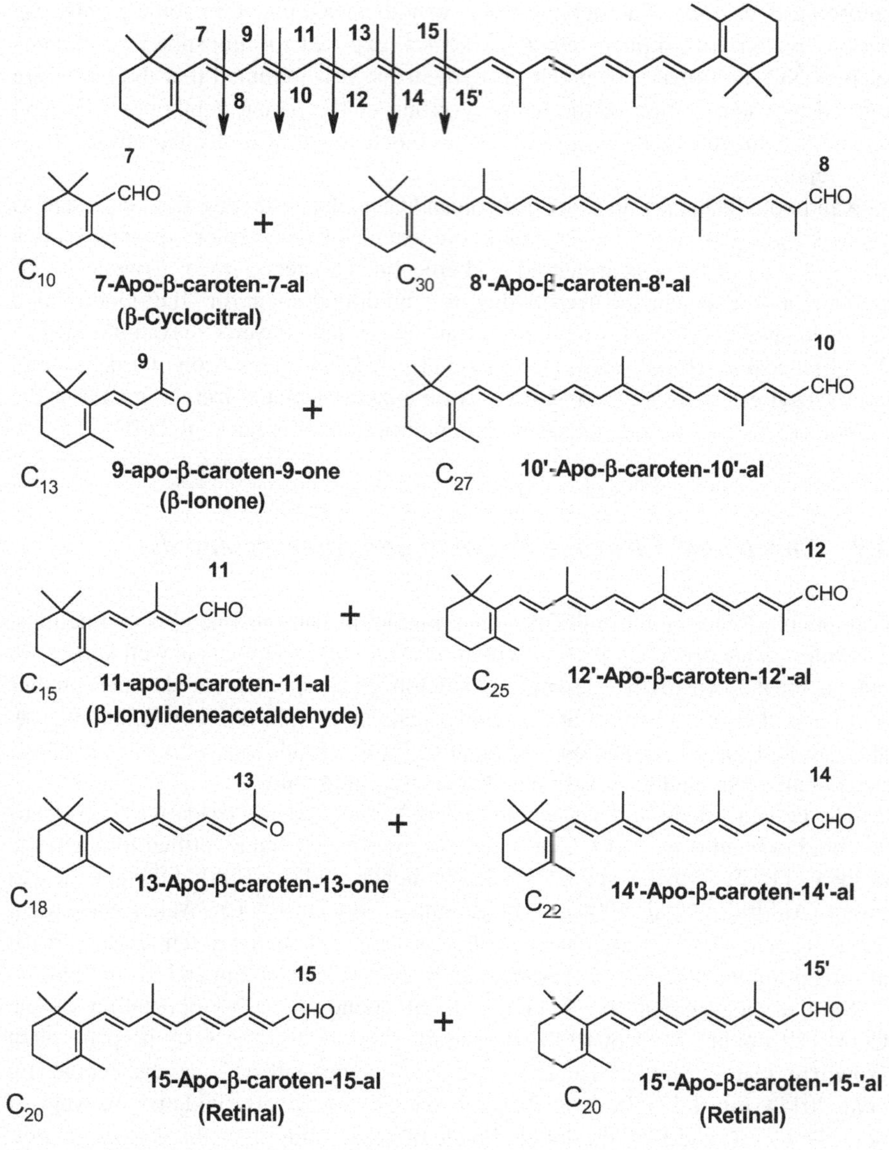

Fig. 1.16 Cleavage of double bonds to form apocarotenoids

oxidised to the dicarboxylic acid norbixin, and esterifed to its monomethyl ester, the food colorant bixin (Fig. 1.19) (Mercadante et al. 1997; Bouvier et al. 2003b).

The enzymes are generally referred to as carotenoid cleavage dioxygenases, CCDs (earlier name carotenases), though, in most cases, a dioxygenase mechanism has not been demonstrated. Molecular oxygen is required for the reactions, and

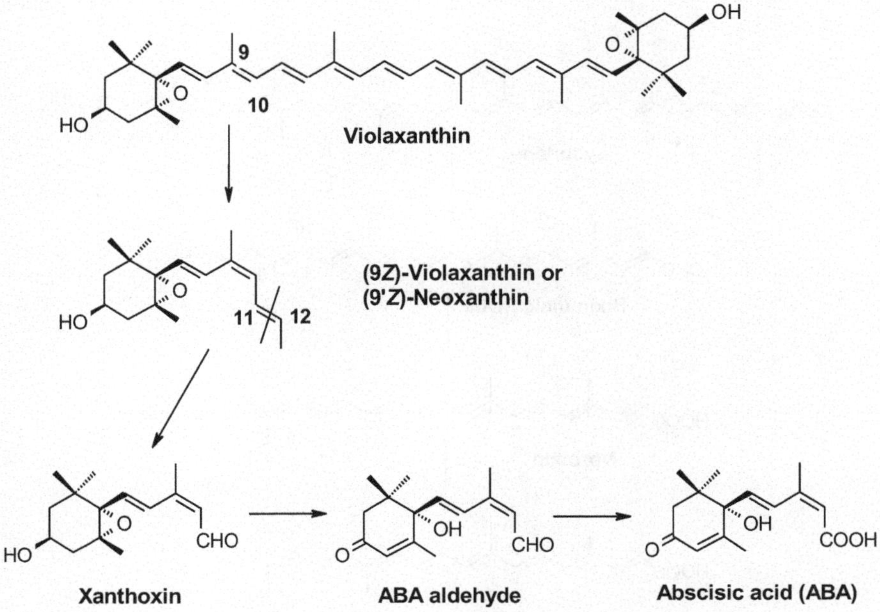

Fig. 1.17 Formation of abscisic acid by cleavage of violaxanthin or neoxanthin

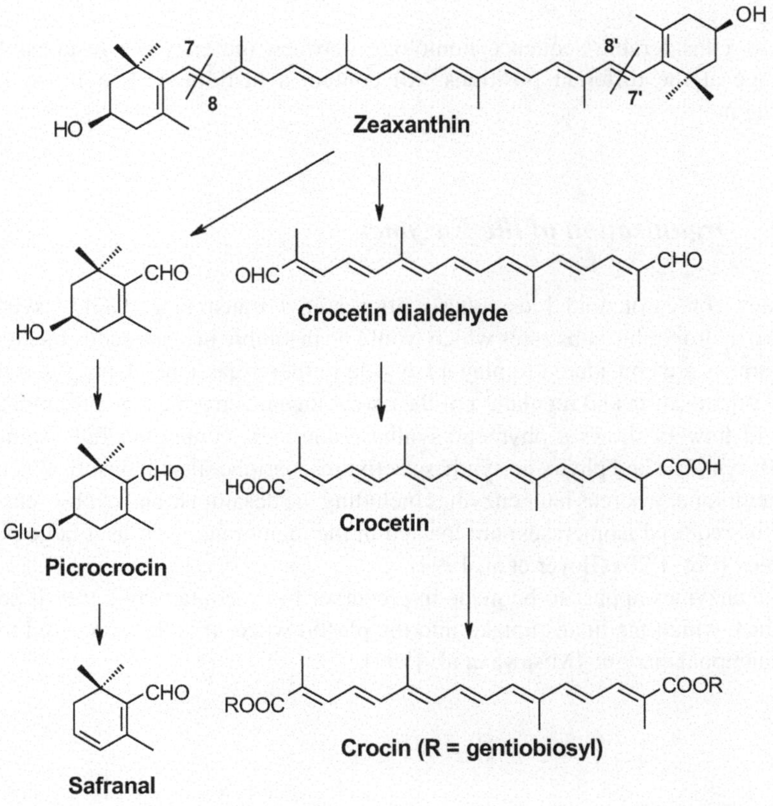

Fig. 1.18 Biosynthesis of saffron apocarotenoids in *Crocus sativus*

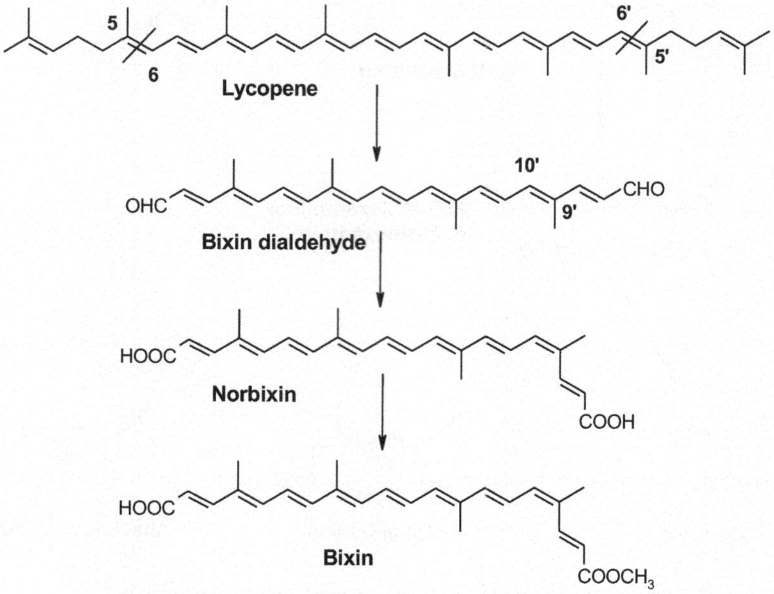

Fig. 1.19 Biosynthesis of bixin by cleavage of lycopene

there is considerable sequence homology between the enzymes responsible for cleavage at the different positions. All contain a histidine-rich non-haem-iron-binding motif.

3.10 Organisation of the Enzymes

All steps of carotenoid biosynthesis after prenyl transferase (GGDP synthase) require hydrophobic substrates which would be insoluble in an aqueous medium, so membranes are considered to play a key role in these reactions. A proposed model of the organisation and regulation of the carotenogenic enzymes in chromoplasts of daffodil flowers shows a phytoene synthase complex, containing IDP isomerase, GGDP synthase and phytoene synthase activities, peripheral and loosely attached to the membrane, whereas later enzymes including the desaturase and cyclase enzymes (and the required isomerases) are located in the membrane itself as a large protein complex (Fig. 1.20) (Beyer et al. 1994).

The enzymes appear to be made in precursor form, containing a transit peptide sequence which facilitates uptake into the plastid where it is then removed to give the functional enzyme (Misawa et al. 1994).

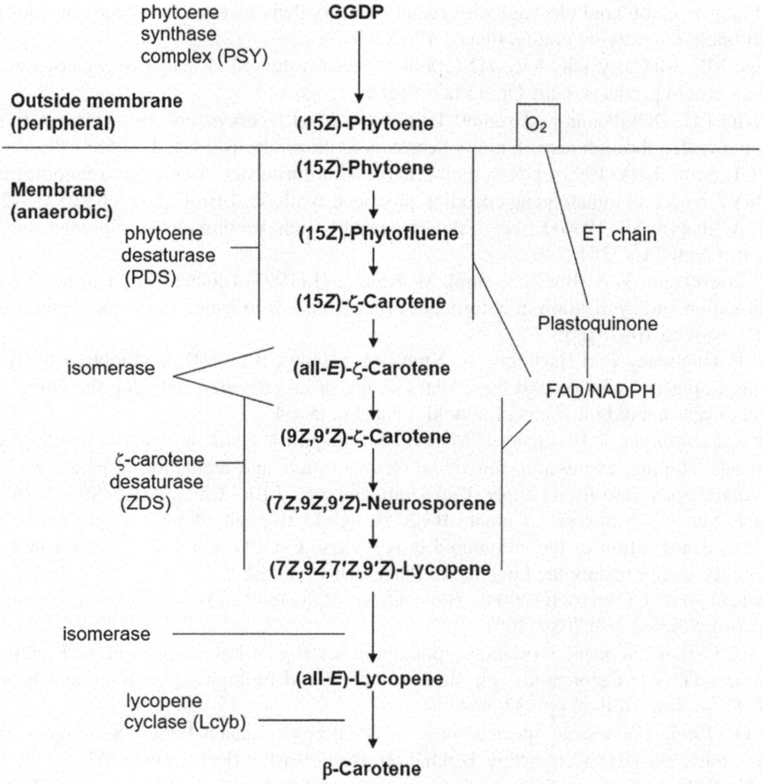

Fig. 1.20 Model of the organisation of carotenogenic enzymes in or associated with a membrane

References

Albrecht M, Klein A, Hugueney P, Sandmann G, Kuntz M (1995) Molecular cloning and functional expression in E. coli of a novel plant enzyme mediating ζ-carotene desaturation. FEBS Lett 372:199–202

Alpha-Tocopherol Beta-Carotene Cancer Prevention Study Group (1994) The effect of vitamin E and beta-carotene on the incidence of lung cancer and other cancers in male smokers. N Engl J Med 330:1029–1035

Altman LJ, Ash L, Kowerski RCW, Epstein W, Larson BR, Rilling HC, Muscio F, Gregonis DE (1972) Prephytoene pyrophosphate: a new intermediate in the biosynthesis of carotenoids. J Am Chem Soc 94:3257–3259

Armstrong GA (1994) Eubacteria show their true colors: genetics of carotenoid pigment biosynthesis from microbes to plants. J Bacteriol 176:4795–4802

Armstrong GA, Alberti M, Leach F, Hearst JE (1989) Nucleotide sequence, organisation, and nature of the protein products of the carotenoid biosynthesis gene cluster of Rhodobacter capsulatus. Mol Gen Genet 216:254–268

Auldridge ME, Block A, Vogel JT, Dabney-Smith C, Mila I, Bouzayen M, Magallanes-Lundback M, DellaPenna D, McCarty DR, Klee HJ (2006a) Characterization of three members of the

Arabidopsis carotenoid cleavage dioxygenase family demonstrates the divergent roles of this multifunctional enzyme family. Plant J 45:982–993

Auldridge ME, McCarty DR, Klee HJ (2006b) Plant carotenoid cleavage oxygenases and their apocarotenoid products. Curr Opin Plant Biol 9:315–321

Bai L, Kim EH, DellaPenna D, Brutnell TP (2009) Novel lycopene epsilon cyclase activities in maize revealed through perturbation of carotenoid biosynthesis. Plant J 59:588–599

Bartley GE, Scolnik PA (1993) cDNA cloning, expression during development and genome mapping of PSY2, a second tomato gene encoding phytoene synthase. J Biol Chem 268:25718–25721

Bendich A, Shapiro SS (1986) Effect of β-carotene and canthaxanthin on the immune responses of the rat. J Nutr 116:2254–2262

Beyer P, Nievelstein V, Al-Babili S, Bonk M, Kleinig H (1994) Biochemical aspects of carotene desaturation and cyclization in chromoplast membranes fron Narcissus pseudonarcissus. Pure Appl Chem 66:1047–1056

Bouvier F, Hugueney P, d'Harlingue A, Kuntz M, Camara B (1994) Xanthophyll biosynthesis in chromoplasts: isolation and moleculat cloning of an enzyme catalyzing the conversion of 5,6-epoxycarotenoid into ketocarotenoid. Plant J 6:45–64

Bouvier F, d'Harlingue A, Hugueney P, Marin E, Marionpoll A, Camara B (1996) Xanthophyll biosynthesis: cloning, expression, functional reconstitution and regulation of beta-cyclohexenyl carotenoid epoxidase from pepper (Capsicum annuum). J Biol Chem 271:28861–28867

Bouvier F, Suire C, Mutterer J, Camara B (2003a) Oxidative remodeling of chromoplast carotenoids: identification of the carotenoid dioxygenase CsCCD and CsZCD genes involved in crocus secondary metabolite biogenesis. Plant Cell 15:47–62

Bouvier F, Dogbo O, Camara B (2003b) Biosynthesis of the food and cosmetic plant pigment bixin (annatto). Science 300:2089–2091

Bowen PE (2010) Lycopene oxidation, uptake and activity in human prostate cell cultures. In: Landrum JT (ed) Carotenoids: physical, chemical and biological functions and properties. CRC Press, Boca Raton, pp 437–464

Britton G (1996) UV/visible spectroscopy. In: Britton G, Liaaen-Jensen S, Pfander H (eds) Carotenoids Vol 1B: spectroscopy. Birkhäuser, Basel/Boston/Berlin, pp 13–62

Britton G (1998) Overview of carotenoid biosynthesis. In: Britton G, Liaaen-Jensen S, Pfander H (eds) Carotenoids Vol 3: biosynthesis and metabolism. Birkhäuser, Basel/Boston/Berlin, pp 13–147

Britton G, Helliwell JR (2008) Carotenoid-protein interactions. In: Britton G, Liaaen-Jensen S, Pfander H (eds) Carotenoids, Vol 4: natural functions. Birkhäuser, Basel/Boston/Berlin, pp 99–118

Britton G, Liaaen-Jensen S, Pfander H (eds) (1995) Carotenoids Vol 1A: isolation and analysis. Birkhäuser, Basel/Boston/Berlin

Britton G, Liaaen-Jensen S, Pfander H (eds) (1996a) Carotenoids Vol 1B: spectroscopy. Birkhäuser, Basel/Boston/Berlin

Britton G, Liaaen-Jensen S, Pfander H (eds) (1996b) Carotenoids Vol 2: synthesis. Birkhäuser, Basel/Boston/Berlin

Britton G, Liaaen-Jensen S, Pfander H (eds) (1998) Carotenoids Vol 3: biosynthesis and metabolism. Birkhäuser, Basel/Boston/Berlin

Britton G, Gambelli L, Dunphy P, Pudney P, Gidney M (2002) Physical state of carotenoids in chromoplasts of tomato and carrot: consequences for bioavailability. In: Empis JA (ed) Functionalities of pigments in food. Sociedade Portuguesa de Quimica, Lisbon, pp 151–154

Britton G, Liaaen-Jensen S, Pfander H (eds) (2004) Carotenoids: Handbook. Birkhäuser, Basel/Boston/Berlin

Britton G, Liaaen-Jensen S, Pfander H (eds) (2008) Carotenoids Vol 4: natural functions. Birkhäuser, Basel/Boston/Berlin

Britton G, Liaaen-Jensen S, Pfander H (eds) (2009) Carotenoids Vol 5: nutrition and health. Birkhäuser, Basel/Boston/Berlin

Britton G, Liaaen-Jensen S, Pfander H (2017) Carotenoids: a colourful history. Fagtrykk, Trondheim

Burton GW, Ingold KU (1984) β-Carotene, an unusual type of lipid antioxidant. Science 224:569–573

Canene-Adams K, Erdman JW Jr (2009) Absorption, transport, distribution in tissues and bioavailability. In: Britton G, Liaaen-Jensen S, Pfander H (eds) Carotenoids, Vol 5: nutrition and health. Birkhäuser, Basel, Boston, Berlin, pp 115–148

Chen Y, Li F, Wurtzel ET (2010) Isolation and characterization of the Z-iso gene encoding a missing component of carotenoid biosynthesis. Plant Physiol 153:66–79

Chew BP, Park JS (2009) The immune system. In: Britton G, Liaaen-Jensen S, Pfander H (eds) Carotenoids, Vol 5: nutrition and health. Birkhäuser, Basel/Boston/Berlin, pp 363–382

Cunningham FX, Gantt E (1998) Genes and enzymes of carotenoid biosynthesis in plants. Ann Rev Plant Physiol Plant Mol Biol 49:557–583

Cunningham FX, Gantt E (2001) One ring or two? Determination of ring number in carotenoids by lycopene ε-cyclases. Proc Natl Acad Sci USA 98:2905–2910

Cunningham FX, Chamovitz D, Misawa N, Gantt E, Hirschberg J (1993) Cloning and functional expression in Escherichia coli of a cyanobacterial gene for lycopene cyclase, the enzyme that catalyzes the biosynthesis of beta-carotene. FEBS Lett 328:130–138

Cunningham FX, Sun ZR, Chamovitz D, Hirschberg J, Gantt E (1994) Molecular structure and enzymatic function of lycopene cyclase from the cyanobacterium Synechococcus sp strain PCC7942. Plant Cell 6:1107–1121

Cunningham FX, Pogson B, Sun ZR, McDonald KA, DellaPenna D, Gantt E (1996) Functional analysis of the β and ε lycopene cyclase enzymes of Arabidopsis reveals a mechanism for control of cyclic carotenoid formation. Plant Cell 8:1613–1626

DellaPenna D, Pogson BJ (2006) Vitamin synthesis in plants: tocopherols and carotenoids. Ann Rev Plant Biol 57:711–738

Dogbo O, Camara B (1987) Purification of isopentenyl pyrophosphate isomerase and geranylgeranyl pyrophosphate synthase from Capsicum annuum by affinity chromatography. Biochim Biophys Acta 920:140–148

Edge R, Truscott G (2010) Properties of carotenoid radicals and excited states and their potential role in biological systems. In: Landrum JT (ed) Carotenoids: physical, chemical and biological functions and properties. CRC Press, Boca Raton, pp 283–308

Eisenreich W, Bacher A, Arigoni D, Rohdich F (2004) Biosynthesis of isoprenoids via the nonmevalonate pathway. Cell Mol Life Sci 61:1401–1426

El-Agamey A, McGarvey DJ (2008) Carotenoid radicals and radical ions. In: Britton G, Liaaen-Jensen S, Pfander H (eds) Carotenoids Vol 4: natural functions. Birkhäuser, Basel/Boston/Berlin, pp 119–154

Fleischmann P, Zorn H (2008) Enzymic pathways for formation of carotenoid cleavage products. In: Britton G, Liaaen-Jensen S, Pfander H (eds) Carotenoids, vol 4: natural Functions. Birkhäuser, Basel/Boston/Berlin, pp 341–366

Fraser PD, Bramley PM (2004) The biosynthesis and nutritional uses of carotenoids. Prog Lipid Res 43:228–265

Fraser PD, Truesdale MR, Bird CR, Schuch W, Bramley PM (1994) Carotenoid biosynthesis during tomato fruit development (evidence for tissue-specific gene expression). Plant Physiol 105:405–413

Fraser PD, Miura Y, Misawa N (1997) In vitro characterization of astaxanthin biosynthetic enzymes. J Biol Chem 272:6128–6135

Fray RG, Grierson D (1993) Molecular genetics of tomato fruit ripening. Trends Genet 9:438–443

Gillaspy G, Ben-David H, Gruissem W (1993) Fruits: a development perspective. Plant Cell 5:1439–1451

Giorio G, Stigliani AL, D'Ambrosio C (2008) Phytoene synthase genes in tomato (Solanum lycopersicon L) – new data on the structures, the deduced amino acid sequences and the expression patterns. FEBS J 275:527–535

Giuliano G (2014) Plant carotenoids: genomics meets multi-gene engineering. Curr Opin Plant Biol 19:111–117

Giuliano G, Bartley GE, Scolnik P (1993) Regulation of carotenoid biosynthesis during tomato development. Plant Cell 5:379–387

Goodwin TW (1952) The comparative biochemistry of the carotenoids. Chapman and Hall, London

Goodwin TW (1980) The biochemistry of the carotenoids, vol I, Plants, 2nd edn. Chapman and Hall, London, New York

Goodwin TW (1984) The biochemistry of the carotenoids, vol II, Animals, 2nd edn. Chapman and Hall, London/New York

Goralczyk R, Wertz K (2009) Skin photoprotection by carotenoids. In: Britton G, Liaaen-Jensen S, Pfander H (eds) Carotenoids, vol 5: nutrition and health. Birkhäuser, Basel/Boston/Berlin, pp 365–362

Gruszecki WI (2010) Carotenoids in lipid membranes. In: Landrum JT (ed) Carotenoids: physical, chemical and biological functions and properties. CRC Press, Boca Raton, pp 19–30

Harker M, Hirschberg J, Oren A (1998) Paracoccus marcusii sp. Nov., an orange gram-negative coccus. Int J Syst Evol Microbiol 48:543–548

Hirschberg J (1998) Molecular biology of carotenoid biosynthesis. In: Britton G, Liaaen-Jensen S, Pfander H (eds) Carotenoids, Vol 3: biosynthesis and metabolism. Birkhäuser, Basel/Boston/Berlin, pp 149–194

Hornero-Mendez D, Britton G (2002) Involvement of NADPH in the cyclisation reaction of carotenoid biosynthesis. FEBS Lett 515:133–136

Hugueney P, Römer S, Kuntz M, Camara B (1992) Characterization and molecular cloning of a flavoprotein catalyzing the synthesis of phytofluene and zeta-carotene in Capsicum chromoplasts. Eur J Biochem 209:399–407

Hugueney P, Badillo A, Chen HC, Klein A, Hirschberg J, Camara B, Kuntz M (1995) Metabolism of cyclic carotenoids: a model for the alteration of this biosynthetic pathway in Capsicum annuum chromoplasts. Plant J 8:417–424

Hundle BS, O'Brien DA, Beyer P, Kleinig H, Hearst JE (1993) In vitro expression and activity of lycopene cyclase and β-carotene hydroxylase from Erwinia herbicola. FEBS Lett 315:329–334

Hundle B, Alberti M, Nievelstein V, Beyer P, Kleinig H, Armstrong GA, Burke DH, Hearst JE (1994) Functional assignment of Erwinia herbicola Eho10 carotenoid genes expressed in Escherichia coli. Mol Gen Genet 245:406–416

Hunter WN (2007) The non-mevalonate pathway of isoprenoid precursor biosynthesis. J Biol Chem 282:21573–21577

Ilg A, Bruno M, Beyer P, Al-Babili S (2014) Tomato carotenoid cleavage dioxygenases 1A and 1B. Relaxed double bond specificity leads to a plenitude of dialdehydes, monoapocarotenoids and isoprenoid volatiles. FEBS Open Bio 4:584–593

Isaacson T, Ronen G, Zamir D, Hirschberg J (2002) Cloning of tangerine from tomato reveals a carotenoid isomerase essential for the production of β-carotene and xanthophylls in plants. Plant Cell 14:333–342

Isaacson T, Ohad I, Beyer P, Hirschberg J (2004) Analysis in vitro of the enzyme CRTISO establishes a poly-cis-carotenoid biosynthetic pathway in plants. Plant Physiol 136:4246–4255

Johnson EJ (2011) Lutein and zeaxanthin relationships to cognitive function in the elderly. Acta Biologica Cracovienska 53(Suppl. 1):17

Johnson EJ, Krinsky NI (2009) Carotenoids and coronary heart disease. In: Britton G, Liaaen-Jensen S, Pfander H (eds) Carotenoids, Vol 5: nutrition and health. Birkhäuser, Basel, Boston, Berlin, pp 287–300

Joyard J, Ferro M, Masselon C, Seigneurin-Berny D, Salvi D, Garin J, Rolland N (2009) Chloroplast proteomics and the compartmentation of plastidial isoprenoid biosynthetic pathways. Mol Plant 2:1154–1180

Kajiwara S, Kakizono T, Saito T, Kondo K, Ohtani T, Nishio N, Nagai S, Misawa N (1995) Isolation and functional identification of a novel cDNA for astaxanthin biosynthesis from

Haematococcus pluvialis, and astaxanthin synthesis in Escherichia coli. Plant Mol Biol 29:343–352

Karrer P, Jucker E (1950) Carotenoids (Braude EA tr). Elsevier, Amsterdam

Kim J, Smith JJ, Tian L, DellaPenna D (2009) The evolution and function of carotenoid hydroxylases in Arabidopsis. Plant Cell Physiol 50:463–479

Köhn S, Kolbe H, Korger M, Köpsel C, Mayer B, Auweter H, Lüddecke E, Bettermann H, Martin H-D (2008) Aggregation and interface behaviour of carotenoids. In: Britton G, Liaaen-Jensen S, Pfander H (eds) Carotenoids, Vol 4: natural functions. Birkhäuser, Basel/Boston/Berlin, pp 53–98

Krinsky NI, Mayne ST, Sies H (eds) (2004) Carotenoids in health and disease. Marcel Dekker, New York

Kuntz M, Romer S, Suire C, Hugueney P, Weil JH, Schantz R, Camara B (1992) Identification of a cDNA for the plastid-located geranylgeranyl pyrophosphate synthase from Capsicum annuum: correlative increase in enzyme activity and transcript level during fruit ripening. Plant J 2:25–34

Liaaen-Jensen S, Lutnaes BF (2008) E/Z isomers and isomerization. In: Britton G, Liaaen-Jensen S, Pfander H (eds) Carotenoids, Vol 4: natural functions. Birkhäuser, Basel, Boston, Berlin, pp 15–36

Linden H, Misawa N, Chamovitz D, Pecker I, Hirschberg J, Sandmann G (1991) Functional complementation in Escherichia coli of different phytoene desaturase genes and analysis of accumulated carotenes. Z Naturforsch 46c:1045–1051

Lotan T, Hirschberg J (1995) Cloning and expression in Escherichia coli of the gene encoding the conversion of β-carotene to the ketocarotenoid canthaxanthin in Haematococcus pluvialis. FEBS Lett 364:125–128

Marin E, Nussaume L, Quesada A, Gonneau M, Sotta B, Hugueney P, Frey A, Marionpoll A (1996) Molecular identification of zeaxanthin epoxidase of Nicotiana plumbaginifolia, a gene involved in abscisic acid biosynthesis and corresponding to the ABA locus of Arabidopsis thaliana. EMBO J 15:2331–2341

Masamoto K, Wada H, Kaneko T, Takaichi S (2001) Identification of a gene required for cis-to-trans carotene isomerization in carotenogenesis of the cyanobacterium Synechocystis sp. PCC6803. Plant Cell Physiol 42:1398–1402

Mathews-Roth MM (1985) Carotenoids and cancer prevention – experimental and epidemiological studies. Pure Appl Chem 47:717–722

Mathews-Roth MM, Pathak MA, Fitzpatrick TB, Harber LC, Kass EH (1970) β-Carotene as a photoprotective agent in erythropoietic protoporphyria. N Engl J Med 282:1231–1234

Matthews PD, Luo R, Wurtzel ET (2003) Maize phytoene desaturase and zeta-carotene desaturase catalyze a poly-Z desaturation pathway: implications for genetic engineering of carotenoid content among cereal crops. J Exp Bot 54:2215–2230

Mayer MP, Beyer P, Kleinig H (1990) Quinone compounds are able to replace molecular oxygen as terminal electron acceptor in phytoene desaturation in chromoplasts of Narcissus pseudonarcissus L. Eur J Biochem 191:359–363

Mayne ST, Wright ME, Cartmel B (2009) Epidemiology and intervention trials. In: Britton G, Liaaen-Jensen S, Pfander H (eds) Carotenoids Vol. 5: nutrition and health. Birkhäuser, Basel, Boston, Berlin, pp 191–210

McCollum EV, Davis M (1915) The essential factors in the diet during growth. J Biol Chem 23:231–254

McLaren DS (2004) The control of Xerophthalmia; a century of contributions and lessons. Sight and Life, Basel

McLaren DS, Frigg M (1997) Sight and life manual on vitamin A deficiency disorders (VADD). Sight and Life, Basel

Mercadante AZ, Steck A, Pfander H (1997) Isolation and identification of new apocarotenoids from annatto (Bixa orellana) seeds. J Agric Food Chem 45:1050–1054

Mialoundame AS, Heintz D, Jadid N, Nkeng P, Rahier A, Deli J, Camara B, Bouvier F (2010) Characterization of plant carotenoid cyclases as members of the flavoprotein family functioning with no net redox change. Plant Physiol 153:970–979

Misawa N, Nakagawa M, Kobayashi K, Yamano S, Izawa I, Nakamura K, Harashima K (1990) Elucidation of Erwinia uredovora carotenoid biosynthesis pathway by functional analysis of gene products expressed in Escherichia coli. J Bacteriol 172:6704–6712

Misawa N, Truesdale MR, Sandmann G, Fraser PD, Bird C, Schuch W, Bramley PM (1994) Expression of a tomato cDNA coding for phytoene synthase in Escherichia coli, phytoene formation in vivo and in vitro, and functional analysis of the various truncated gene products. J Biochem Tokyo 116:980–985

Misawa N, Kajiwara S, Kondo K, Yokoyama A, Satomi Y, Saito T, Miki W, Ohtani T (1995a) Canthaxanthin biosynthesis by the conversion of methylene to keto groups in a hydrocarbon β-carotene by a single gene. Biochem Biophys Res Commun 209:867–876

Misawa N, Satomi Y, Kondo K, Yokoyama A, Kajiwara S, Saito T, Ohtani T, Miki W (1995b) Structure and functional analysis of a marine bacterial carotenoid biosynthesis gene cluster and astaxanthin biosynthetic pathway proposed at the gene level. J Bacteriol 177:6575–6584

Moise AR, Al-Babili S, Wurtzel ET (2013) Mechanistic aspects of carotenoid biosynthesis. Chem Rev 114:164–193

Moore T (1930) Vitamin A. Elsevier, Amsterdam

Moraga AR, Nohales PF, Perez JA, Gomez-Gomez L (2004) Glycosylation of the saffron apocarotenoid crocetin by a glucosyltransferase isolated from Crocus sativus stigmas. Planta 219:955–966

Moran NA, Jarvik T (2010) Lateral transfer of genes from fungi underlies carotenoid production in aphids. Science 328:624–627

Nisar N, Li L, Lu S, Khin NC, Pogson BJ (2015) Carotenoid metabolism in plants. Mol Plant 8:68–82

Norris SR, Barrette TR, DellaPenna D (1995) Genetic dissection of carotenoid synthesis in Arabidopsis defines plastoquinone as an essential component of phytoene desaturation. Plant Cell 7:2139–2149

Obermüller-Jevic UC, Packer L (2004) Lycopene and prostate cancer. In: Packer L, Obermüller-Jevic U, Kraemer K, Sies H (eds) Carotenoids and Retinoids: molecular aspects and health issues. AOCS Press, Champaign, pp 295–302

Packer L, Obermüller-Jevic U, Kraemer K, Sies H (eds) (2004) Carotenoids and Retinoids: molecular aspects and health issues. AOCS Press, Champaign

Pandian S, Saengchjan S, Raman TS (1981) An alternative pathway for the biosynthesis of isoprenoid compounds in bacteria. Biochem J 196:675–681

Park H, Kreunen SS, Cuttriss AJ, DellaPenna D, Pogson BJ (2002) Identification of the carotenoid isomerase provides insight into carotenoid biosynthesis, prolamellar body formation, and photomorphogenesis. Plant Cell 14:321–332

Pecker I, Chamovitz D, Linden H, Sandmann G, Hirschberg J (1992) A single polypeptide catalyzing the conversion of phytoene to ζ-carotene is transcriptionally regulated during tomato fruit ripening. Proc Natl Acad Sci USA 89:4962–4965

Pecker I, Gabbay R, Cunningham FX, Hirschberg J (1996) Cloning and characterization of the cDNA for lycopene beta-cyclase from tomato reveals decrease in its expression during fruit ripening. Plant Mol Biol 30:807–819

Peto R, Doll R, Buckley JD, Sporn MB (1981) Can dietary beta-carotene materially reduce human cancer rates? Nature 290:201–208

Pogson B, McDonald KA, Truong M, Britton G, DellaPenna D (1996) Arabidopsis carotenoid mutants demonstrate that lutein is not essential for photosynthesis in higher plants. Plant Cell 8:1627–1639

Porter JW, Lincoln RE (1950) 1. Lycopersicon selections containing a high content of carotenes and colorless polyenes. 2. The mechanism of carotene biosynthesis. Arch Biochem Biophys 27:390–403

Porter JW, Spurgeon SL (eds) (1981) Biosynthesis of isoprenoid compounds, Vol 1. Wiley, New York

Rohmer M (1999) The discovery of a mevalonate-independent pathway for isoprenoid biosynthesis in bacteria, algae and higher plants. Nat Prod Rep 16:565–574

Rohmer M, Knani M, Simonin P, Sutter B, Sahm H (1993) Isoprenoid biosynthesis in bacteria; a novel pathway for the early steps leading to isopentenyl diphosphate. Biochem J 295:517–524
Rohmer M, Seeman M, Horbach S, Bringer-Meyer S, Sahm H (1996) Glyceraldehyde 3-phosphate and pyruvate as precursors of isoprenic units in an alternative non-mevalonate pathway for terpenoid biosynthesis. J Am Chem Soc 118:2564–2566
Ronen G, Cohen M, Zamir D, Hirschberg J (1999) Regulation of carotenoid biosynthesis during tomato fruit development: expression of the gene for lycopene epsilon-cyclase is down-regulated during ripening and is elevated in the Delta mutant. Plant J 17:341–351
Sandmann G (1994) Carotenoid biosynthesis in microorganisms and plants. Eur J Biochem 223:7–24
Schalch W, Landrum JT, Bone RA (2009) The eye. In: Britton G, Liaaen-Jensen S, Pfander H (eds) Carotenoids, Vol 5: nutrition and health. Birkhäuser, Basel/Boston/Berlin, pp 301–334
Schalch W, Bone RA, Landrum JT (2010) The functional role of xanthophylls in the primate retina. In: Landrum JT (ed) Carotenoids: physical, chemical and biological functions and properties. CRC Press, Boca Raton, pp 275–282
Schledz M, Al-Babili S, von Lintig J, Haubruck H, Rabbani S, Kleinig H, Beyer P (1996) Phytoene synthase from Narcissus pseudonarcissus: functional expressiom, galactolipid requirement, topological distribution in chromoplasts and induction during flowering. Plant J 10:781–792
Schwender J, Seeman M, Lichtenthaler HK, Rohmer M (1996) Biosynthesis of isoprenoids (carotenoids, sterols, prenyl sidechains of chlorophylls and plastoquinone) via a novel pyruvate/glyceraldehyde 3-phosphate non-mevalonate pathway in the green alga Scenedesmus obliquus. Biochem J 316:73–80
Shi Y, Wang R, Luo Z, Jin L et al (2014) Molecular cloning and functional characterization of the lycopene ε-cyclase gene via virus-induced gene silencing and expression pattern in Nicotiana tabacum. Int J Mol Sci 8:14766–14785
Shumskaya M, Wurtzel ET (2013) The carotenoid biosynthetic pathway: thinking in all dimensions. Plant Sci 208:58–63
Sommer A, West KP Jr (1996) Vitamin A deficiency: health, survival, and vision. Oxford University Press, New York/Oxford
Spurgeon SL, Porter JW (1983) Biosynthesis of carotenoids. In: Porter JW, Spurgeon SL (eds) Biochemistry of isoprenoid compounds, Vol 2. Wiley, New York, pp 1–122
Stra A, Almarwaey LO, Alagoz Y, Moreno JC, Al-Babili S (2023) Carotenoid metabolism: new insights and synthetic approaches. Front Plant Sci 13:1–11
Sun ZR, Gantt E, Cunningham FX (1996) Cloning and functional analysis of the beta-carotene hydroxylase of Arabidopsis thaliana. J Biol Chem 271:24349–24352
Taylor RF, Davies BH (1974) Triterpenoid carotenoids and related lipids: the triterpenoid carotenes of Streptococcus faecium UNH564P. Biochem J 139:751–760
Tian L, Musetti V, Kim J, Magallanes-Lundback M, DellaPenna D (2004) The Arabidopsis LUT1 locus encodes a member of the cytochrome P450 family that is required for carotenoid-ring hydroxylation activity. Proc Natl Acad Sci USA 101:402–407
To KY, Lai EM, Lee LY, Lin TP, Hung CH, Chen CL, Chang YS, Liu ST (1994) Analysis of the gene cluster encoding carotenoid biosynthesis in Erwinia herbicola Eho13. Microbiology 140:331–339
Vishwanathan R, Kuchan MJ, Johnson EJ (2011) Lutein is the predominant carotenoid in infant brain. Acta Biologica Cracovienska 53(suppl.1):29
von Lintig J, Quadro L (eds) (2020) Special issue on carotenoids. Molecular and cell biology of lipids, Biochim Biophys Acta, vol 1865
Walter MH, Strack D (2011) Carotenoids and their cleavage products: biosynthesis and function. Nat Prod Rep 28:663–692
Wurtzel ET (ed) (2022a) Methods in enzymology, Vol 670 carotenoids: carotenoid and apocarotenoid analysis. Academic Press, Cambridge, MT, and Elsevier, Amsterdam

Wurtzel ET (ed) (2022b) Methods in enzymology Vol 671 carotenoids: carotenoid and apocarotenoid biosynthesis, metabolic engineering and synthetic biology. Academic Press, Cambridge, MT, and Elsevier, Amsterdam

Wurtzel ET (ed) (2022c) Methods in enzymology Vol 674 carotenoids: biological functions of carotenoids and apocarotenoids in natural and artificial systems. Academic Press, Cambridge, MT, and Elsevier, Amsterdam

Yamamoto HY, Chichester CO, Nakayama TOM (1962) Biosynthetic origin of oxygen in the leaf xanthophylls. Arch Biochem Biophys 96:645–649

Yu Q, Beyer P (2012) Reaction specificities of the ε-ionone-forming lycopene cyclase from rice (Oryza sativa) elucidated in vitro. FEBS Lett 586:3415–3420

Yu Q, Ghisla S, Hirschberg J, Mann V, Beyer P (2011) Plant carotene cis-trans isomerase CRTISO: a new member of the FAD(RED)-dependent flavoproteins catalyzing non-redox reactions. J Biol Chem 286:8666–8676

Chapter 2
Biosynthesis of Carotenoids in Plants and Microorganisms

Karyme do Socorro de Souza Vilhena, Marcilene Paiva da Silva, Anderson de Santana Botelho, Fábio José Bonfim Cardoso, Oberdan Oliveira Ferreira, Mozaniel Santana de Oliveira, and Eloisa Helena de Aguiar Andrade

1 Introduction

Essential for life, carotenoids are isoprenoid metabolites produced by eukaryotic cells, including plants, algae and fungi, and by prokaryotic cells, such as bacteria. Some animals, including flamingos, accumulate carotenoids in their plumage because their diet is rich in crustaceans that contain these pigments. In plants, these substances are responsible for photosynthesis and photoprotection, playing key roles as pigments in capturing light and as structural components in photosystems (López et al. 2023; Sun et al. 2022).

Chemically, carotenoids can be classified in several ways. Based on whether they have a hydrogen bond in their chemical structure, carotenoids can be divided into carotenes, such as lycopene, α, β, γ-carotene, and torulene, and into xanthophylls, which contain oxygen in addition to carbon and hydrogen, as exemplified by lutein, capsaicin, zeaxanthin, and astaxanthin (Liu et al. 2021; Shen et al. 2023). Based on their number of carbon atoms, carotenoids can be classified as C30, C40, C45, and C50. C40 carotenoids, such as lutein, α-carotene, β-carotene, zeaxanthin,

K. do Socorro de Souza Vilhena (✉) · M. P. da Silva · A. de Santana Botelho
O. O. Ferreira · E. H. de Aguiar Andrade
Laboratório Adolpho Ducke, Museu Paraense Emílio Goeldi, Belém, Brazil
e-mail: karyme@ufpa.br; andersonbotelho@museu-goeldi.br; oberdan@museu-goeldi.br; eloisa@museu-goeldi.br

M. S. de Oliveira
Programa de Pós-Graduação em Ciências Biológicas - Botânica Tropical Universidade Federal Rural da Amazônia/Museu Paraense Emílio Goeldi, Belém, Brazil
e-mail: mozanieloliveira@museu-goeldi.br

F. J. B. Cardoso
Laboratório de Química Multidisciplinar, Universidade Federal do Pará, Tucuruí, Brazil

R. Campos Chisté et al. (eds.), *Carotenoids*,
https://doi.org/10.1007/978-3-031-75322-0_2

canthaxanthin, and astaxanthin, are the most abundant naturally occurring carotenoids (Progress in Carotenoid Research 2018; Gottl et al. 2023).

Thanks to their chemical and structural characteristics, carotenoids display diverse biological activities. Among these activities, antioxidant activity stands out for the potential to deactivate free radicals and singlet oxygen. As such, these compounds can be used to prevent some cancers, including leukemia, prostate cancer, hepatocellular carcinoma and skin cancer, in addition to cardiovascular diseases. Furthermore, carotenoids can also be used to treat *diabetes mellitus* and to reduce macular degeneration (Saini et al. 2015; Hussain et al. 2022; Koklesova et al. 2020; Van Hoang et al. 2018). Vitamin A precursors, in turn, have been widely studied for their essential activities related to tissue growth and maintenance, photoreceptor regeneration and immune response (González-Peña et al. 2023; Amengual 2019).

Considering their biological activities and functions and their benefits for organisms, carotenoids can be used as dietary supplements, food colorants, nutraceuticals, cosmetics and even incorporated into poultry, fish, and crustacean feed (Nabi et al. 2020; Saini and Keum 2019; Pereira da Costa and Campos Miranda-Filho 2020). Approximately 80–90% carotenoids used in the industry, healthcare, and food are chemically synthesized. However, due to their harmful health effects, the demand for natural carotenoids has been increasingly markedly (Saini and Keum 2019; Bogacz-Radomska et al. 2020).

This Chapter aims at gathering recent data on carotenoids, highlighting their classification and nomenclature, the reactions involved in the biosynthesis of these compounds, the main carotenoids found in plants and microorganisms and their main biological activities, as well as the formation of vitamin A derivatives from carotenoids. In addition, this Chapter reports on the main positive effects of these compounds on human health.

2 Carotenoids: Classification and Applications

Carotenoids are a class of secondary organic compounds. These fat-soluble tetraterpenes play key roles in photosynthetic organisms, such as plants, bacteria, fungi, algae, and animals. These substances are usually yellow, orange, or red pigments, but when they bind to proteins, they become green, purple, or blue. In plants, these pigments are found alongside chlorophyll, helping in the photosynthesis process, and protecting against damage from oxidative reactions, due to their antioxidant activity. Furthermore, carotenoids are associated with several human health benefits (Bohn et al. 2021; Britton 2022).

Carotenoids consist of a conjugated system of hydrocarbon bonds with eight isoprenoid units connected in a pattern in which the arrangement of the isoprenoid units is inverted at the center of the molecule. Such a pattern provides the molecule with symmetry so that the two central methyl groups are in a 1,6 positional relationship, and the remaining, non-terminal methyl groups are in a 1,5 positional relationship (Fig. 2.1). All carotenoids can be derived from the acyclic structure $C_{40}H_{56}$,

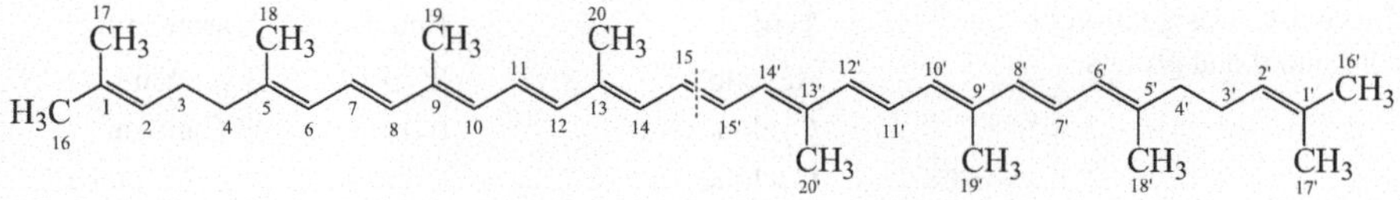

Fig. 2.1 Basic structure of carotenoid $C_{40}H_{56}$

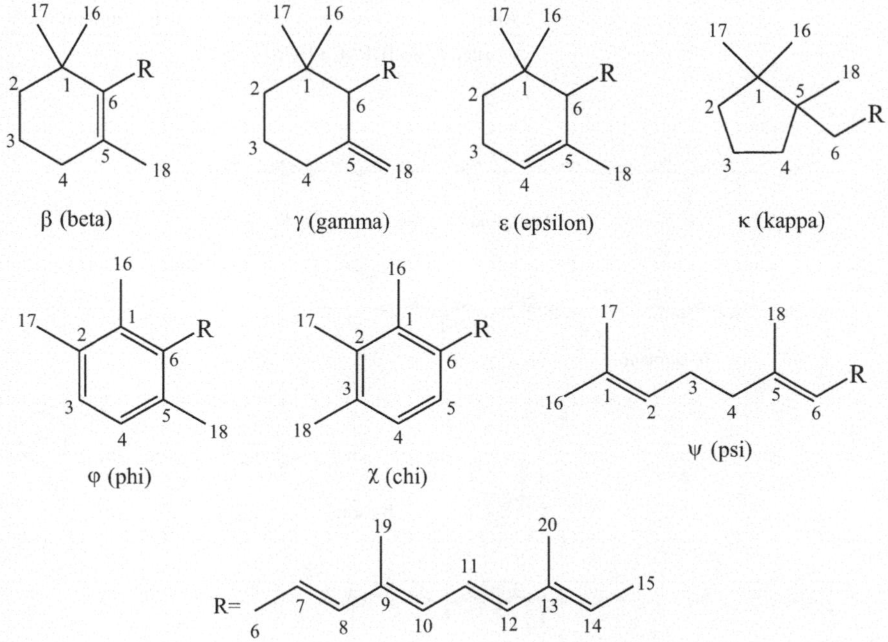

Fig. 2.2 End groups present in carotenoids

with a long central chain of conjugated double bonds commonly known as a polyene chain (Fernandes et al. 2018; Meléndez-Martínez et al. 2019).

Based on whether carotenoids contain rings, they can be classified into two main groups, namely cyclic or acyclic carotenoids. The C_9 end groups of carotenoids are shown in Fig. 2.2, and additional data are outlined in Table 2.1. The name of a specific carotenoid is defined by adding two Greek letters as prefixes to the main name "carotene". These prefixes are characteristic of the two end groups, as in α-carotene, β-carotene, β,ψ-carotene, γ,γ-carotene and others (Maoka 2020; Meléndez-Martínez et al. 2019).

Transformations in the polyene chain give yield acyclic carotenoids, which exclusively contain carbon and hydrogen atoms, such as lycopene. Each ψ-acyclic end of lycopene can be cyclized into α-, β-, or γ-ionone rings, as shown in Fig. 2.3. The types of end groups and ring substitutions found in carotenoids are important for their function and determine their fate (Maoka 2020; Meléndez-Martínez et al. 2019; Moise et al. 2014).

Table 2.1 Designation of carotenoid end groups

Type	End group	Carotenoid
Acyclic	Ψ (psi)	Lycopene
Cyclohexene	β (beta)	β-Carotene
Cyclohexene	ε (epsilon)	ε,ε- Carotene
Methylenecyclohexane	γ (gamma)	γ,γ- Carotene
Cyclopentane	κ (kappa)	Capsorubin
Aryl	χ (chi)	Renierapurpurin
Aryl	χ (chi)	Isorenieratene

Adapted from IUPAC (1975)

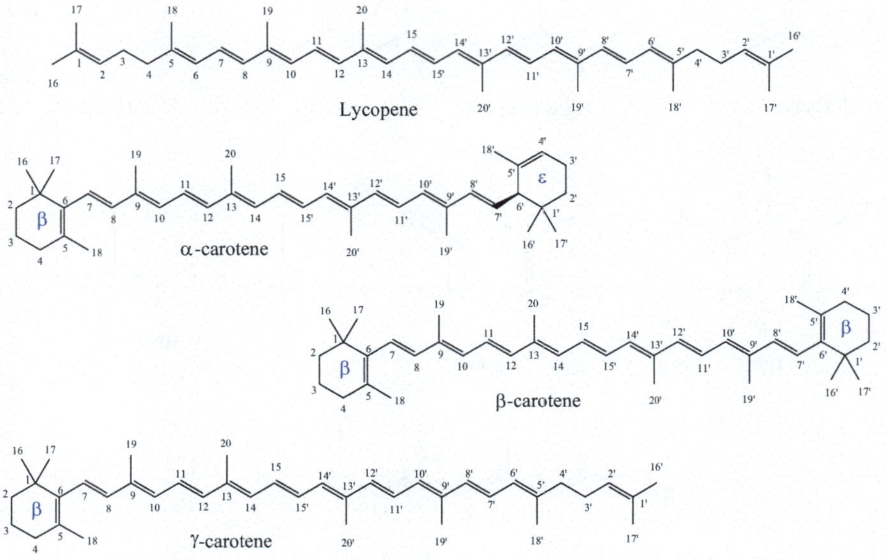

Fig. 2.3 Structure of the acyclic (Lycopene) and cyclic (α-, β- and γ-carotene) carotenoid

Carotenoids can also be classified based on whether they contain oxygen, such as xanthophylls, the oxygenated derivatives of carotenes. The oxygenated functional groups most commonly found in carotenoids include hydroxyl groups, for example, in β-cryptoxanthin, lutein, and zeaxanthin (Fig. 2.4), and oxygenated epoxy groups, in antheraxanthin and violaxanthin (Fig. 2.5). These xanthophylls are formed depending on some types of specific reactions. The main types of reactions that occur in the ionone ring are cyclization, hydroxylation, and epoxidation (Meléndez-Martínez et al. 2019; Moise et al. 2014).

Naturally occurring carotenoids are produced by eukaryotic cells, such as plants, algae, and fungi, and by prokaryotic cells, such as bacteria. Humans and animals cannot synthesize carotenoids. Therefore, these compounds must be included in their diets.

In this context, oxygenated carotenoids, primarily lutein and zeaxanthin (Fig. 2.4), are essential dietary sources through natural products, with several health benefits for their ability to bind to physiological proteins in humans (López et al.

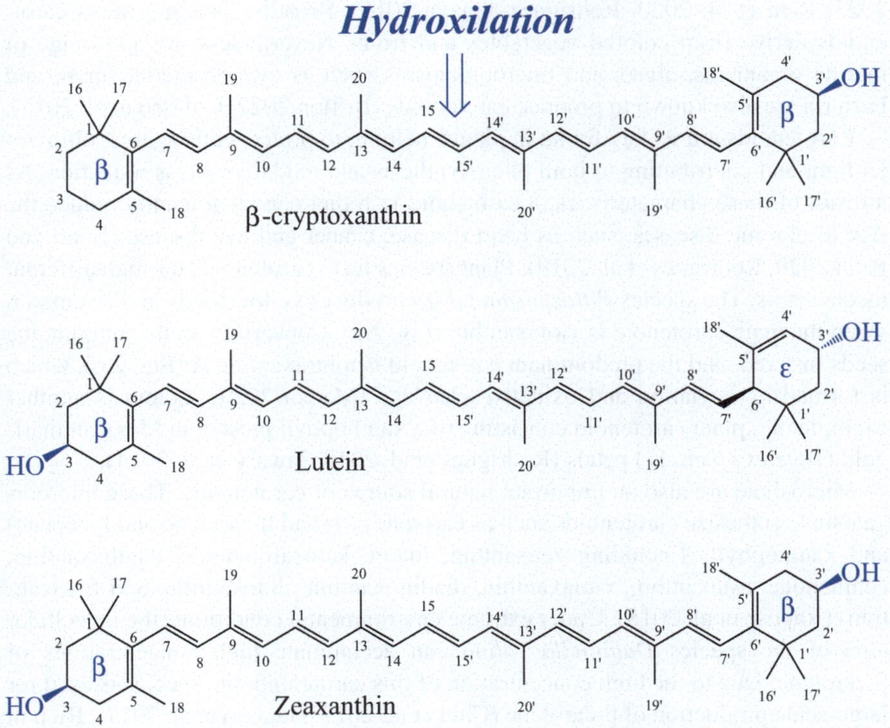

Fig. 2.4 Chemical structures of xanthophylls with hydroxyl groups

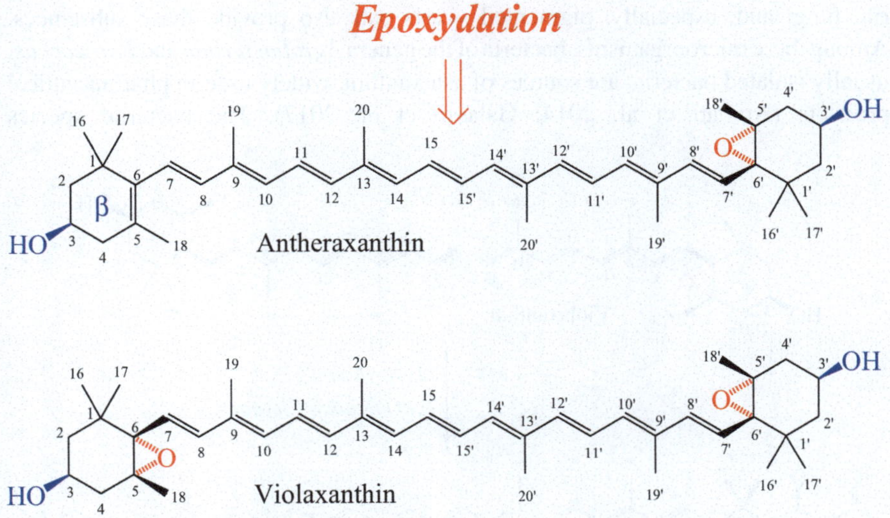

Fig. 2.5 Chemical structures of xanthophylls with epoxy groups

2023; Ram et al. 2020; Rodriguez-Amaya 2015). Broadly speaking, most carotenoids derive from colored vegetables and fruits. Nevertheless, a wide range of marine organisms, algae, and microorganisms such as cyanobacteria, fungi, and bacteria are also known to produce carotenoids (Britton 2022; Galasso et al. 2017).

Carotenoids are widely found in plants, helping to protect cells against ultraviolet light and contributing to both photosynthesis and oxidative stress reduction. As a result of these characteristics, a carotenoid-rich diet can significantly reduce the risk of chronic diseases, such as heart disease, cancer and eye disease (Bhatt and Patel 2020; Rodrigues et al. 2019). Plant species form carotenoids through different mechanisms. The species *Pittosporum tobira* produces yellow seeds in the summer, when the main carotenoid is violaxanthin (Fig. 2.5). Conversely, in the autumn, the seeds turn red, and the predominant carotenoid is tobiraxanthin A (Fig. 2.6), which is formed by acylation and oxidative cleavage (Maoka 2020). Lutein is another example of a plant carotenoid consisting of a xanthophyll present in Mexican marigold (*Tagetes erecta* L.) petals (Rodrigues et al. 2019; Tiwary et al. 2014).

Microalgae are also an important natural source of carotenoids. These microorganisms synthesize carotenoids such as carotenes (α- and β-carotene and lycopene) and xanthophylls (including zeaxanthin, lutein, ketocarotenoids, canthaxanthin, echinenone, astaxanthin, violaxanthin, diadinoxanthin, dinoxanthin, and fucoxanthin) (Raposo et al. 2015). Under extreme environmental conditions, the unicellular alga of the species *Dunaliella salina* can accumulate high concentrations of β-carotene. Due to the high concentration of this carotenoid, the species is used for large-scale production of β-carotene (Chiu et al. 2017; Galasso et al. 2017). Rich in carotenoids, especially tocopherol, the microalga *Tetraselmis suecica* has been suggested as a food supplement in human and animal diets (Sansone et al. 2017).

As with algae, which are sources of carotenoids, microorganisms such as bacteria, fungi and, especially, pigmented yeasts can also provide these substances. Among these microorganisms, bacteria of the genera *Agrobacterium* and *Paracoccus*, usually isolated bacteria, are sources of astaxanthin, widely used in pharmaceutical products (Ambati et al. 2014; Galasso et al. 2017). The bacterial species

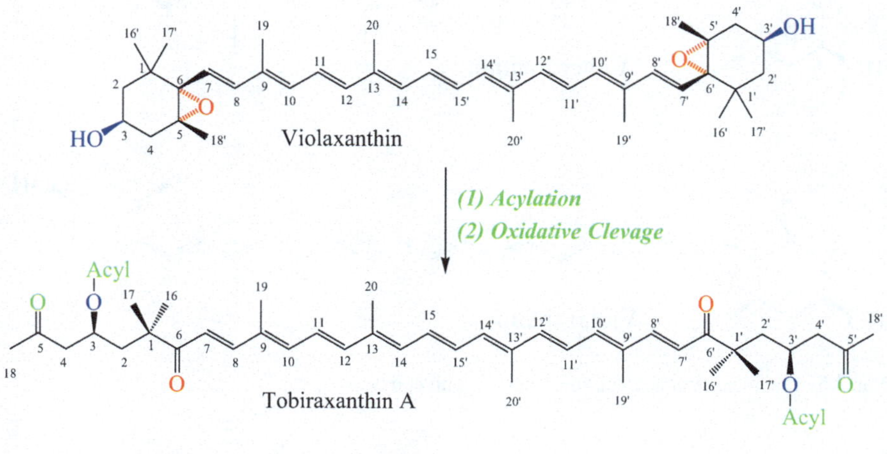

Fig. 2.6 Tobiraxanthin A from *Pittosporum tobira*

Flavobacterium spp., *Agrobacterium* spp., *Micrococcus* spp., *Pseudomonas aeruginosa*, *Serratia marcescens*, *Chromobacterium* spp., *Rheinheimera* spp., and *Arthrobacter* spp. also produce carotenoids such as astaxanthin, β-carotene, zeaxanthin, canthaxanthin, and lycopene, which are commonly used in food, feed and prophylactic agents against various diseases (Ram et al. 2020).

Various yeast species isolated from the environment synthesize carotenoids. IN particular, the genera *Rhodotorula mucilaginosa, Sporidiobolus pararoseus*, and *Pichia fermentans* are used to produce astaxanthin, lutein, and β-carotene. These microorganisms are a promising new source of biocompounds and demonstrate biotechnological potential for agro-industrial applications (Cipolatti et al. 2019; Allahkarami et al. 2021; Liu et al. 2019).

Marine animals such as oysters, clams and scallops accumulate carotenoids from foods, including algae and other animals, and modify them through metabolic reactions. Many carotenoids found in marine animals are β -carotene, fucoxanthin, peridinin, diatoxanthin, alloxanthin, and astaxanthin metabolites (Cardoso et al. 2017; Maoka 2020). The main carotenoid of diatoms is fucoxanthin, and the main carotenoids of shellfish include β-carotene, lutein A, zeaxanthin, diatoxanthin, alloxanthin, astaxanthin, pectenolone, pectenol A, mytiloxanthin, and mactraxanthin (Galasso et al. 2017; Maoka 2020).

Crustaceans such as crabs, shrimp and lobsters accumulate astaxanthin, which is synthesized from β-carotene ingested from algae. The purple, blue or yellow colors of crustaceans derive from proteins in the crustacyanin (Saini et al. 2018; Wade et al. 2017). Various fish species, such as salmonids, accumulate carotenoids and can synthesize astaxanthin from crustaceans, convert astaxanthin into zeaxanthin, and accumulate astaxanthin in their muscles, which accounts for their pinkish-red pigmentation (Maoka 2020; Vo et al. 2023). The carotenoids included in the diet of these organisms not only contribute to their color but also play a key role in their health, providing potentially beneficial properties for their survival.

Table 2.2 outlines some of the most common microorganism-based sources of natural carotenoids for food, pharmaceutical, cosmetic, and other purposes.

3 Carotenoid Biosynthesis in Plants

Carotenoid biosynthesis in plants involves two biochemically active isoprene structures, dimethylallyl diphosphate and isopentenyl diphosphate (Fig. 2.7). These units can be formed through two biosynthetic routes named after the corresponding intermediates mevalonic acid and methylerythritol phosphate (2-*C*-methyl-D-erythritol 4-phosphate) (Krause et al. 2020). Both biosynthetic routes occur in plants, but the enzymes of the mevalonic acid pathway are found in the cytosol, while the enzymes of the methylerythritol pathway are found in chloroplasts (Chen et al. 2018; Tetali 2019; Vranová et al. 2019).

The mevalonic acid pathway starts with the condensation of three acetyl-coenzyme A molecules involving an aldol reaction, yielding 3-hydroxy-3-methylglutaryl-coenzyme A. The thioester is then reduced to alcohol, forming mevalonic acid, which in

Table 2.2 Main sources of carotenoids

Carotenoid	Microorganisms	Industrial applications	References
Torulene and Torularhodin	*Rhodotorula mucilaginosa* (Yeasts)	Production of carotenoids for the treatment of various agro-industrial wastes.	Ghilardi et al. (2020)
Astaxanthin and β-Carotene	*Xanthophyllomyces dendrorhous* (Fungi)	Application in the agricultural industry.	Gervasi et al. (2018)
Astaxanthin	*Escherichia coli* (Bacteria)	Applications in the aquaculture, food, cosmetics, and pharmaceutical industries	Lu et al. (2017)
β-carotene	*Arthrospira platensis* Spirulina (Cyanobacteria)	Production of dietary supplements, animal nutrition and cosmetic products	Foong et al. (2021)
Fucoxanthin	*Phaeodactylum tricornutum* (Microalgae)	Dietary supplement production	
β-Carotene and Lycopene	*Fusarium* spp. (Fungi)	Production of natural pigments for applications in food, feed, and cosmetics.	Gmoser et al. (2017)
β-Carotene	*Dunaliella salina* (Chlorophyceae) (Microalgae)	Production in the food, pharmaceutical and cosmetics industries.	Chiu et al. (2017)
Lutein	*Chlorella protothecoides* (Chlorophyceae) (Microalgae)	Production of natural pigments for use in the food, cosmetics, and pharmaceutical industries.	Lin et al. (2015)

Dimethylallyl diphosphate Isopentenyl diphosphate

Fig. 2.7 Biochemically active structures of isoprene—dimethylallyl diphosphate and isopentenyl diphosphate

turn undergoes double phosphorylation in the primary alcohol, thereby generating mevalonic acid diphosphate. Subsequently, mevalonic acid diphosphate undergoes decarboxylation and dehydration of the tertiary alcohol, yielding isopentenyl diphosphate, whose isomerization gives rise to dimethylallyl diphosphate (Miziorko 2011; Dewick 2009). This biosynthetic route is shown in Fig. 2.8:

The methylerythritol phosphate pathway starts with pyruvic acid decarboxylation and its addition to the D-glyceraldehyde 3-phosphate. Both reactions are mediated by thiamine diphosphate, yielding 1-deoxy-D-xylulose 5-phosphate. After a sequence of rearrangements involving aldol and reduction reactions, 1-deoxy-D-xylulose 5-phosphate is converted into 2-C-methyl-D-erythritol 4-phosphate. The resulting intermediate then undergoes addition, phosphorylation, elimination and rearrangement reactions, which produce 4-hydroxy-3-methylbut-2-enyl

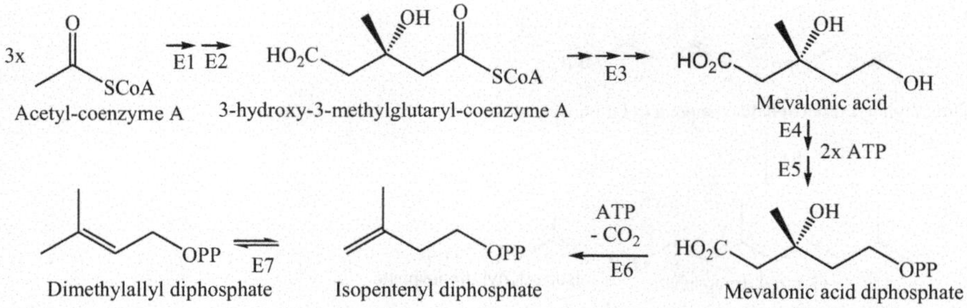

Fig. 2.8 Biosynthetic route of dimethylallyl diphosphate and isopentenyl diphosphate via the mevalonic acid pathway. E1: acetoacetyl-CoA synthase; E2: 3-hydroxy-3-methylglutaryl-CoA synthase; E3: 3-hydroxy-3-methylglutaryl-CoA reductase; E4: mevalonate kinase; E5: phosphomevalonate kinase; E6: mevalonate 5-diphosphate decarboxylase; E7: isopentenyl diphosphate isomerase

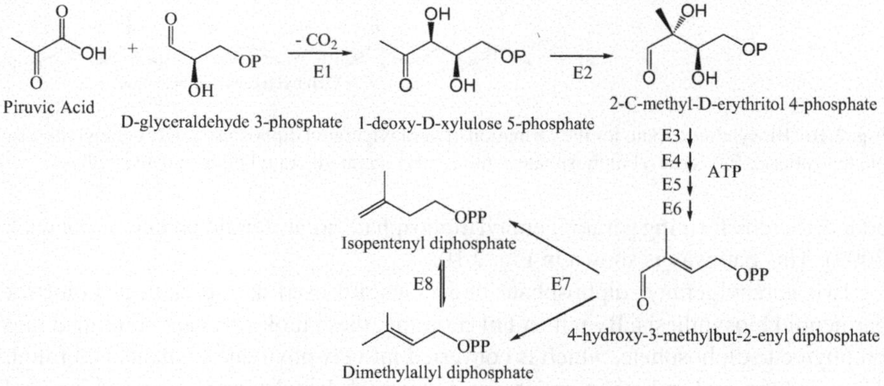

Fig. 2.9 Biosynthetic route of dimethylallyl diphosphate and isopentenyl diphosphate via the 2-C-methyl-D-erythritol 4-phosphate pathway. E1: 1-deoxy-D-xylulose 5-phosphate synthase; E2: 2-C-methyl-D-erythritol 4-phosphate synthase; E3: 4-diphosphocytidyl-2-C-methyl-D-erythritol synthase; E4: 4-diphosphocytidyl-2-C-methyl-D-erythritol kinase; E5: 2-C-methyl-D-erythritol-2,4-cyclodiphosphate synthase; E6: 4-hydroxy-3-methylbut-2-enyl diphosphate synthase; E7: 4-hydroxy-3-methylbut-2-enyl diphosphate reductase; E8: isopentenyl diphosphate isomerase

diphosphate. Ultimately, 4-hydroxy-3-methylbut-2-enyl diphosphate is reduced, yielding isopentenyl diphosphate and dimethylallyl diphosphate (Movahedi et al. 2022; Dewick 2009). This biosynthetic route is shown in Fig. 2.9:

Once the active isoprene units (isopentenyl diphosphate and dimethylallyl diphosphate) are formed, carotenoids can be synthesized via a sequence of combinations of these units. First, dimethylallyl diphosphate ionization generates an allylic cation, in addition to the leaving group diphosphate. Ionization is followed by addition to the double bond of isopentenyl diphosphate and by proton loss, forming geranyl diphosphate. The resulting molecule receives isopentenyl diphosphate, yielding farnesyl diphosphate to which a new isopentenyl diphosphate molecule is

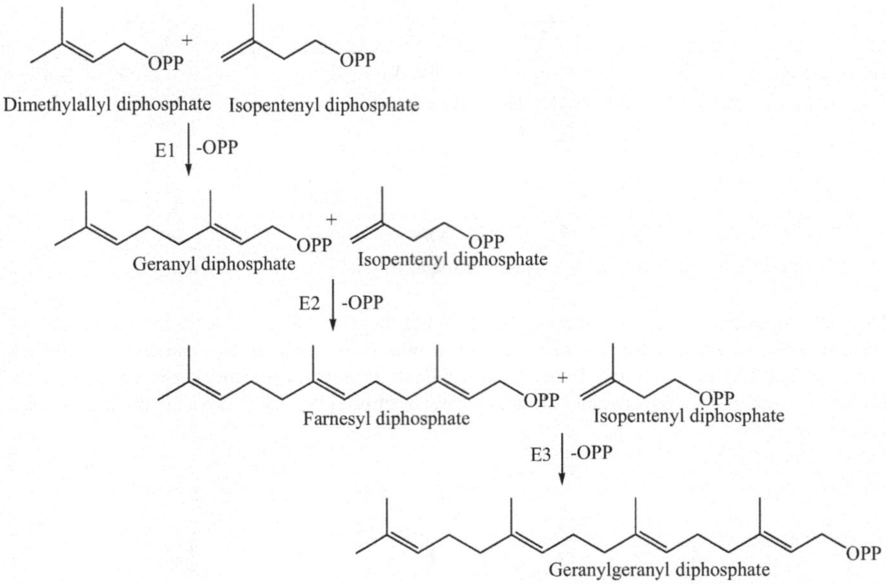

Fig. 2.10 Biosynthetic route for the formation of geranylgeranyl diphosphate. E1: geranyl diphosphate synthase; E2: farnesyl diphosphate synthase; E3: geranylgeranyl diphosphate synthase

added, thereby forming geranylgeranyl diphosphate, a carotenoid precursor (Dewick 2009). This pathway is shown in Fig. 2.10:

Two geranylgeranyl diphosphate molecules are used as the starting point for carotenoid biosynthesis. By tail-to-tail coupling, these molecules are combined into prephytoene diphosphate, which is converted into *cis*-phytoene in plants and fungi. The resulting molecule then undergoes a series of dehydrogenation reactions, and *cis-trans* isomerization of the central double bond yields the carotenoid lycopene, a pigment that is responsible for the red color of tomatoes (Dewick 2009). This biosynthetic route is shown in Fig. 2.11:

From lycopene, several carotenoids can be produced. Among them, the most commonly known carotenoids in plants are β-carotene, α-carotene, β-cryptoxanthin, zeaxanthin, lutein, and capsanthin (Fig. 2.12) (Quian-Ulloa and Stange 2021). The biosynthesis of these carotenoids involves cyclization and oxidation reactions at the end groups of lycopene promoted by cyclases, hydroxylases, ketolases, and epoxidases (Maoka 2020). The biosynthetic route of these and other carotenoids is shown in Fig. 2.13:

In plants, these carotenoids have various functions and are found in both photosynthetic and non-photosynthetic organs (Maoka 2020; Zia-Ul-Haq 2021). These essential compounds participate in photosynthesis and photoprotection, capturing light, transferring energy to chlorophylls and absorbing excess energy (Dhami et al. 2018; Sun et al. 2022). In addition, they also act as color attractants for pollinators by providing flowers and fruits with pigmentation. Therefore, carotenoids derived from lycopene play a key role in reproduction (Maoka 2020; Dhami and Cazzonelli 2020).

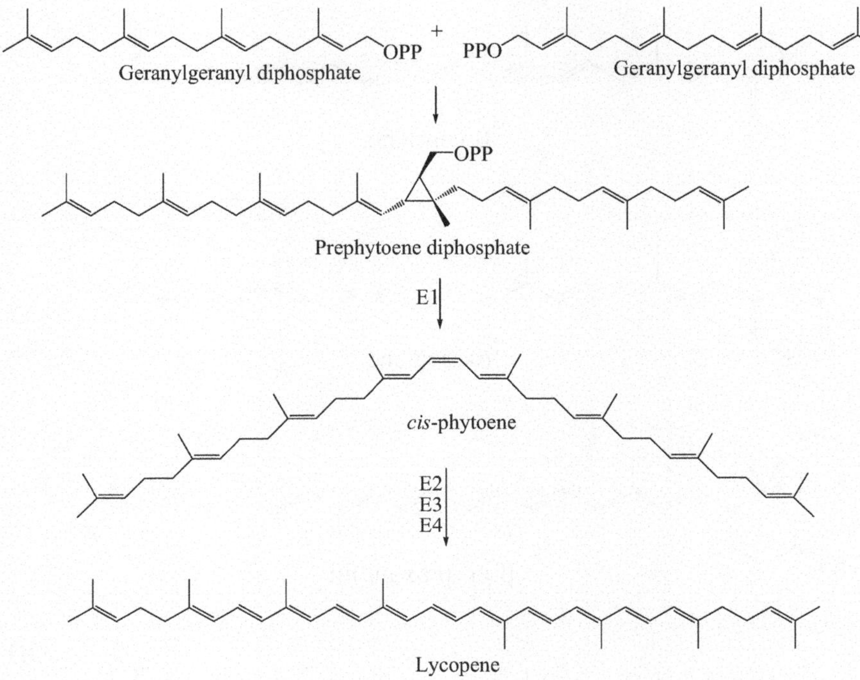

Fig. 2.11 Biosynthetic route for the formation of lycopene. E1: phytoene synthase; E2: phytoene desaturase; E3: ζ-carotene desaturase; E4: carotene isomerase

4 Biosynthesis of Carotenoids in Microorganisms: Bacteria and Fungi

Based on their chemical characteristics, carotenoids are defined as water-insoluble, lipophilic compounds with chromophoric groups. The main starting molecule for carotenoid biosynthesis is phytoene, a colorless carotenoid with 40 carbon atoms (Martínez-Cámara et al. 2021). Carotenoids are produced by phototrophic and non-phototrophic organisms, such as archaebacteria, cyanobacteria, algae, fungus, and plants and such substances provides these organisms with vibrant colors (Ashokkumar et al. 2023; González-Peña et al. 2023).

Currently, more than 1200 carotenoids and carotenoid precursors are known, deriving from 722 different organisms according to the Carotenoid Database (http://carotenoiddb.jp/index.html). Among them, 324 derive from bacterial sources, of which 251 are exclusively produced by these microorganisms (Yabuzaki 2017). Based on their chemical structure, carotenoids are classified into two main groups, namely carotenes, which contain only carbon and hydrogen, for example, lycopene, α-carotene, β-carotene, and torulene, and xanthophylls, which contain oxygen in addition to carbon and hydrogen, for example, torularhodin, lutein, fucoxanthin, and violaxanthin (Liu et al. 2021; Progress in Carotenoid Research 2018; Rodriguez-Concepcion et al. 2018).

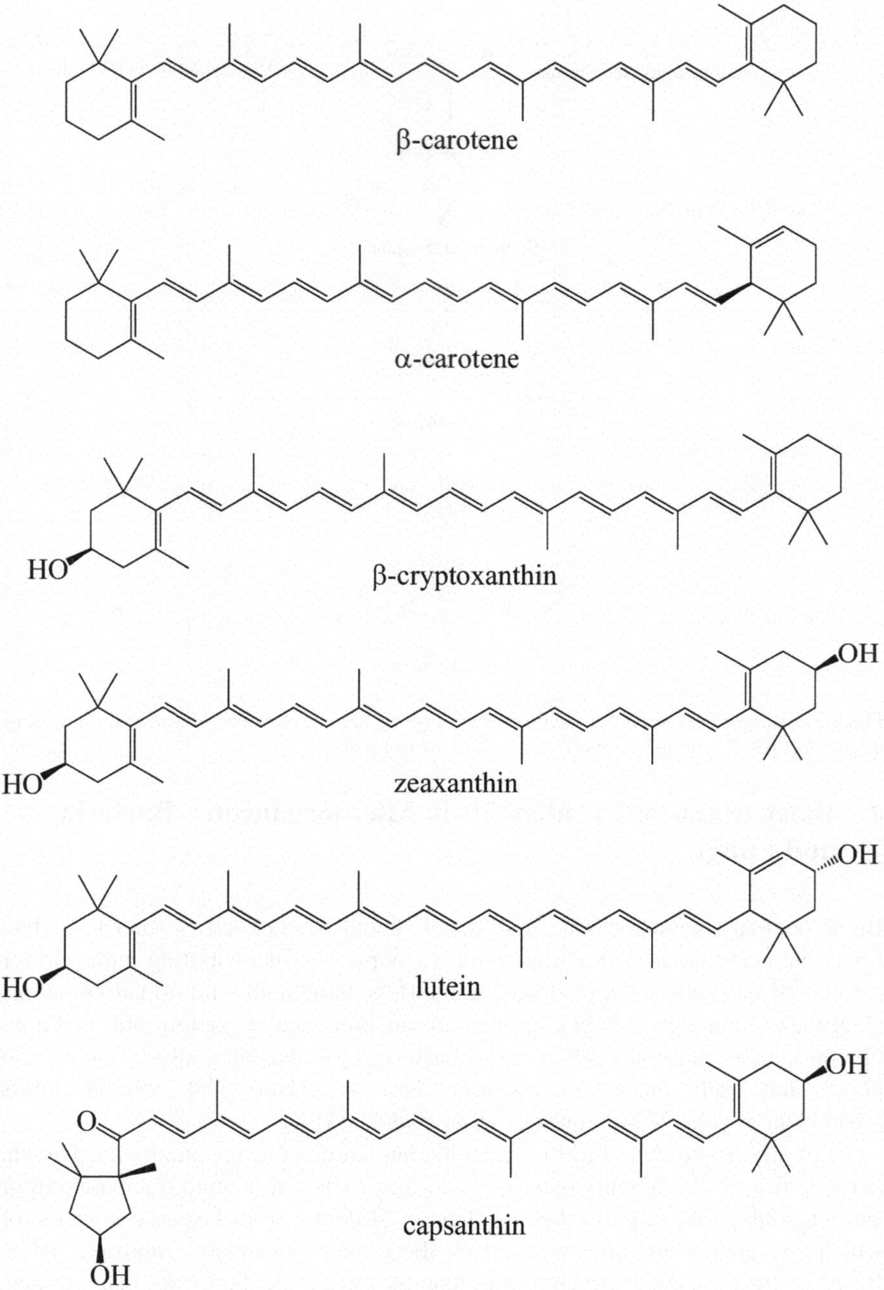

Fig. 2.12 Some of the most common carotenoids found in plants

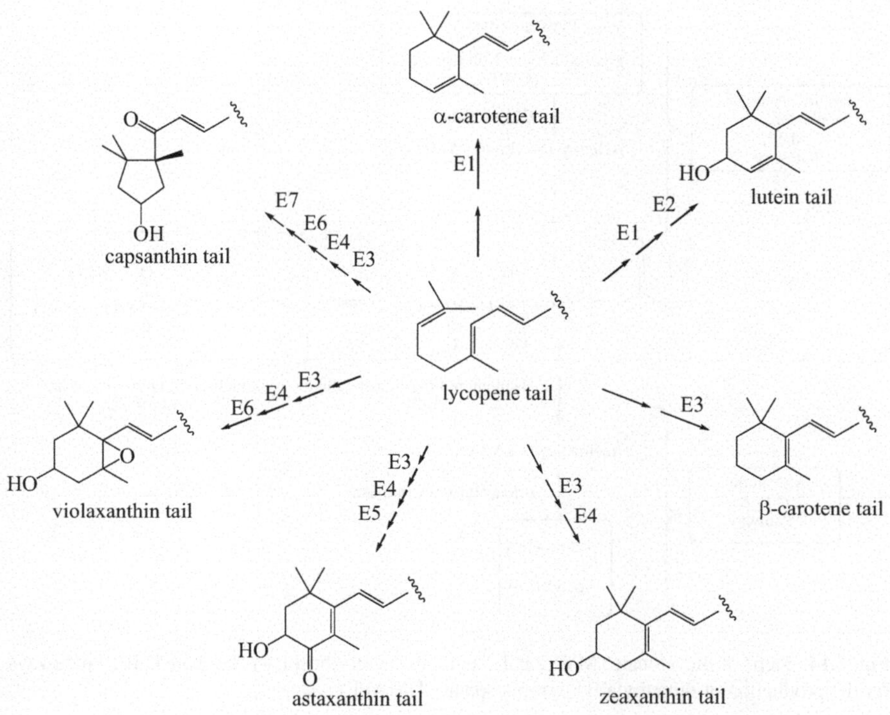

Fig. 2.13 Biosynthesis of carotenoids from lycopene tail. E1: ε-cyclase; E2: ε-ring hydroxylase; E3: β-cyclase; E4: β-ring hydroxylase; E5: β-carotene oxygenase; E6: ε-ring hydroxylase: E7: capsanthin synthase

Carotenoids can also be classified based on the number of carbon atoms found in their structure, most often C30, C40, C45 and C50. C40 are the most abundant naturally occurring carotenoids, and their chemical structure consists of a wide variety of end groups. C30 and C50 carotenoids are usually synthesized by archaebacteria and bacteria, and they contain 6 and 10 C5 isoprene units, respectively. In turn, only bacteria can synthesize C45 carotenoids, which are formed by 9 C5 isoprene units (Progress in Carotenoid Research 2018).

Broadly speaking, in carotenoid-producing organisms, biosynthesis involves the precursor isopentenyl pyrophosphate (IPP, C5) and its allylic isomer dimethylallyl pyrophosphate (DMAPP). Both dimethylallyl pyrophosphate and isopentenyl pyrophosphate are produced via 2-C-methyl-D-erythritol 4-phosphate (MEP), that is, through the mevalonate (MVA) pathway (Fig. 2.14). Thus, photosynthetic organisms, such as microalgae and plants, predominantly use the mevalonate pathway, whereas bacteria and fungi synthesize dimethylallyl pyrophosphate and isopentenyl pyrophosphate primarily through the 2-C-methyl-d-erythritol 4-phosphate (MEP) pathway (Saini and Keum 2019; Liu et al. 2021; Moise et al. 2014).

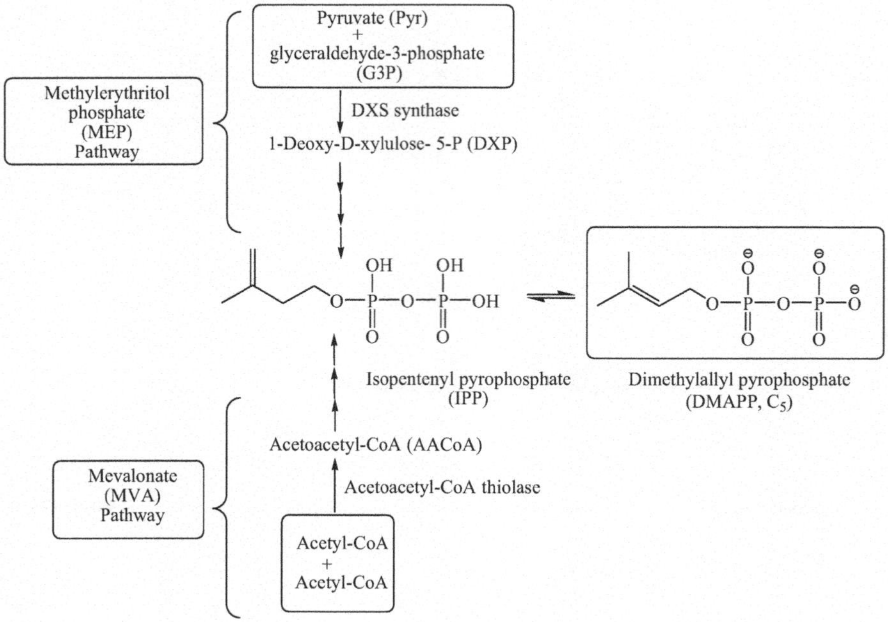

Fig. 2.14 Steps of mevalonate (MVA) and 2-C-methyl-d-erythritol 4-phosphate (MEP) pathways for the production of dimethylallyl pyrophosphate (DMAPP)

4.1 Bacterial Carotenoids

Bacteria can synthesize C40 derivates of β-carotene (Fig. 2.12) and γ-carotene, in addition to acyclic carotenoids, including spheroidene and spirilloxanthin (Fig. 2.15). C30 carotenoids, such as diaponeurosporene, are synthesized only in non-phototropic bacteria because they can condense two C15 isoprenoid units. C45 and C50 carotenoids, such as decaprenoxanthin (Fig. 2.15), are produced by some gram-positive Eubacteria and archaea (Liang et al. 2018). The main biosynthetic routes used to produce carotenoids in these microorganisms are the 2-C-methyl-D-erythritol-4-phosphate pathway and the mevalonate pathway. But while archaea use the mevalonate pathway to produce the isoprenoid precursor necessary for carotenoid production, many bacteria use the 2-C-methyl-D-erythritol-4-phosphate pathway. A small minority of bacteria use the mevalonate pathway, and only a few bacteria, including some *Streptomyces* species use both pathways (Rodriguez-Concepcion et al. 2018).

In bacteria, carotenoid biosynthesis starts from the C5 isoprenoid unit known as isopentenyl pyrophosphate. In the mevalonate pathway, isopentenyl pyrophosphate isomerizes, yielding dimethylallyl pyrophosphate. Initially, an isopentenyl pyrophosphate unit condenses with a dimethylallyl pyrophosphate unit, forming a C10-geranyl pyrophosphate unit, as shown in Fig. 2.10 (López et al. 2023). From C10-geranyl pyrophosphate, other units can be added, resulting in C15-farnesyl

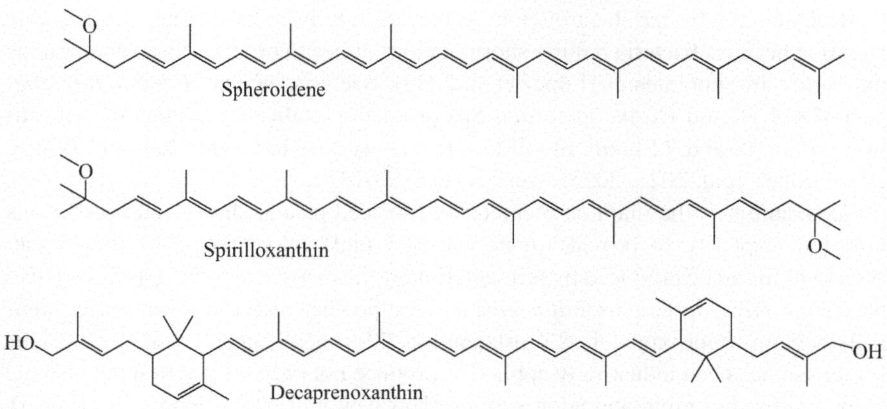

Fig. 2.15 Bacterial carotenoids Spheroidene, Spirilloxanthin, and Decaprenoxanthin

pyrophosphate and C20-geranylgeranyl pyrophosphate. After a sequence of desaturation reactions, catalyzed by the enzyme phytoene desaturase, and phytoene isomerization, the first physiologically relevant carotenoid, lycopene, is synthesized, as shown in Fig. 2.11 (Saini and Keum 2019; Liang et al. 2018; Quian-Ulloa and Stange 2021).

Carotenoids derived from C30 and C40 chains follow different biosynthetic routes involving specific bacterial Crt gene clusters (Takemura et al. 2021). C30 carotenoids are synthesized by addition of two C15-farnesyl pyrophosphate units catalyzed by the enzyme CrtM, bacterial carotenoid synthase. Conversely, C40 carotenoids are synthesized by condensation of two C20-geranylgeranyl pyrophosphate units catalyzed by the bacterial enzyme geranylgeranyl pyrophosphate synthase (CrtE). The precursor of the C30 carotenoids is dehydrosqualene, whose synthesis is catalyzed by the bacterial enzyme carotenoid desaturase (CrtN), whereas the precursor of C40 carotenoids is 15-cis-phytoene, whose synthesis is catalyzed by the bacterial enzyme carotenoid synthase (CrtB—phytoene synthase). During the biosynthesis of other carotenoids, desaturations are mediated by the enzymes CrtN and CrtI, bacterial desaturases that act specifically on the biosynthetic routes of C30 and C40 carotenoids, respectively (Liang et al. 2018; Ashokkumar et al. 2023; López et al. 2023).

Lycopene (ψ,ψ-carotene) is a common intermediate for the biosynthesis of almost all other C40 carotenoids. After its synthesis, the ends of the lycopene chain cyclized, yielding β or ε rings (Fig. 2.13). Some enzymes are responsible for cyclizing the chain ends of lycopene, including lycopene β-cyclase and lycopene ε-cyclase (Martínez-Cámara et al. 2021; Saini and Keum 2019). Depending on the type of bacteria, such as those of the *Streptomyces* genus, C5 units may be added through the 2-C-methyl-D-erythritol-4-phosphate pathway. The mevalonate and 2-C-methyl-D-erythritol-4-phosphate units are synthesized during bacterial glycolysis (Moise et al. 2014; Liang et al. 2018).

Research on bacterial carotenoid synthesis has been attracting considerable attention because bacteria require shorter culture times to produce these compounds than other microorganisms (López et al. 2023). Bacteria such as *Halobacillus trueperi* MXM-16 and *Exiguobacterium* Sps produce significant amounts of astaxanthin within 48 and 72 hours of culture, respectively (Kharangate-Lad and Bhosle 2016; López et al. 2021, 2023; Arulselvi et al. 2013).

Astaxanthin is the main carotenoid used to feed trout, salmon and crustaceans raised in captivity to provide them with the pink color typical of their meat. Astaxanthin can be produced by fermentation by the marine, aerobic, Gram-negative bacterium *Paracoccus carotinifaciens*, and the powder resulting from drying their cells contains approximately 2% astaxanthin (Martínez-Cámara et al. 2021). This bacterium has been industrially applied to produce not only astaxanthin but also the carotenoids adonirubin and adonixanthin (Hayashi et al. 2021; Honda et al. 2020, 2022). Structures of carotenoids derived from *Paracoccus carotinifaciens* are shown in Fig. 2.16.

Other bacterial species produce significant amounts of carotenoids, especially canthaxanthin. This substance holds a considerable share of the carotenoid market because canthaxanthin has been widely used as a colorant in dietary supplements and food additives, in hygiene and cleaning products and in pharmaceutical applications (Gaur and Bera 2023). Marine bacteria have been extensively studied as a source of the carotenoid canthaxanthin, especially the species *Paracoccus* sp., *Dietzia* sp., *Gordonia* sp., *Brevibacterium* sp., *Agrobacterium* sp. and *Micrococcus* sp., (Choi et al. 2021; Venil et al. 2021; Nidhi et al. 2023).

The following bacterial strains also stand out in carotenoid production: *Micrococcus roseus*, *Brevibacterium* spp., *Corynebacterium michiganense*, and *Bradyrhizobium* spp. (Watcharawipas and Runguphan 2023). Concurrently, bacteria have emerged as an alternative source of carotenoids because different species produce a wide variety of compounds. Moreover, bacteria can be considered a renewable source of carotenoids (Ashokkumar et al. 2023; López et al. 2021, 2023).

4.2 Fungal Carotenoids

In fungi and yeasts, carboxylic and monocyclic carotenoids, such as torulene and torularhodin, are the most commonly found carotenoids (Liang et al. 2018). In these organisms, carotenoid production is photo-regulated, which highlights the antioxidant importance of these compounds as their production is associated with a light-dependent process (Sandmann 2022).

Several fungal and yeast species can synthesize and intracellularly accumulate high amounts of carotenoids. Among yeasts commercially explored for carotenoid bioproduction, the following stand out: *Xanthophyllomyces dendrorhous*, *Rhodotorula glutinis*, *Rhodotorula mucilaginosa*, *Sporobolomyces*, and *Phaffia rhodozyma*. Asexual fungi of the genera *Zygomycetes*, *Ascomycetes*, *Basidiomycetes*

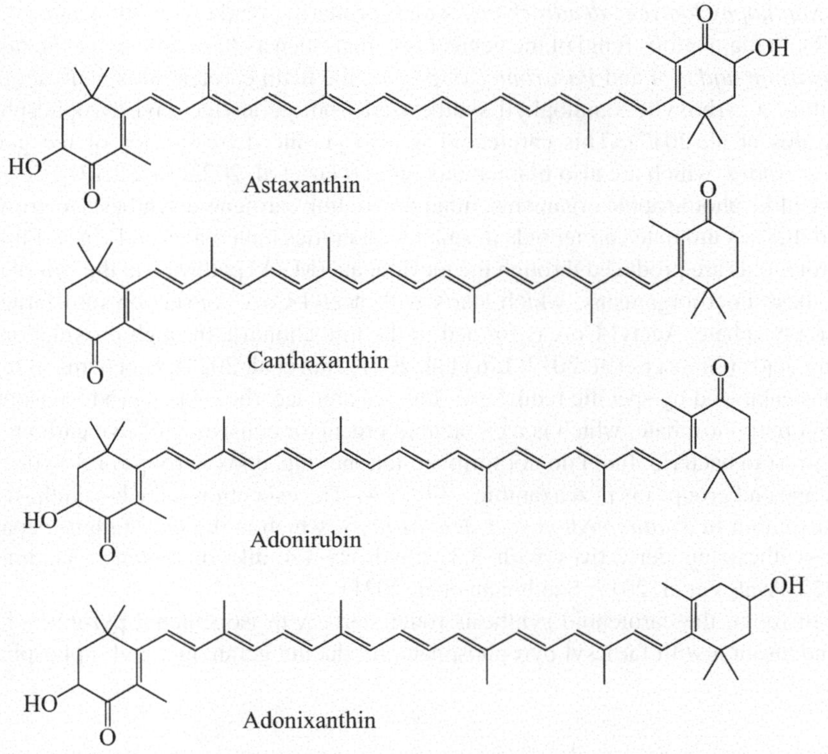

Fig. 2.16 Bacterial carotenoids from *Paracoccus carotinifaciens*

and *Deuteromycetes* also produce larger amounts of carotenoids than other organisms (López et al. 2023).

The genus *Rhodotorula* encompasses yeasts that produce carotenoids, primarily β-carotene (Fig. 2.12), torulene and torularhodin. The amount of carotenoids that these yeasts produce depends on the strain and culture conditions. In *Rhodotula glutinis* and *Rhodotula graminis*, γ-carotene accounts for 11–15% of the total carotenoids produced by these species (Liu et al. 2021).

Rhodosporidium species are also considered promising strains for producing commercial carotenoids, including β-carotene, torulene, and torularhodin. Species such as *R. toruloides, R. diobovatum*—known as red yeast strains—and *R. glutinis* have been studied for their high yield of these substances (Liu et al. 2021; Zheng et al. 2021; Landolfo et al. 2018). A study using *R. glutinis* strains and a bioreactor under adapted aeration and agitation conditions showed a 25.8, 11.88 and 24.5% (m/v) increase in β-carotene, torularhodin, and torulene yield, respectively (Mussagy et al. 2021; Landolfo et al. 2018).

One of the most important commercially produced carotenoids is astaxanthin for its prominent antioxidant potential resulting from the 13 conjugated double bonds and hydroxyl groups of the β-ionone rings. The main fungal source of astaxanthin is *Xanthophyllomyces dendrorhous*, which primarily synthesizes the isomer (3R, 3'R). In filamentous fungi of the genus *Fusarium*, such as *Fusarium aquaeductuum*, *Fusarium fujikuroi* and *Fusarium oxysporum*, the main carotenoid is neurosporaxanthin, a carboxylic xanthophyll synthesized from geranylgeranyl pyrophosphate (Avalos et al. 2017). This carotenoid is also produced by species of the genus *Neurospora*, which are also filamentous fungi (Naz et al. 2023; Fig. 2.17).

Unlike phototrophic organisms, fungi do not link carotenoid synthesis to growth, and they accumulate carotenoids in smaller quantities than plants and algae. Fungal carotenoids are produced through the mevalonate (MVA) pathway, in the cytoplasm of these microorganisms, which starts with acetyl-CoA conversion into farnesyl pyrophosphate. Acetyl-CoA is formed in the mitochondria through β-oxidation of fatty acids (Gmoser et al. 2017; Liu et al. 2021; Naz et al. 2023). Biochemical reactions catalyzed by specific reductases, kinases and decarboxylases produce isopentenyl pyrophosphate, which is a carotenoid precursor consisting of five carbons. In contrast to bacteria, fungi do not form an -ionone ring, epoxy groups or 3-hydroxyionone end groups (as in zeaxanthin—Fig. 2.4). The exception is the biosynthesis of astaxanthin in *Xanthophyllomyces dendrorhous*, which is the only fungus capable of synthesizing derivatives with 3,3′-dihydroxy-4,4′-diketo-carotene (Sandmann 2022; Avalos et al. 2017; Sandmann et al. 2021).

In fungi, the carotenoid synthesis route starts with isopentenyl pyrophosphate condensation with farnesyl pyrophosphate, producing geranylgeranyl diphosphate.

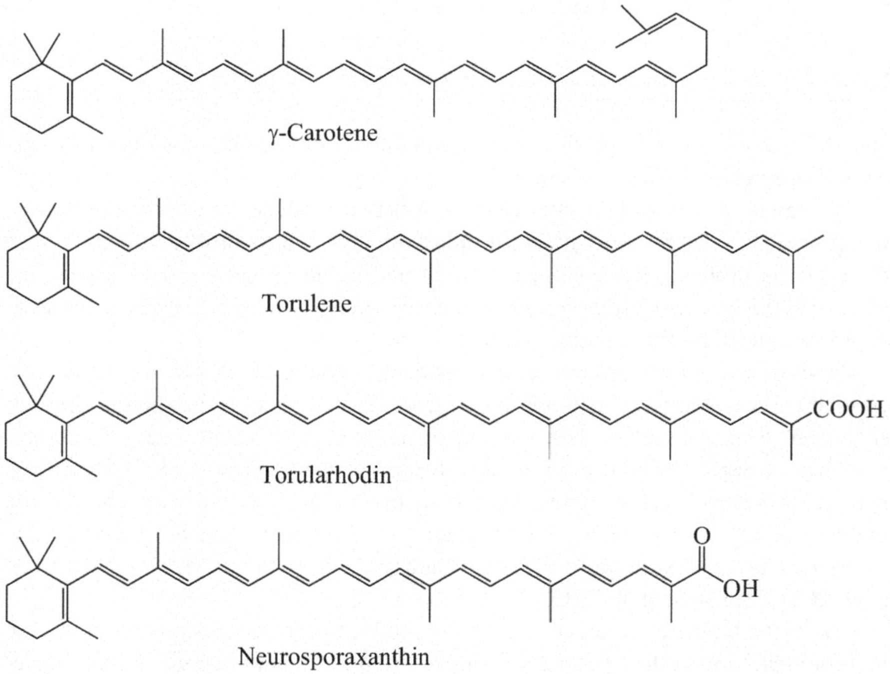

Fig. 2.17 Fungi and yeast carotenoids γ-carotene, torulene, torularhodin and neurosporaxanthin

Two geranylgeranyl diphosphate units are then condensed by the enzyme carotenoid synthase (CrtI). Subsequently, phytoene is dehydrogenated by the bifunctional enzyme phytoene synthase-lycopene cyclase (CrtYB), yielding lycopene. This enzyme (CrtYB) is responsible for both the dehydrogenation and cyclization of the rings at the end of the chain, producing γ-carotene (one β-ionone ring) and β-carotene (two β-ionone rings). The carotene torulene, for example, is synthesized after γ-carotene dehydrogenation. Torulene hydroxylation and oxidation reactions form a carboxylic acid end group, thus synthesizing torularhodin. Another pathway, exclusive to *X. dendrorhous*, involves astaxanthin synthase (CrtS), a cytochrome P450 enzyme. From the β-carotene, this enzyme forms astaxanthin (Sandmann 2022; Avalos et al. 2017; Watcharawipas and Runguphan 2023; Liang et al. 2018; Naz et al. 2023).

5 Biosynthesis of Carotenoids: Retinol—A_1 and Dehydroretinol—A_2—Metabolites Derived from Carotenoids

Vitamins are essential compounds for human and animal health, as proposed by the Polish biochemist Casimir Funk (Semba, 2012), who stated that as deficiency diseases could be prevented and treated by adding some substances to the diet, which he termed "vitamins". Elmer McCollum, in 1914, added that vitamins were not just a compound but a set of compounds, and he was one of the first scientists to use letters to differentiate these compounds. Among these compounds, vitamin A (retinol), Fig. 2.18, was discovered in 1916. In 1931, vitamin A was isolated, and its chemical structure was determined. The term vitamin A is broadly used to refer to a set of molecules with biological activity based on the compound retinol. These compounds are monocyclic diterpenes often termed retinoids (Bonrath et al. 2023).

Retinol is the main active form of vitamin A (also known as vitamin A1), supports growth and development, promotes collagen regeneration, and maintains the normal shape and functions of bones and epithelial tissues, in addition to improving vision and immunity (Wang et al. 2023). The human body cannot synthesize vitamin A, but vitamin A can be introduced into the diet from animal products, such as eggs, fish, dairy products and meat. Alternatively, provitamin A compounds can be converted into vitamin A in the body, and these compounds are usually found in plants. The most important carotenoids are β-carotene, α-carotene and β-cryptoxanthin, Fig. 2.18 (Maurya et al. 2022).

In nature, vitamin A results from the degradation of carotenoids termed provitamin A, which are naturally synthesized in photosynthetic organisms and can release at least one retinal molecule under catalysis by the carotenoid cleavage enzyme dioxygenase (Hu et al. 2022). Carotenoid cleavage enzymes are a family of enzymes that can metabolize carotenoids and are classified into central and eccentric cleavage enzymes based on the position of the asymmetric cleavage site. β-carotene

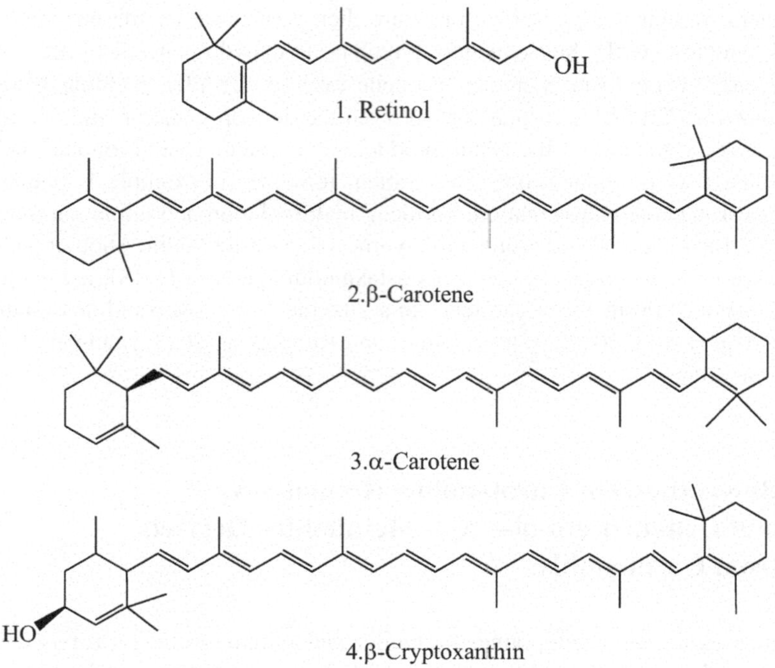

Fig. 2.18 Two-dimensional structures of carotenoids retinol, β-carotene, α-carotene and β-cryptoxanthin

oxygenases (BCOs) play a key role in carotenoid metabolism in vertebrates and are divided into two types, namely β-carotene 15,15′-oxygenases (BCO1, EC 1.13.11.63) and β-carotene 9′,10′-oxygenases (BCO2, EC 1.13.11.71) based on the cleavage position of β-carotene (Shin et al. 2022).

BCO1 are important for retinoid production, acting on the 15,15′ double bond of β-carotene and thus generating all-*trans*-retinal, Fig. 2.19 (route a). All-*trans*-retinal can be converted into all-*trans*-retinol by the enzyme Retinal reductase with cellular retinol-binding protein (CRBP) and subsequently into retinyl esters for storage by the enzyme Lecithin:retinol acyltransferase (LRAT, EC 2.3.1.135) (Redfern 2020).

Retinol can also be oxidized to retinal by the enzyme Retinol dehydrogenase, and then retinal is irreversibly oxidized by the enzyme Retinal dehydrogenase, yielding retinoic acid. Subsequently, retinoic acid is oxidized to more water-soluble hydroxy and oxo forms by one of several cytochromes of the P450 enzyme (O'Byrne and Blaner 2013).

BCO2 act on the double bond 9,10′ of β-carotene and produce ionone and β-apo-10′-carotenal compounds, Fig. 2.19 (route b), which can subsequently cleaved by the enzyme BCO1, yielding other retinoids, including retinol and retinoic acid (Jin et al. 2020).

Traditionally, research and nutrition have focused on vitamin A1 and its derivatives, whereas vitamin A2 (dehydroretinol), Fig. 2.20, have been overlooked because

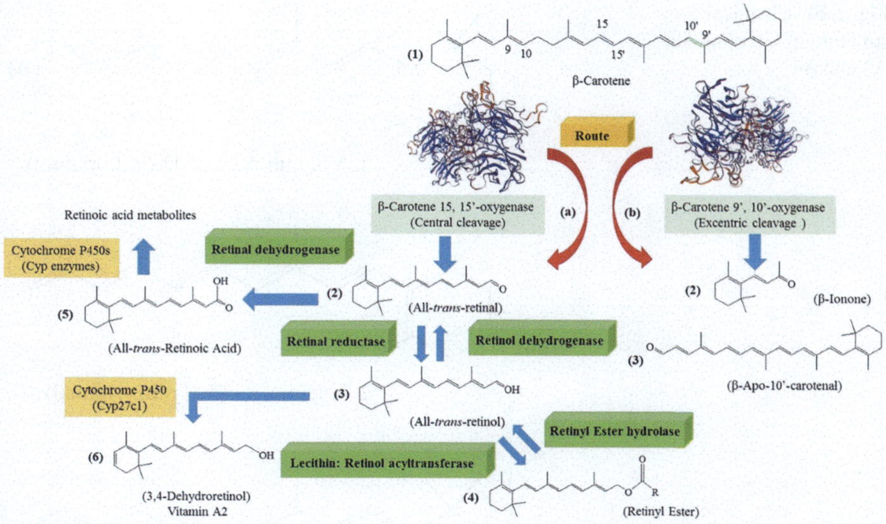

Fig. 2.19 Metabolic cleavage and conversion of β-carotene (1). Through pathway (a), the formation of retinal (2) is due to the enzyme β-Carotene 15,15′-oxygenase (BCO1, Uniprot code: Q9HAY6). Retinol (3) is formed by the reversible reduction of retinal due to the action of the enzyme Retinal reductase. The enzyme Lecithin: Retinol acyltransferase synthesizes retinyl esters (4). The enzyme Retinal dehydrogenase is responsible for converting retinal (2) to retinoic acid (5). 3,4-dehydroretinol (6) comes from retinol and is produced by the enzyme Cyp27c1. In pathway (b) the compounds β-Ionone (2) and β-Apo-10'carotenal (3) are formed by the action of the enzyme β-Carotene 9′,10′-oxygenase (BCO2, Uniprot code: Q9BYV7)

they are less physiologically and nutritionally relevant for the human organism, in contrast to birds and fish. For these animals, vitamin A2 is highly physiologically important and nutritionally relevant. Vitamins A3 and A4, Fig. 2.20, are retinal analogs and serve as visual pigments for arthropods or crustaceans (Bohn et al. 2023).

In humans, a considerable fraction of the retinoid pool in the skin derives from vitamin A2, which can be produced locally by keratinocytes. Keratinocytes can convert vitamin A1 into A2 in cell cultures, and vitamin A2 derivatives are also used as a visual chromophore in cold-blooded vertebrates and some reptiles. The use of this chromophore is associated with freshwater environments because some animals that migrate from freshwater to marine environments dynamically change from vitamin A1- to vitamin A2-based chromophores (Kramlinger et al. 2016).

Vitamin A2 is a constituent of the visual pigment porphyropsin and a partner of the protein rhodopsin in the retina of the eyes of teleost fish. Vitamin A2 is usually the retinoid and predominant visual pigment of freshwater fish, whereas retinol is found in marine fish (Green and Fascetti 2016).

In some vertebrate species (*Petromyzon marinus*, *Oncorhynchus kisutch*, *Lithobates pipiens*) vitamin A1 is converted into vitamin A2, Fig. 2.18, by a member of the cytochrome P450 family, Cyp27c1. This conversion enables these species to extend their visual sensitivity beyond the range of human vision, which improves

Fig. 2.20 Chemical structure of vitamins A2, A3 and A4

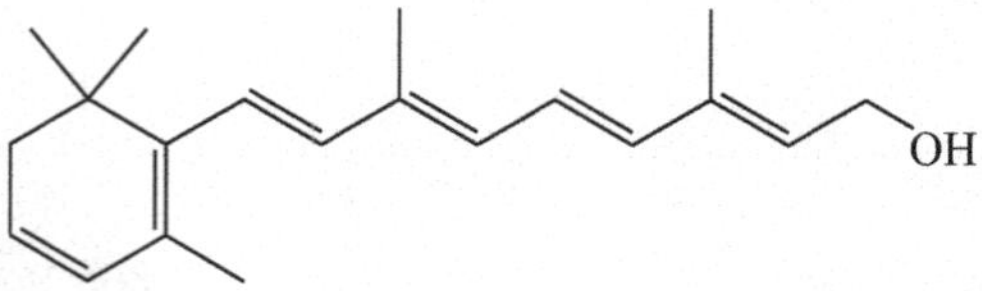

1. Vitamin A2 (3,4-Dehydroretinol)

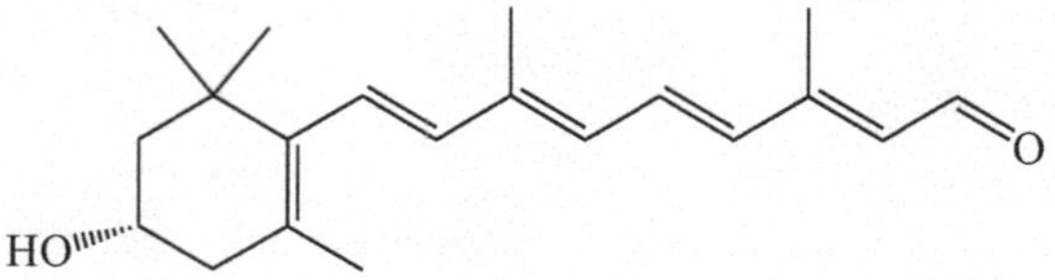

2. Vitamin A3 (3-Hydroxyretinal)

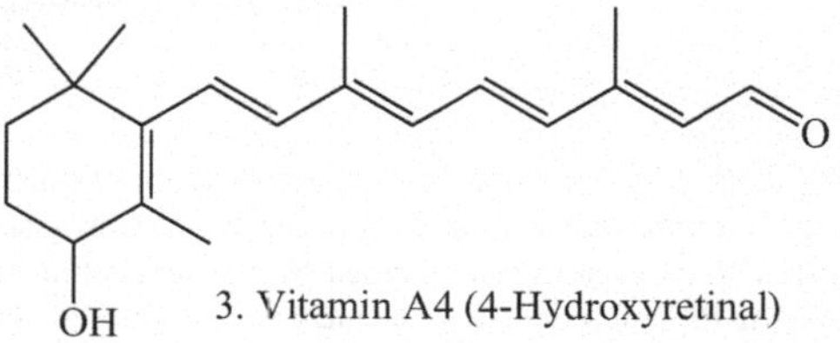

3. Vitamin A4 (4-Hydroxyretinal)

their perceptual capacity (Enright Jennifer et al. 2015). Cytochrome P450 (P450 or Cyp) 27c1 stands out as a mitochondrial enzyme present in human skin that catalyzes all-trans-retinoid desaturation to 3,4-dehydroretinoids, which serve as a potential source of biologically active retinoids after exposure to ultraviolet (UV) rays. This is the main function of skin 3,4-dehydroretinoids (Glass et al. 2023).

References

Allahkarami S, Sepahi AA, Hosseini H, Razavi MR (2021) Isolation and identification of carotenoid-producing Rhodotorula sp. from Pinaceae forest ecosystems and optimization of in vitro carotenoid production. Biotechnol Rep 32:e00687. https://doi.org/10.1016/j.btre.2021.e00687

Ambati RR, Phang S-M, Ravi S, Aswathanarayana RG (2014) Astaxanthin: sources, extraction, stability, biological activities and its commercial applications: a review. Mar Drugs 12(1):128–152

Amengual J (2019) Bioactive properties of carotenoids in human health. Nutrients 11(10):2388. https://doi.org/10.3390/nu11102388

Arulselvi I, Sasidharan P, Raja R, Karthik C, Gurumayum RS (2013) Isolation and characterization of yellow pigment producing Exiguobacterium sps. J Biochem Technol 4(4):632–635

Ashokkumar V, Flora G, Sevanan M, Sripriya R, Chen WH, Park J-H, Rajesh Banu J, Kumar G (2023) Technological advances in the production of carotenoids and their applications– a critical review. Bioresour Technol 367:128215. https://doi.org/10.1016/j.biortech.2022.128215

Avalos J, Pardo-Medina J, Parra-Rivero O, Ruger-Herreros M, Rodríguez-Ortiz R, Hornero-Méndez D, Carmen Limón M (2017) Carotenoid biosynthesis in fusarium. J Fungi 3(3):39. https://doi.org/10.3390/jof3030039

Bhatt T, Patel K (2020) Carotenoids: potent to prevent diseases review. Nat Prod Bioprospect 10:109–117

Bogacz-Radomska L, Harasym J, Piwowar A (2020) 10 – Commercialization aspects of carotenoids. In: Galanakis CM (ed) Carotenoids: properties, processing and applications. Academic, pp 327–357. https://doi.org/10.1016/B978-0-12-817067-0.00010-5

Bohn T, Bonet ML, Borel P, Keijer J, Landrier J-F, Milisav I, Ribot J, Riso P, Winklhofer-Roob B, Sharoni Y, Corte-Real J, van Helden Y, Loizzo MR, Poljesak B, Porrini M, Roob J, Trebese P, Tundis R, Wawrzyniak A, Ruhl R, Dulienska-Litewka J (2021) Mechanistic aspects of carotenoid health benefits – where are we now? Nutr Res Rev 34(2):276–302. https://doi.org/10.1017/S0954422421000147

Bohn T, Hellman-Regen J, de Lera AR, Böhm V, Rühl R (2023) Human nutritional relevance and suggested nutritional guidelines for vitamin A5/X and provitamin A5/X. Nutr Metab 20(1):34. https://doi.org/10.1186/s12986-023-00750-3

Bonrath W, Gao B, Houston P, McClymont T, Müller M-A, Schäfer C, Schweiggert C, Schütz J, Medlock JA (2023) 75 Years of vitamin a production: a historical and scientific overview of the development of new methodologies in chemistry, formulation, and biotechnology. Org Process Res Dev 27(9):1557–1584. https://doi.org/10.1021/acs.oprd.3c00161

Britton G (2022) Chapter one – getting to know carotenoids. In: Wurtzel ET (ed) Methods in enzymology, vol 670. Academic, Cambridge, pp 1–56. https://doi.org/10.1016/bs.mie.2022.04.005

Cardoso LAC, Karp SG, Vendruscolo F, Kanno KYF, Zoz LIC, JlC C (2017) Biotechnological production of carotenoids and their applications in food and pharmaceutical products. Carotenoids 8:125–141

Carotenoids Database (2015). http://carotenoiddb.jp/

Chen N, Wang P, Li C, Wang Q, Pan J, Xiao F, Wang Y, Zhang K, Li C, Yang B, Sun C, Deng X (2018) A single nucleotide mutation of the IspE gene participating in the MEP pathway for isoprenoid biosynthesis causes a green-revertible yellow leaf phenotype in rice. Plant Cell Physiol 59(9):1905–1917. https://doi.org/10.1093/pcp/pcy108

Chiu HF, Liao JY, Lu YY, Han YC, Shen YC, Venkatakrishnan K, Golovinskaia O, Wang CK (2017) Anti-proliferative, anti-inflammatory and pro-apoptotic effects of Dunaliella salina on human KB oral carcinoma cells. J Food Biochem 41(3):e12349

Choi SS, Seo YB, Nam S-W, Kim G-D (2021) Enhanced production of astaxanthin by co-culture of Paracoccus haeundaensis and lactic acid bacteria. Front Mar Sci 7. https://doi.org/10.3389/fmars.2020.597553

Cipolatti EP, Remedi RD, dos Santos SC, Rodrigues AB, Ramos JMG, Burkert CAV, Furlong EB, de Medeiros Burkert JF (2019) Use of agroindustrial byproducts as substrate for production of carotenoids with antioxidant potential by wild yeasts. Biocatal Agric Biotechnol 20:101208

Dewick PM (2009) The mevalonate and methylerythritol phosphate pathways: terpenoids and steroids. In: Medicinal natural products, pp 187–310. https://doi.org/10.1002/9780470742761.ch5

Dhami N, Cazzonelli CI (2020) Environmental impacts on carotenoid metabolism in leaves. Plant Growth Regul 92(3):455–477. https://doi.org/10.1007/s10725-020-00661-w

Dhami N, Tissue DT, Cazzonelli CI (2018) Leaf-age dependent response of carotenoid accumulation to elevated CO2 in Arabidopsis. Arch Biochem Biophys 647:67–75. https://doi.org/10.1016/j.abb.2018.03.034

Enright Jennifer M, Toomey Matthew B, Sato S-y, Temple Shelby E, Allen James R, Fujiwara R, Kramlinger Valerie M, Nagy Leslie D, Johnson Kevin M, Xiao Y, How Martin J, Johnson Stephen L, Roberts Nicholas W, Kefalov Vladimir J, Guengerich FP, Corbo Joseph C (2015) Cyp27c1 red-shifts the spectral sensitivity of photoreceptors by converting vitamin A1 into A2. Curr Biol 25(23):3048–3057. https://doi.org/10.1016/j.cub.2015.10.018

Fernandes AS, Do Nascimento TC, Jacob-Lopes E, De Rosso VV, Zepka LQ (2018) Carotenoids: a brief overview on its structure, biosynthesis, synthesis, and applications, vol 1. pp 1–17. https://doi.org/10.5772/intechopen.179542

Foong LC, Loh CWL, Ng HS, Lan JC-W (2021) Recent development in the production strategies of microbial carotenoids. World J Microbiol Biotechnol 37(1):12. https://doi.org/10.1007/s11274-020-02967-3

Galasso C, Corinaldesi C, Sansone C (2017) Carotenoids from marine organisms: biological functions and industrial applications. Antioxidants 6(4):96

Gaur V, Bera S (2023) Microbial canthaxanthin: an orange-red keto carotenoid with potential pharmaceutical applications. Biotechnologia 104(3):315–328. https://doi.org/10.5114/bta.2023.130733

Gervasi T, Pellizzeri V, Benameur Q, Gervasi C, Santini A, Cicero N, Dugo G (2018) Valorization of raw materials from agricultural industry for astaxanthin and β-carotene production by Xanthophyllomyces dendrorhous. Nat Prod Res 32(13):1554–1561. https://doi.org/10.1080/14786419.2017.1385024

Ghilardi C, Sanmartin Negrete P, Carelli AA, Borroni V (2020) Evaluation of olive mill waste as substrate for carotenoid production by Rhodotorula mucilaginosa. Bioresour Bioprocess 7(1):52. https://doi.org/10.1186/s40643-020-00341-7

Glass SM, Tateishi Y, Guengerich FP, Wang H-J (2023) 3,4-Desaturation of retinoic acid by cytochrome P450 27C1 prevents P450-mediated catabolism. Arch Biochem Biophys 743:109669. https://doi.org/10.1016/j.abb.2023.109669

Gmoser R, Ferreira JA, Lennartsson PR, Taherzadeh MJ (2017) Filamentous ascomycetes fungi as a source of natural pigments. Fungal Biol Biotechnol 4(1):4. https://doi.org/10.1186/s40694-017-0033-2

González-Peña MA, Ortega-Regules AE, Anaya de Parrodi C, Lozada-Ramírez JD (2023) Chemistry, occurrence, properties, applications, and encapsulation of carotenoids–a review. Plants 12(2):313. https://doi.org/10.3390/plants12020313

Gottl VL, Pucker B, Wendisch VF, Henke NA (2023) Screening of structurally distinct lycopene b-Cyclases for production of the cyclic C40 carotenoids b-carotene and astaxanthin by corynebacterium glutamicum. J Agric Food Chem 71(20):7765–7776. https://doi.org/10.1021/acs.jafc.3c01492

Green AS, Fascetti AJ (2016) Meeting the vitamin A requirement: the efficacy and importance of β-carotene in animal species. Sci World J 2016:7393620. https://doi.org/10.1155/2016/7393620

Hayashi M, Ishibashi T, Kuwahara D, Hirasawa K (2021) Commercial production of astaxanthin with Paracoccus carotinifaciens. In: Misawa N (ed) Carotenoids: biosynthetic and biofunctional approaches. Springer Singapore, Singapore, pp 11–20. https://doi.org/10.1007/978-981-15-7360-6_2

Honda M, Kageyama H, Hibino T, Sowa T, Kawashima Y (2020) Efficient and environmentally friendly method for carotenoid extraction from Paracoccus carotinifaciens utilizing naturally occurring Z-isomerization-accelerating catalysts. Process Biochem 89:146–154. https://doi.org/10.1016/j.procbio.2019.10.005

Honda M, Murakami K, Osawa Y, Kawashima Y, Wasai M, Hirasawa K, Kuroda I (2022) Supercritical CO2 extraction of carotenoids (astaxanthin, adonirubin, and adonixanthin) from Paracoccus carotinifaciens: improved Z-isomer ratio and recovery of carotenoids via high-temperature extraction. Eur J Lipid Sci Technol 124(7):2200021. https://doi.org/10.1002/ejlt.202200021

Hu Q, Zhang T, Yu H, Ye L (2022) Selective biosynthesis of retinol in S. cerevisiae. Bioresour Bioprocess 9(1):22. https://doi.org/10.1186/s40643-022-00512-8

Hussain Y, Abdullah, Alsharif KF, Aschner M, Theyab A, Khan F, Saso L, Khan H (2022) Therapeutic role of carotenoids in blood cancer: mechanistic insights and therapeutic potential. Nutrients 14(9):1949. https://doi.org/10.3390/nu14091949

IUPAC (1975) Nomenclature of carotenoids (Rules approved 1974). Pure Appl Chem 41. https://doi.org/10.1351/pac197541030405

Jin Y, Yu Y, Zhang C, Li S, Zhang X, Li F (2020) Characterization and function analysis of the beta-carotene oxygenase-like genes in carotenoids metabolism of the Ridgetail white prawn Exopalaemon carinicauda. Front Physiol 11:745. https://doi.org/10.3389/fphys.2020.00745
Kharangate-Lad A, Bhosle S (2016) Studies on siderophore and pigment produced by an adhered bacterial strain Halobacillus trueperi MXM-16 from the mangrove ecosystem of Goa, India. Indian J Microbiol 56(4):461–466. https://doi.org/10.1007/s12088-016-0591-7
Koklesova L, Liskova A, Samec M, Zhai K, Abotaleb M, Ashrafizadeh M, Brockmueller A, Shakibaei M, Biringer K, Bugos O, Najafi M, Golubnitschaja O, Busselberg D, Kubatka P (2020) Carotenoids in cancer metastasis—status Quo and outlook. Biomol Ther 10(12):1653. https://doi.org/10.3390/biom10121653
Kramlinger VM, Nagy LD, Fujiwara R, Johnson KM, Phan TTN, Xiao Y, Enright JM, Toomey MB, Corbo JC, Guengerich FP (2016) Human cytochrome P450 27C1 catalyzes 3,4-desaturation of retinoids. FEBS Lett 590(9):1304–1312. https://doi.org/10.1002/1873-3468.12167
Krause T, Reichelt M, Gershenzon J, Schmidt A (2020) Analysis of the isoprenoid pathway intermediates, dimethylallyl diphosphate and isopentenyl diphosphate, from crude plant extracts by liquid chromatography tandem mass spectrometry. Phytochem Anal 31(6):770–777. https://doi.org/10.1002/pca.2941
Landolfo S, Ianiri G, Camiolo S, Porceddu A, Mulas G, Chessa R, Zara G, Mannazzu I (2018) CAR gene cluster and transcript levels of carotenogenic genes in Rhodotorula mucilaginosa. Microbiology 164(1):78–87. https://doi.org/10.1099/mic.0.000588
Liang M-H, Zhu J, Jiang J-G (2018) Carotenoids biosynthesis and cleavage related genes from bacteria to plants. Crit Rev Food Sci Nutr 58(14):2314–2333. https://doi.org/10.1080/10408398.2017.1322552
Lin J-H, Lee D-J, Chang J-S (2015) Lutein production from biomass: Marigold flowers versus microalgae. Bioresour Technol 184:421–428. https://doi.org/10.1016/j.biortech.2014.09.099
Liu C, Cheng Y, Du C, Lv T, Guo Y, Han M, Pi F, Zhang W, Qian H (2019) Study on the wall-breaking method of carotenoids producing yeast Sporidiobolus pararoseus and the antioxidant effect of four carotenoids on SK-HEP-1 cells. Prep Biochem Biotechnol 49(8):767–774. https://doi.org/10.1080/10826068.2019.1608448
Liu C, Hu B, Cheng Y, Guo Y, Yao W, Qian H (2021) Carotenoids from fungi and microalgae: a review on their recent production, extraction, and developments. Bioresour Technol 337:125398. https://doi.org/10.1016/j.biortech.2021.125398
López G-D, Álvarez-Rivera G, Carazzone C, Ibañez E, Leidy C, Cifuentes A (2021) Carotenoids in bacteria: biosynthesis, extraction, characterization and applications. Preprints. https://doi.org/10.20944/preprints202108.0383.v1
López G-D, Álvarez-Rivera G, Carazzone C, Ibáñez E, Leidy C, Cifuentes A (2023) Bacterial carotenoids: extraction, characterization, and applications. Crit Rev Anal Chem 53(6):1239–1262. https://doi.org/10.1080/10408347.2021.2016366
Lu Q, Bu Y-F, Liu J-Z (2017) Metabolic engineering of Escherichia coli for producing astaxanthin as the predominant carotenoid. Mar Drugs 15(10):296. https://doi.org/10.3390/md15100296
Maoka T (2020) Carotenoids as natural functional pigments. J Nat Med 74(1):1–16
Martínez-Cámara S, Ibañez A, Rubio S, Barreiro C, Barredo J-L (2021) Main carotenoids produced by microorganisms. Encyclopedia 1(4):1223–1245. https://doi.org/10.3390/encyclopedia1040093
Maurya VK, Shakya A, Bashir K, Kushwaha SC, McClements DJ (2022) Vitamin A fortification: recent advances in encapsulation technologies. Compr Rev Food Sci Food Saf 21(3):2772–2819. https://doi.org/10.1111/1541-4337.12941
Meléndez-Martínez AJ, Mapelli-Brahm P, Hornero-Méndez D, Vicario IM (2019) Structures, nomenclature and general chemistry of carotenoids and their esters. In: Mercadante AZ (ed) Carotenoid esters in foods: physical, chemical and biological properties. The Royal Society of Chemistry, Cambridge. https://doi.org/10.1039/9781788015851-00001
Miziorko HM (2011) Enzymes of the mevalonate pathway of isoprenoid biosynthesis. Arch Biochem Biophys 505(2):131–143. https://doi.org/10.1016/j.abb.2010.09.028

Moise AR, Al-Babili S, Wurtzel ET (2014) Mechanistic aspects of carotenoid biosynthesis. Chem Rev 114(1):164–193. https://doi.org/10.1021/cr400106y
Movahedi A, Wei H, Pucker B, Ghaderi-Zefrehei M, Rasouli F, Kiani-Pouya A, Jiang T, Zhuge Q, Yang L, Zhou X (2022) Isoprenoid biosynthesis regulation in poplars by methylerythritol phosphate and mevalonic acid pathways. Front Plant Sci 13:968780
Mussagy CU, Guimarцëes AAC, Rocha LVF, Winterburn J, VrdC S-E, Pereira JFB (2021) Improvement of carotenoids production from Rhodotorula glutinis CCT-2186. Biochem Eng J 165:107827. https://doi.org/10.1016/j.bej.2020.107827
Nabi F, Arain MA, Rajput N, Alagawany M, Soomro J, Umer M, Soomro F, Wang Z, Ye R, Liu J (2020) Health benefits of carotenoids and potential application in poultry industry: a review. J Anim Physiol Anim Nutr 104(6):1809–1818. https://doi.org/10.1111/jpn.13375
Naz T, Ullah S, Nazir Y, Li S, Iqbal B, Liu Q, Mohamed H, Song Y (2023) Industrially important fungal carotenoids: advancements in biotechnological production and extraction. J Fungi 9(5):578. https://doi.org/10.3390/jof9050578
Nidhi S, Pinakin D, Janki NT (2023) Carotenoids from marine bacteria: a natural antioxidant and UV protectant. Int J Adv Health Sci (IJHS) 10(5):49–62
O'Byrne SM, Blaner WS (2013) Retinol and retinyl esters: biochemistry and physiology: thematic review series: fat-soluble vitamins: vitamin A. J Lipid Res 54(7):1731–1743. https://doi.org/10.1194/jlr.R037648
Pereira da Costa D, Campos Miranda-Filho K (2020) The use of carotenoid pigments as food additives for aquatic organisms and their functional roles. Rev Aquac 12(3):1567–1578. https://doi.org/10.1111/raq.12398
Progress in Carotenoid Research (2018) IntechOpen, Rijeka. https://doi.org/10.5772/intechopen.73775
Quian-Ulloa R, Stange C (2021) Carotenoid biosynthesis and plastid development in plants: the role of light. Int J Mol Sci 22(3):1184. https://doi.org/10.3390/ijms22031184
Ram S, Mitra M, Shah F, Tirkey SR, Mishra S (2020) Bacteria as an alternate biofactory for carotenoid production: a review of its applications, opportunities and challenges. J Funct Foods 67:103867
Raposo MFJ, Morais AMMB, Morais RMSC (2015) Carotenoids from marine microalgae: a valuable natural source for the prevention of chronic diseases. Mar Drugs 13(8):5128–5155
Redfern CPF (2020) Chapter one – vitamin A and its natural derivatives. In: Pohl E (ed) Methods in enzymology, vol 637. Academic, Cambridge, pp 1–25. https://doi.org/10.1016/bs.mie.2020.02.002
Rodrigues DB, Mercadante AZ, Mariutti LRB (2019) Marigold carotenoids: much more than lutein esters. Food Res Int 119:653–664
Rodriguez-Amaya DB (2015) Carotenes and xanthophylls as antioxidants. In: Handbook of antioxidants for food preservation. Elsevier, Sawston, Cambridge, pp 17–50
Rodriguez-Concepcion M, Avalos J, Bonet ML, Boronat A, Gomez-Gomez L, Hornero-Mendez D, Limon MC, Meléndez-Martínez AJ, Olmedilla-Alonso B, Palou A, Ribot J, Rodrigo MJ, Zacarias L, Zhu C (2018) A global perspective on carotenoids: metabolism, biotechnology, and benefits for nutrition and health. Prog Lipid Res 70:62–93. https://doi.org/10.1016/j.plipres.2018.04.004
Saini RK, Keum Y-S (2019) Microbial platforms to produce commercially vital carotenoids at industrial scale: an updated review of critical issues. J Ind Microbiol Biotechnol 46(5):657–674. https://doi.org/10.1007/s10295-018-2104-7
Saini RK, Nile SH, Park SW (2015) Carotenoids from fruits and vegetables: chemistry, analysis, occurrence, bioavailability and biological activities. Food Res Int 76:735–750. https://doi.org/10.1016/j.foodres.2015.07.047
Saini RK, Moon SH, Keum Y-S (2018) An updated review on use of tomato pomace and crustacean processing waste to recover commercially vital carotenoids. Food Res Int 108:516–529
Sandmann G (2022) Carotenoids and their biosynthesis in fungi. Molecules 27:1431. https://doi.org/10.3390/molecules27041431

Sandmann G, Pollmann H, Gassel S, Breitenbach J (2021) Xanthophyllomyces dendrorhous, a versatile platform for the production of carotenoids and other acetyl-CoA-derived compounds. In: Misawa N (ed) Carotenoids: biosynthetic and biofunctional approaches. Springer Singapore, Singapore, pp 137–151. https://doi.org/10.1007/978-981-15-7360-6_11

Sansone C, Galasso C, Orefice I, Nuzzo G, Luongo E, Cutignano A, Romano G, Brunet C, Fontana A, Esposito F (2017) The green microalga Tetraselmis suecica reduces oxidative stress and induces repairing mechanisms in human cells. Sci Rep 7:41215

Semba RD (2012) The discovery of the vitamins. Int J Vitam Nutr Res 82(5):310–315. https://doi.org/10.1024/0300-9831/a000124

Shen N, Ren J, Liu Y, Sun W, Li Y, Xin H, Cui Y (2023) Natural edible pigments: a comprehensive review of resource, chemical classification, biosynthesis pathway, separated methods and application. Food Chem 403:134422. https://doi.org/10.1016/j.foodchem.2022.134422

Shin K-C, Seo M-J, Kim Y-S, Yeom S-J (2022) Molecular properties of β-carotene Oxygenases and their potential in industrial production of vitamin A and its derivatives. Antioxidants 11(6):1180. https://doi.org/10.3390/antiox11061180

Sun T, Rao S, Zhou X, Li L (2022) Plant carotenoids: recent advances and future perspectives. Mol Hortic 2(1):3. https://doi.org/10.1186/s43897-022-00023-2

Takemura M, Sahara T, Misawa N (2021) Violaxanthin: natural function and occurrence, biosynthesis, and heterologous production. Appl Microbiol Biotechnol 105(16):6133–6142. https://doi.org/10.1007/s00253-021-11452-2

Tetali SD (2019) Terpenes and isoprenoids: a wealth of compounds for global use. Planta 249(1):1–8. https://doi.org/10.1007/s00425-018-3056-x

Tiwary BK, Kumar A, Nanda AK, Chakraborty R (2014) A study on optimization of marigold petal yield, pure lutein, and formulation of free-flowing lutein esters. J Crop Sci Biotechnol 17:175–181

Van Hoang D, Pham NM, Lee AH, Tran DN, Binns CW (2018) Dietary carotenoid intakes and prostate cancer risk: a case-control study from Vietnam. Nutrients 10(1):70. https://doi.org/10.3390/nu10010070

Venil CK, Malathi M, Velmurugan P, Renuka Devi P (2021) Green synthesis of silver nanoparticles using canthaxanthin from Dietzia maris AURCCBT01 and their cytotoxic properties against human keratinocyte cell line. J Appl Microbiol 130(5):1730–1744. https://doi.org/10.1111/jam.14889

Vo TTM, Tran TD, Amoroso G, Ventura T, Elizur A (2023) Analysis of carotenoids and fatty acid compositions in Atlantic salmon exposed to elevated temperatures and displaying flesh color loss. Food Chem 417:135867

Vranová E, Kopcsayová D, Kosuth J, Colinas M (2019) Mutant-based model of two independent pathways for carotenoid-mediated chloroplast biogenesis in Arabidopsis embryos. Front Plant Sci 10:1034

Wade NM, Gabaudan J, Glencross BD (2017) A review of carotenoid utilisation and function in crustacean aquaculture. Rev Aquac 9(2):141–156

Wang X, Xu X, Liu J, Liu Y, Li J, Du G, Lv X, Liu L (2023) Metabolic engineering of Saccharomyces cerevisiae for efficient retinol synthesis. J Fungi 9(5):512. https://doi.org/10.3390/jof9050512

Watcharawipas A, Runguphan W (2023) Red yeasts and their carotenogenic enzymes for microbial carotenoid production. FEMS Yeast Res 23:foac063. https://doi.org/10.1093/femsyr/foac063

Yabuzaki J (2017) Carotenoids Database: structures chemical fingerprints and distribution among organisms. *Database*, Volume 2017, 2017, bax004. https://doi.org/10.1093/database/bax004

Zheng X, Hu R, Chen D, Chen J, He W, Huang L, Lin C, Chen H, Chen Y, Zhu J, Qi F, Xue T (2021) Lipid and carotenoid production by the Rhodosporidium toruloides mutant in cane molasses. Bioresour Technol 326:124816. https://doi.org/10.1016/j.biortech.2021.124816

Zia-Ul-Haq M (2021) Historical and introductory aspects of carotenoids. In: Zia-Ul-Haq M, Dewanjee S, Riaz M (eds) Carotenoids: structure and function in the human body. Springer International Publishing, Cham, pp 1–42. https://doi.org/10.1007/978-3-030-46459-2_1

Chapter 3
Biosynthesis of Carotenoids in Insects: What We Currently Know?

Norihiko Misawa, Takashi Maoka, Tetsuyuki Ueda, and Miho Takemura

1 Introduction

Carotenoids are natural pigments of yellow—orange—red colors with long conjugated double bonds, and play important physiological roles in organisms. For examples as their crucial functions, carotenoids are essential for photosynthesis as the light-harvesting pigments (Frank and Cogdell 1993; Maoka 2020), and protect organisms from photooxidative damage caused by excessive light or other oxidative damage (Krinsky 2001; Stahl and Sies 2003). Furthermore, several animals (the kingdom Animalia) use carotenoids as a precursor of retinal (vitamin A) as well as an enhancer of immunity, a contributor to reproduction, and as signals for intraspecies (sexual signaling, social status signaling, and parent-offspring signaling) and inter-species (species recognition, warning coloration, mimicry, and crypsis) communications (Blout and McGraw 2008; Frank and Cogdell 1993; Stahl and Sies 2003).

Carotenoid pigments are de novo synthesized not only in photosynthetic organisms, which contain plants such as higher plants, ferns, mosses and liverworts, and algae such as green algae, brown algae, red algae and diatoms (eukaryotes; the

N. Misawa (✉) · M. Takemura
Research Institute for Bioresources and Biotechnology, Ishikawa Prefectural University, Suematsu, Nonoichi-shi, Japan
e-mail: n-misawa@ishikawa-pu.ac.jp; mtake@ishikawa-pu.ac.jp

T. Maoka
Research Institute for Production Development, Sakyo-ku, Kyoto, Japan
e-mail: maoka@mbox.kyoto-inet.or.jp

T. Ueda
Ishikawa Prefectural University, Suematsu, Nonoichi-shi, Japan
e-mail: uzu@spacelan.ne.jp

R. Campos Chisté et al. (eds.), *Carotenoids*,
https://doi.org/10.1007/978-3-031-75322-0_3

domain Eukarya), and contain cyanobacteria and photosynthetic bacteria (the domain Bacteria), but also in some species of non-photosynthetic bacteria, archaea (the domain Archae), and fungi (the kingdum Fungi) (Britton et al. 2004; Misawa 2010; Moise et al. 2014). On the other hand, animals, which contain a variety of structurally diverse carotenoids, ordinarily utilize exogeneous carotenoids acquired from their diet, by metabolizing them to other carotenoid forms or apocarotenoids as needed, for supporting their health or life, since they generally have no ability to synthesize carotenoids de novo.

2 Arthropods That Can Synthesize Carotenoids de Novo

All of the organisms that belong to the kingdom Animalia (animals) had been believed not to synthesize carotenoids de novo. However, several species of insects and arachnids (the class Insecta and Arachnida, respectively) belonging to arthropods (the phylum Arthropoda) have unexpectedly been shown to possess carotenoid biosynthesis genes of fungal origin in the genomes, since 2010. Carotenoid biosynthesis (carotenogenic) genes homologous to fungal ones, which contain lycopene β-cyclase/phytoene synthase (*CrtYB*) and phytoene desaturase (*CrtI*) genes, were first found to exist in the pea aphid (*Acyrthosiphon pisum*; Hemiptera: Aphididae) of the class Insecta (Moran and Jarvik 2010). In 2011, similar carotenoid biosynthesis genes were discovered in the two spotted spider mite (*Tetranychus urticae*; Acari: Tetranychidae) of the class Arachnida (Grbić et al. 2011). Aphids and spider mites are both widespread and persistent agricultural pests. Further, the carotenogenic genes have been found in flies of the family Cecidomyiidae known as gall midges, i.e., the goldenrod gall midge (*Asteromyia carbonifera*) that lives around tall goldenrod (*Solidago altissma*), and the Hessian fly (*Mayetiola destructor*) (the order Diptera; the class Insecta) (Cobbs et al. 2013).

The *CrtYB* and *CrtI* genes appearing in the genomes of the above-mentioned arthropods (insects and arachnids) are thought to have been acquired with ancient lateral gene transfer from fungal donors, since only carotenogenic fungi have possessed the *CrtYB* gene, which codes for a fusion protein of lycopene β-cyclase (CrtY) and phytoene synthase (CrtB) (Misawa et al. 1995; Verdoes et al. 2003). The fungal two genes, *CrtYB* and *CrtI*, have been shown to mediate the biosynthesis of β-carotene (IUPAC semi-systematic name: β,β-Carotene) and torulene (3′,4′-Didehydro-β,ψ-carotene) from geranylgeranyl diphosphate (GGPP) (Sandmann and Misawa 2002; Verdoes et al. 2003). The *CrtYB* and *CrtI* genes derived of aphids, spider mites, and gall midges are considered to retain the same or similar function to that of the fungal corresponding genes.

3 Carotenoid Biosynthesis in Spider Mites

Spider mite *Tetranychus urticae* that sucks plant leaf contents was reported to contain typical higher plant-type carotenoids, β-carotene (main), α-carotene (β,ε-Carotene), lutein, violaxanthin, and neoxanthin, in addition to ketocarotenoids, esterified 3-hydroxyechinenone (3-Hydroxy-β,β-caroten-4-one), adonirubin (3-Hydroxy-β,β-carotene-4,4′-dione) and astaxanthin (3,3′-Dihydroxy-β,β-carotene-4,4′-dione), half a century ago (Veeman 1974). These carotenoids have been shown to exist widely in spider mite species (Goodwin 1984). Veeman (1974) also found that ketocarotenoid contents were increased in the diapausing females that harbor red body color. Disruption of the *CrtI* gene in *T. urticae* as well as in the citrus red mite *Panonychus citri* was shown to abolish carotenoid accumulation and result in complete albinism. This result indicates that the laterally transferred *CrtI* gene (Altincicek et al. 2012) is required for normal pigmentation (Bryon et al. 2017). It was further shown that the *CrtI* gene was required for the induction of diapause. These results means that some spider mites can forgo dietary carotenoids.

Wybouw et al. (2019) demonstrated through knock out experiments that CYP384A1 retains carotenoid 4,4′-ketolase activity, which is involved in the synthesis of ketocarotenoids such as astaxanthin and echinenone (β,β-Caroten-4-one) from β-carotene in *T. urticae*. This cytochrome P450 gene was shown to have been aquired by the convergent evolution of an ancient endogenous gene. Recently, Huo et al. (2021) suggested through comparative transcriptome analysis that the endogenous CYP389B1 gene of *T. urticae* is also involved in ketocarotenoid biosynthesis from β-carotene.

4 Carotenoid Biosynthesis in Aphids

The carotenoid biosynthesis genes (*CrtYB* and *CrtI*) of aphids (the Aphididae family) were discovered first in the pea aphid (*Acyrthosiphon pisum*) (Moran and Jarvik 2010), and found to exist in all of 34 aphid species examined (Nováková and Moran 2012). Further, the genes proved to inherit aphid relatives, adelgids (the Adelgidae family) and phylloxerids (the Phylloxeridae family) (Zhao and Nabity 2017). Sequence analysis of these insect genes and fungal genes from databases supported diversification of ancestral genes transferred from a fungus (Nováková and Moran 2012). They further reported that the *CrtI* genes of aphids and aphid relatives are closest to that of the β-carotene-producing fungus *Phycomyces blakesleeanus* (Mucoromycotina) (Murillo et al. 1978) among fungal *CrtI* sequences available in databases. The *CrtYB* genes of the pea aphid were also found to be closest to that of the same fungus species (Moran and Jarvik 2010).

Aphid *Macrosiphum liliodendri* had been reported to contain β,γ-carotene and γ,γ-carotene that were uncommon carotenoids with the γ-ring (called here "γ-carotenoids") [attention, "γ-carotenoids" are entirely distinct from γ-carotene (β,ψ-Carotene)] in ordinal green morphs, and contained lycopene (ψ,ψ-Carotene), γ-carotene, and 3,4-didehydrolycopene (3,4-Didehydro-ψ,ψ-carotene) in pink variants, until 4 decade ago (Andrewes et al. 1971; Goodwin 1984). "γ-Carotenoids" had also been isolated from the ladybird beetle *Coccinella septempunctata*, which fed on aphids preferably (Britton et al. 1977). Recently, we analyzed carotenoids contained in 15 aphid species including the pea aphids, the broad bean aphids (*Megoura crassicauda*), Japanese radish aphid (*Brevicoryne brassicae*), Taiwan bear aphid (*Uroleucon formosanum*), and celery aphids (*Semiaphis heraclei*) (Maoka et al. 2021; Takemura et al. 2021). Consequently, we were able to confirm the production of the uncommon carotenoids with the γ-ring ("γ-carotenoids"), i.e., (6'*S*)-β,γ-carotene (as a main carotenoid component), γ,γ-carotene and γ,ψ-carotene, and typical carotenoids with only the β-ring, i.e., β-carotene (β,β-Carotene main), often along with γ-carotene (β,ψ-Carotene) and β-zeacarotene. Exceptionally, the large chestnut aphid (*Lachnus tropicalis*) produced no dedectable carotenoids. Torulene was found in the red morphs (Moran and Jarvik 2010). Carotenoid species contained in the aphids and their chemical structures are shown in Fig. 3.1. These carotenoids are not observed in the leaf, stem, and sap of higher plants. Thus, aphids, as sap-feeding insects, are very likely to biosynthesize all these carotenoids by their own genes.

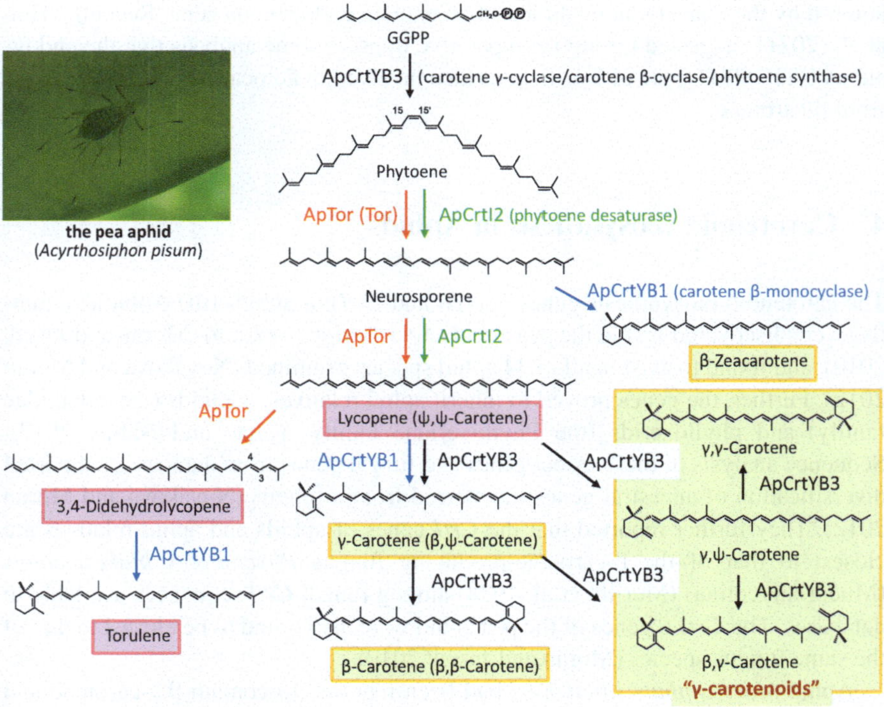

Fig. 3.1 Carotenoid biosynthetic pathway in the pea aphid (*Acyrthosiphon pisum*) and functions of its carotenoid biosynthesis genes

5 Functional Assignment of Aphid Carotenoid Biosynthesis Genes

As described above, a few insects and arachnids, representatively including aphids and spider mites, respectively, possess active carotenoid biosynthesis genes of fungal origin. The genome of the pea aphid (*Acyrthosiphon pisum*) harbored 4 *CrtI* and 4 *CrtYB* gene sequences, which were named *ApCrtI1*, *ApCrtI2*, *ApCrtI3* and *ApTor* (*ApCrtI4*), and *ApCrtYB1*, *ApCrtYB2*, *ApCrtYB3* and *ApCrtYB4* (Takemura et al. 2021). These aphid genes and relevant information are shown in Table 3.1.

Analysis for corroborating the catalytic functions of the aphid carotenogenic gene products (proteins) had scarcely been performed. Thus, we carried out functional complementation analysis on all the carotenoid biosynthesis gene sequences of the pea aphid, using recombinant *Escherichia coli* cells that synthesized carotenoid substrates. Consequently, 4 carotenoid biosynthesis genes, *ApCrtI2*, *ApTor* (*ApCrtI4*), *ApCrtYB1*, and *ApCrtYB3*, were identified to synthesize all carotenoids accumulated in the aphids from GGPP, as shown in Fig. 3.1 (Takemura et al. 2021). It was interestingly discovered that the *ApCrtYB3* gene can convert lycopene not only into carotenoids with the β-ring such as β-carotene, but also into the uncommon carotenoids with the γ-ring such as (6′*S*)-β,γ-carotene and γ,γ-carotene

Table 3.1 The carotenoid biosynthesis genes that are present in the genome of the pea aphid (*Acyrthosiphon pisum*) and their functional assignment

Gene	Accession	Accession	Gene	Assigned enzyme name
Our name[a]	Our study	Protein	Other name[b]	‘Our study
ApCrtI1	LC517091	XP_001943225.2	*CdeA*	‘
ApCrtI2	LC517092	XP_001950764	*CdeD*	phytoene desaturase Phytoene → Lycopene
ApCrtI3	LC517093	XP_001946689.2	*CdeC*	
ApTor (*ApCrtI4*)	LC517094	XP_001943938	*tor* *CdeB*	phytoene desaturase Phytoene → Lycopene → 3,4-Didehydrolycopene
ApCrtYB1	LC517095	XP_001943170	*CscA*	acyclic carotene β-monocyclase acyclic carotene → β-monocyclic carotene
ApCrtYB2	LC517096	XP_001950787	*CscC*	
ApCrtYB3	LC517097	XP_003241668.1	*CscB*	carotene β- and γ-cyclase / phytoene synthase carotene → β- or γ-cyclic carotene / GGPP → phytoene
ApCrtYB4	LC517098	XP_008178846		

[a] Takemura et al. (2021)

[b] *tor*, Moran and Jarvik (2010); *Cde* and *Csc*, Ding et al. (2020)

("γ-carotenoids"). Carotenogenic fungi can ordinarily biosynthesize not "γ-carotenoids" but carotenoids with the β-ring only, e.g., β-carotene, torulene, and torularhodin (Misawa 2021). The only exception is the orange peel fungus (the discomycete fungus; *Caloscypha fulgens*; the phylum Ascomycota) (discomycete was renamed ascomycete), which can produce β,γ-carotene as a minor carotenoid (Arpin et al. 1971). Ascomycota is taxonomically distant from Mucoromycotina that is closest to the carotenogenic gene source of aphids. Except for the orange peel fungus and aphids, no carotenogenic organisms have been reported to produce "γ-carotenoids" to our knowledge. Thus, *ApCrtYB3* was considered to have functionally evolved in aphids, after the lateral transfer.

The *ApTor* gene, named *tor* (Moran and Jarvik 2010), was found to mediate desaturation reaction from phytoene to 3,4-didehydrolycopene via lycopene. 3,4-Didehydrolycopene was then converted to torulene with *ApCrtYB1*. It was reported that the *ApTor* (*tor*) gene was significantly up-regulated in the red morph of the pea aphid (*A. pisum*) (Ding et al. 2020; Zhang et al. 2018). Trissi et al. (2023) recently showed that the *tor* ortholog exists in aphid *Myzus persicae*. Thus, the phytoene desaturase gene that mediates 3,4-didehydrolycopene generation is likely to be the key enzyme for the biosynthesis of torulene in aphids, which has longer conjugated double bonds, to turn into the red morph.

Plasmids pAC-HIEBI-ApCrtYB3 and pAC-HIEBI-ApCrtYB1 are gene constructs to let *E. coli* produce (6'*S*)-β,γ-carotene and torulene, which are available from Riken BRC as RDB 19,502 and RDB 19,503, respectively (Misawa et al. 2022).

6 Aphid-Mediated Food Chain

Maoka et al. (2020) conducted comprehensive carotenoid analysis on 20 dragonfly species (adults) and consequently showed that they all possessed "γ-carotenoids", which were considered to be derived from aphids. Among them, akiakane (*Sympetrum frequens*) (Odonata: Libellulidae), commonly known as "akatombo" in Japan, which means the red dragonfly, has been developed along with the long history of the climate and culture of Japanese wet-paddy rice cultivation that began 3000 years ago (Asahina 1974; Corbet 1999). The red dragonfly had been one of the most familiar insects for Japanese until the rapid decrease of the populations that started from the beginning of this century. Its unique life cycle is well known, as illustrated in Fig. 3.2.

We considered that the uncommon "γ-carotenoids" can work as ecological indicators for estimating the food chain from aphids to the red dragonfly and other predatory arthropods. The content rate of "γ-carotenoids" that exist in several aphids averaged 40.2% (Takemura et al. 2021). We carried out quantitative carotenoid analysis on the red dragonfly in the several life stages. As expected, the immature and mature adults were found to accumulate β,γ-carotene and γ,γ-carotene (Takemura et al. 2021). Content rate of these "γ-carotenoids" and contribution rate of the carotenoids derived from aphids (content rate per 40.2) in the individual life stages of the red dragonfly (*S. frequens*) were calculated as follows (shown as

Fig. 3.2 Life cycle of "akatombo" (the red dragonfly) (akiakane; *Sympetrum frequens*) and contribution rate of the carotenoids derived from aphids (%)

percentage, respectively): the aquatic larvae (0%, 0%), teneral adults (0%, 0%), the immature adults (5%, 12.4%), and the mature adults (17.8%, 44.3%), as illustrated in Fig. 3.2. The aquatic larvae and teneral adults are unlikely to feed on aphids. The contribution rate of the aphid carotenoids (44.3%) in the mature adults was considered at a much higher level than anticipated. It is feasible that the red-dragonfly adults prey on the winged form of aphids from "aerial plankton" (also called "aeroplankton") composed of small drifting insects that are present high in the air (Johnson 1969). The contribution of aphids to the red dragonfly was also supported by the findings that the gut microflora of the red-dragonfly mature adults included endosymbiotic bacteria such as *Serratia symbiotica* specific to the black bean aphid (*Aphis fabae*) (Takemura et al. 2021).

We further carried out quantitative carotenoid analysis on other predatory insects and arachnids (Takemura et al. 2021). Content rate of "γ-carotenoids" and contribution rate of the aphid carotenoids in the individuals were calculated as follows (shown as percentage, respectively): the larvae of the seven-spot ladybird beetle (*Coccinella septempunctata*) (24.1%, 60%), the adults of the beetle (24.7%, 61.4%), sasa-spider (*Oxyopes sertatus*) (11%, 27.4%), and joro spider (*Nephila clavata*) (1.9%, 4.7%). These rate values were considered at high levels, as anticipated.

This result indicated that aphids made significant contributions to the food chain of several predatory arthropods including the red-dragonfly adults. Aphids are likely to be positioned as an important "phytochemicals" source for some predatory insects and arachnids, which are often active under bright sunlight.

7 Postscript

In this chapter we clarified current results about functions of the carotenoid biosynthesis genes contained in the carotenogenic arthropods, especially insect aphids. We further introduced food chain from aphids to the red dragonfly and other predatory insects and arachnids, which was seen from aspects of the uncommon carotenoids.

Acknowledgements The authors thank to Naoki Kawase for his supply of the picture of teneral adults in Fig. 3.2.

References

Altincicek B, Kovacs JL, Gerardo NM (2012) Horizontally transferred fungal carotenoid genes in the two-spotted spider mite *Tetranychus urticae*. Biol Lett 8:253–257

Andrewes AG, Kjosen H, Liaaen-Jensen S, Weisgraber KH, Lousberg RJ, Weiss U (1971) Animal carotenoids. 7. Carotenes of two colour variants of the aphid *Macrosiphum liliodendri*—identification of natural γ,γ-carotene. Acta Chem Scand 25:3878–3880

Arpin N, Fiasson JL, Bouchez-Dangye-Caye MP, Francis GW, Liaaen-Jensen S (1971) A new $C_{40}H_{56}$ carotene with a terminal methylene group. Phytochemistry 10:1595

Asahina S (1974) The development of odonatology in the Far East. Odonatologica 3:5–12

Blout JD, McGraw KJ (2008) Signal function of carotenoid coloration. In: Britton G, Liaaen-Jensen S, Pfander H (eds) Carotenoids vol 4 Natural functions. Birkhäuser, Basel, pp 212–236

Britton G, Goodwin TW, Harriman GE, Lockley WJS (1977) Carotenoids of the ladybird beetle; *Coccinella septempunctata*. Insect Biochem 7:337–345

Britton G, Liaaen-Jensen S, Pfander H (2004) Carotenoids handbook. Birkhäuser Verlag, Basel, Boston, Berlin

Bryon A, Kurlovs AH, Dermauw W, Greenhalgh R, Riga M, Grbić M, Tirry L et al (2017) Disruption of a horizontally transferred phytoene desaturase abolishes carotenoid accumulation and diapause in *Tetranychus urticae*. PNAS 114:E5871–E5880

Cobbs C, Heath J, Stireman JO III, Abbot P (2013) Carotenoids in unexpected places: gall midges, lateral gene transfer, and carotenoid biosynthesis in animals. Mol Phylogenet Evol 68:221–228

Corbet PS (1999) Portrait: *Sympetrum frequens*. In: Corbet PS (ed) Dragonflies: behavior and ecology of Odonata. Cornell University Press, New York, pp 390–392

Ding BY, Niu J, Shang F, Yang L, Zhang W, Smagghe G, Wang JJ (2020) Parental silencing of a horizontally transferred carotenoid desaturase gene causes a reduction of red pigment and fitness in the pea aphid. Pest Manag Sci 76:2423–2433

Frank HA, Cogdell RJ (1993) The photochemistry and function of carotenoids in photosynthesis. In: Young AJ, Britton G (eds) Carotenoids in photosynthesis. Springer, Dordrecht, pp 252–326

Goodwin TW (1984) Arachnids and insects. In: Goodwin TW (ed) The biochemistry of the carotenoids, Animals, vol II. Chapman and Hall, London and New York, pp 97–121

Grbić M, Van Leeuwen T, Clark RM, Rombauts S, Rouzé P, Grbić V et al (2011) The genome of *Tetranychus urticae* reveals herbivorous pest adaptations. Nature 479:487–492

Huo SM, Zhang YY, Song ZR, Xiong XH, Hong XY (2021) The potential pigmentation-related genes in spider mites revealed by comparative transcriptomes of the red form of *Tetranychus urticae*. Insect Mol Biol 30:580–593

Johnson CG (1969) The composition of aerial populations and the vertical distribution of insects in the air. In: Johnson CG (ed) Migration and dispersal of insects by flight. Methuen & Co Ltd, London, pp 294–360

Krinsky NI (2001) Carotenoids as antioxidants. Nutrition 17:815–817

Maoka T (2020) Carotenoids as natural functional pigments. J Nat Med 74:1–16. https://doi.org/10.1007/s11418-019-01364-x

Maoka T, Kawase N, Ueda T, Nishida R (2020) Carotenoids of dragonflies, from the perspective of comparative biochemical and chemical ecological studies. Biochem Systematics Ecol 89:104001

Maoka T, Kawase N, Hironaka M, Nishida R (2021) Carotenoids of hemipteran insects, from the perspective of chemo-systematic and chemical ecological studies. Biochem Systematics Ecol 95:104241

Misawa N (2010) Carotenoids. In: Mander L, Lui H-W (eds) Comprehensive natural products II chemistry and biology, vol 1. Elsevier, Oxford, pp 733–753

Misawa N (ed) (2021) Carotenoids: biosynthetic and biofunctional approaches, Advances in experimental medicine and biology, vol 1261. Springer, Singapore

Misawa N, Satomi Y, Kondo K, Yokoyama A, Kajiwara S, Saito T, Ohtani T, Miki W (1995) Structure and functional analysis of a marine bacterial carotenoid biosynthesis gene cluster and astaxanthin biosynthetic pathway proposed at the gene level. J Bacteriol 177:6575–6584

Misawa N, Maoka T, Takemura M (2022) Carotenoids: carotenoid and apocarotenoid analysis — use of *E. coli* to produce carotenoid standards. Methods Enzymol 670:87–137

Moise AR, Al-Babili S, Wutzel ET (2014) Mechanistic aspects of carotenoid biosynthesis. Chem Rev 114:164–193

Moran NA, Jarvik T (2010) Lateral transfer of genes from fungi underlies carotenoid production in aphids. Science 328:624–627

Murillo FJ, Calderón IL, López-Diaz I, Cerdá-Olmedo E (1978) Carotene-superproducing strains of Phycomyces. Appl Environ Microbiol 36:639–642

Nováková E, Moran NA (2012) Diversification of genes for carotenoid biosynthesis in aphids following an ancient transfer from a fungus. Mol Biol Evol 29:313–323

Sandmann G, Misawa N (2002) Fungal carotenoids. In: Osiewacz HD (ed) The Mycota X industrial applications. Springer, Berlin, Heidelberg, pp 247–262

Stahl W, Sies H (2003) Antioxidant activity of carotenoids. Mol Asp Med 24:345–351. ISSN 0098-2997. https://doi.org/10.1016/S0098-2997(03)00030-X

Takemura M, Maoka T, Koyanagi T, Kawase N, Nishida R, Tsuchida T, Hironaka M, Ueda T, Misawa N (2021) Elucidation of the whole carotenoid biosynthetic pathway of aphids at the gene level and arthropodal food chain involving aphids and the red dragonfly. BMC Zool 6(1):1–13

Trissi N, Troczka BJ, Ozsanlav-Harris L, Singh KS, Mallott M, Aishwarya V, O'Reilly A, Bass C, Wilding CS (2023) Differential regulation of the *tor* gene homolog drives the red/green pigmentation phenotype in the aphid *Myzus persicae*. Insect Biochem Mol Biol 153:103896. https://doi.org/10.1016/j.ibmb.2022.103896

Veeman A (1974) Carotenoid metabolism in *Tetranychus urticae* Koch (Acari; Tetranychidae). Comp Biochem Physiol 47B:101–116

Verdoes JC, Sandmann G, Visser H, Diaz M, van Mossel M, van Ooyen AJJ (2003) Metabolic engineering of the carotenoid biosynthetic pathway in the yeast *Xanthophyllomyces dendrorhous* (*Phaffia rhodozyma*). Appl Environ Microbiol 69:3728–3738

Wybouw N, Kurlos AH, Greenhalgh R, Bryon A, Kosterlitz O, Manabe Y, Osakabe M et al (2019) Convergent evolution of cytochrome P450s underlies independent origins of keto-carotenoid pigmentation in animals. Proc R Soc B Biol Sci. https://doi.org/10.1098/rspb.2019.1039

Zhang L, Wang MY, Li XP, Wang XT, Jia CL, Yang XZ, Feng RQ, Yuan ML (2018) A small set of differentially expressed genes was associated with two color morphs in natural populations of the pea aphid *Acyrthosiphon pisum*. Gene 651:23–32

Zhao C, Nabity PD (2017) Phylloxerids share ancestral carotenoid biosynthesis genes of fungal origin with aphids and adelgids. PLoS One 12:e0185484

Chapter 4
C_{30} Carotenoids: Biosynthesis, Chemical and Biological Aspects

Norihiko Misawa, Miho Takemura, Wataru Matsumoto, Paul D. Fraser, and Kazutoshi Shindo

1 Introduction

Carotenoids are biosynthesized in all photosynthetic organisms, which extend from photosynthetic bacteria and cyanobacteria (Domain Bacteria) to plants and algae [Kingdom Plantae or Chromista in Domain Eukarya (eukaryotes)], since carotenoids are essential for photosynthesis acting as the light-harvesting pigments (Maoka 2020). The pigments are also biosynthesized in some of non-photosynthetic organisms, which contain bacteria (chemoheterotrophs), archaea (Domain Archaea) and fungi (Kingdom Fungi), and even in a few arthropods, e.g., aphids and spider mites (Kingdom Animalia) (Misawa 2021). These carotenoids are considered to prevent themselves from having (photo)oxidative stress as a main role (Krinsky 2001; Stahl and Sies 2003).

Major bacterial species and all eukaryotes that have carotenoid biosynthetic (carotenogenic) capability, *de novo* synthesize carotenoids that retain typical basic structures composed of 40 carbons (Britton et al. 2004). On the other hand, some non-photosynthetic bacterial species with carotenogenic capability, de novo biosynthesize acyclic carotenoids with a basic 30 carbon skeleton (described

N. Misawa (✉) · M. Takemura · W. Matsumoto
Research Institute for Bioresources and Biotechnology, Ishikawa Prefectural University, Suematsu, Nonoichi-shi, Japan
e-mail: n-misawa@ishikawa-pu.ac.jp; mtake@ishikawa-pu.ac.jp

P. D. Fraser
School of Biological Sciences, Royal Holloway University of London, Surrey, UK
e-mail: p.fraser@rhul.ac.uk

K. Shindo
Department of Food and Nutrition, Japan Women's University, Mejirodai, Bunkyo-ku, Tokyo, Japan
e-mail: kshindo@fc.jwu.ac.jp

R. Campos Chisté et al. (eds.), *Carotenoids*,
https://doi.org/10.1007/978-3-031-75322-0_4

C_{30}-carotenoids or C_{30} carotenoids) (Siziya et al. 2022a). Recently, studies on the biosynthesis and biological function of C_{30}-carotenoids has advanced carotenoid biosynthesis and the potential exploitation of these molecules (Elmesseri et al. 2022; Xue et al. 2019). In this chapter, we note such studies of C_{30}-carotenoids and report their biosynthesis as a main topic along with their biological function, in a comprehensive manner.

2 Distribution of C_{30}-Carotenoid Biosynthetic Bacteria

Carotenogenic-bacterial species that belong to the phylum Firmicutes (usually class Bacilli), so-called low GC Gram-positive bacteria, have been reported to biosynthesize only C_{30} carotenoids. This category includes, commencing with a human pathogen *Staphylococcus aureus* (Marshall and Wilmoth 1981), various bacterial species *Streptococcus faecium* (Taylor and Davies 1974), *Bacillus indicus* (reclassified to *Metabacillus indicus*), *Bacillus firmus* (reclassified to *Cytobacillus firmus*) (Steiger et al. 2012a), and *Sporosarcina aquimarina* (Steiger et al. 2012b). Further the category involves, as stress-tolerant bacteria, the genus *Planococcus*, e.g., *Planococcus maritimus* (Shindo et al. 2008b, 2014), *Planococcus faecalis* (Kim et al. 2018) and *Planococcus* sp. CP5-4 (Moyo et al. 2022), and *Halobacillus halophilus* (Osawa et al. 2010, 2014). The category also involves lactic acid bacteria *Lactobacillus plantarum* (reclassified to *Lactiplantibacillus plantarum*) (Garrido-Fernández et al. 2010), *Leuconostoc mesenteroides* (Matsumoto et al. 2024), and *Enterococcus gilvus* (Hagi et al. 2018).

C_{30} Carotenoids have also been found in bacterial species that belong to other phyla, e.g., *Rubritalea squalenifaciens* of the phylum Verrucomicrobia (Shindo et al. 2008a), and *Methylomonas* sp. strain 16a (Tao et al. 2005) and *Methylobacterium extorquens* (Mo et al. 2023), which belong to the phylum Proteobacteria.

3 Structures of C_{30}-Carotenoid Final Products

Several C_{30}-carotenogenic bacteria have been shown to produce a C_{30} carotenoid that forms a complex glycosylated structure, i.e., a mono-glycosylated C_{30} carotenoid, which adds a sugar D-glucose (or D-xylose, rarely) with/without an acylated aliphatic acid to the one terminus region. Figure 4.1 exhibits chemical structures of five examples of such C_{30}-carotenoids. As detailed in the Sect. 4, we here found the principle that the chemical structures of their aglycones converge on three types, i.e., 4,4′-diaponeurosporen-4-oic acid (staphyloxanthin-type), 4,4′-diapolycopene-4,4′-dioic acid (dioic acid-type), and 5-hydroxy-5,6-dihydro-4,4′-diapolycopen-4′-oic acid (methyl ester) (high modification-type). Figure 4.1 contains staphyloxanthin [β-D-glucosyl 1-*O*-(4,4′-diaponeurosporen-4-oate)-6-*O*-(12-methyltetradecanoate)] of *S. aureus* (Pelz et al. 2005) (Fig. 4.1a;

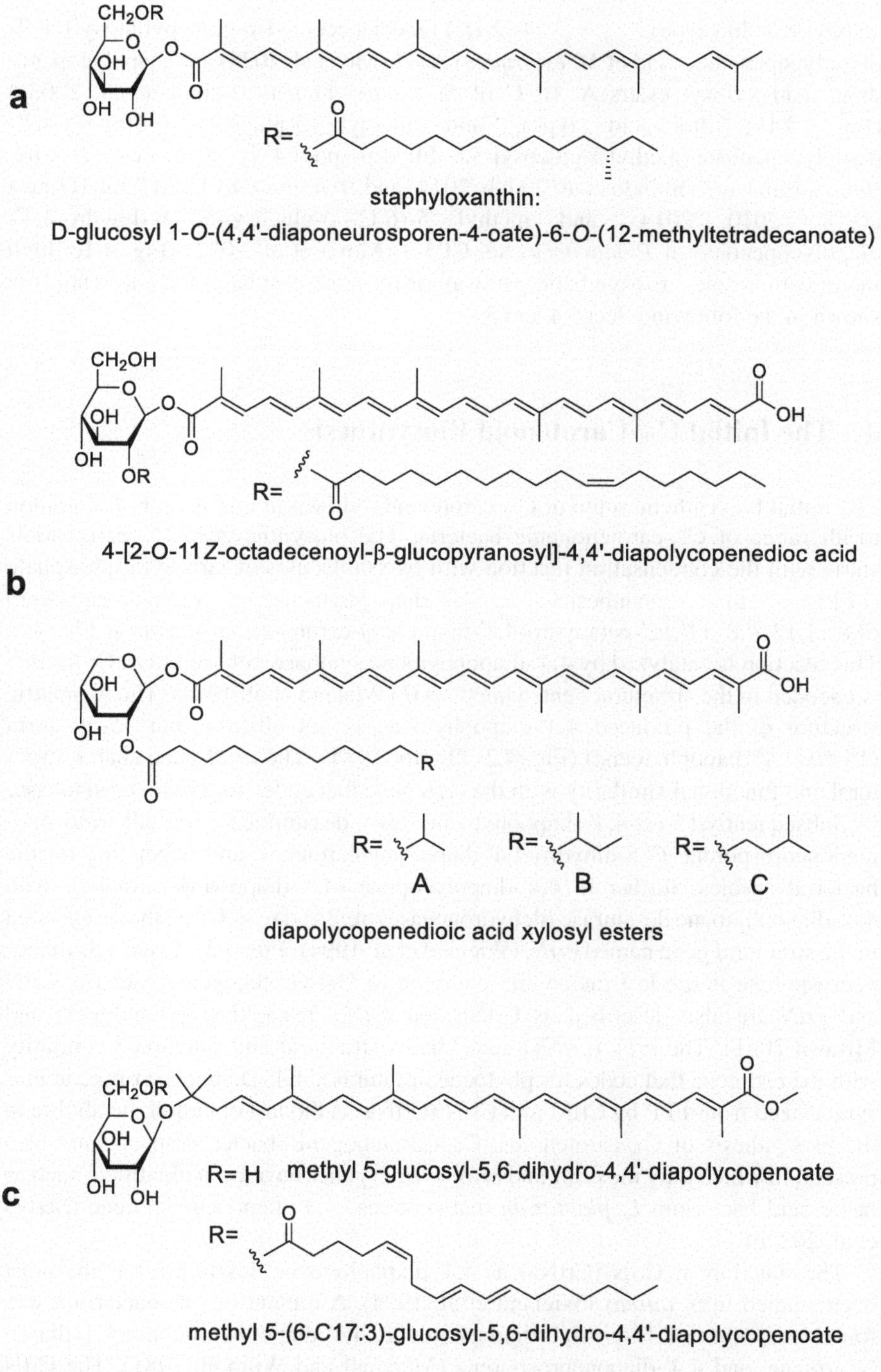

Fig. 4.1 Structures of glycosylated C_{30} carotenoids, biosynthesized as the final products by several C_{30}-carotenogenic bacteria. (**a**) carotenoid glucosyl ester with staphyloxanthin-type aglycone (staphyloxanthin); (**b**) carotenoid glycosyl ester with dioic acid-type aglycone; and (**c**) carotenoid glucoside with high modification-type aglycone

staphyloxanthin-type), 4-[2-*O*-11Z-octadecenoyl-β-glucopyranosyl]-4,4′-diapolycopenedioc acid of *M. extorquens* AM1 (Mo et al. 2023) and diapolycopene-dioic acid xylosyl esters A, B, C of *R. squalenifaciens* (Shindo et al. 2008a) (Fig. 4.1b; dioic-acid type), and methyl 5-glucosyl-5,6-dihydro-4,4′-diapolycopenoate (methyl 5-glucosyl-5,6-dihydro-apo-4,4′-lycopenoate) of *P. maritimus* strain iso-3 (Shindo et al. 2008b, 2014) and *H. halophilus* DSM 2266 (Osawa et al. 2010, 2014), and methyl 5-(6-$C_{17:3}$)-glucosyl-5, 6-dihydro-4,4′-diapolycopenoate of *Planococcus* sp. CP5-4 (Moyo et al. 2022) (Fig. 4.1c; high modification-type). Biosynthetic pathway of these glycosylated C_{30} carotenoids is shown in the following Sects. 4 and 5.

4 The Initial C_{30}-Carotenoid Biosynthesis

The initial biosynthetic route of C_{30}-carotenoids, shown in this section, is common to all range of C_{30}-carotenogenic bacteria. The biosynthesis of C_{30}-carotenoids starts with the condensation reaction with two molecules of farnesyl diphosphate (FPP) to synthesize 4,4′-diapophytoene (dehydrosqualene) (7,8,11,12,7′,8′,11′,12′-octahydro-4,4′-diapo-ψ,ψ-carotene), as shown in Fig. 4.2. This reaction is catalyzed by 4,4′-diapophytoene synthase, referred to as CrtM, that is encoded in the structural gene named *crtM* (Wieland et al. 1994). The geometric structure of the produced 4,4′-diapophytoene is not all-*trans* but 15-*cis* form (15-*cis*-4,4′-diapophytoene) (Fig. 4.2) (Taylor 1984). The *crtM* gene shares structural and functional similarity with the *crtB* gene that codes for phytoene synthase.

Subsequently, 15-*cis*-4,4′-diapophytoene is desaturated to all-*trans*-4,4′-diaponeurosporene (7,8-dihydro-4,4′-diapo-ψ,ψ-carotene), and depending on the bacterial species, further to 4,4′-diapolycopene (4,4′-diapo-ψ,ψ-carotene), with 4,4′-diapophytoene desaturase (dehydrogenase), referred to as CrtN, that is encoded in the structural gene named *crtN*, (Wieland et al. 1994) (Fig. 4.2). Thus, 4,4′-diaponeurosporene is the last metabolite common to C_{30}-carotenogenic bacteria. CrtN and *crtN* are also described as CrtNa and *crtNa*, respectively (Sandmann and Misawa 2021). The *crtN* (*crtNa*) gene shares structural and functional similarity with the *crtI* gene that codes for phytoene desaturase. 4,4′-Diaponeurosporene biosynthesized from FPP by CrtM and CrtN (CrtNa) is the last common metabolite in the biosynthesis of C_{30}-carotenoids. C_{30}-Carotenogenic bacterial species are also present, in which only the *crtM* and *crtN* (*crtNa*) genes have been identified, such as lactic acid bacterium *L. plantarum* that produces 4,4′-diaponeurosporene (Siziya et al. 2022b).

The function of CrtN (CrtNa) as 4,4′-diapophytoene desaturase has the most been studied in *S. aureus* (Wieland et al. 1994). A mutant of this bacterium was found to synthesize 15-*cis*-4,4′-diapophytoene, 4,4′-diapophytofluene, 4,4′-diapo-ζ-carotene, and 4,4′-diaponeurosporene (Marshall and Wilmoth 1981). The CrtN enzyme was shown to convert 4,4′-diapophytoene to 4,4′-diaponeurosporene by

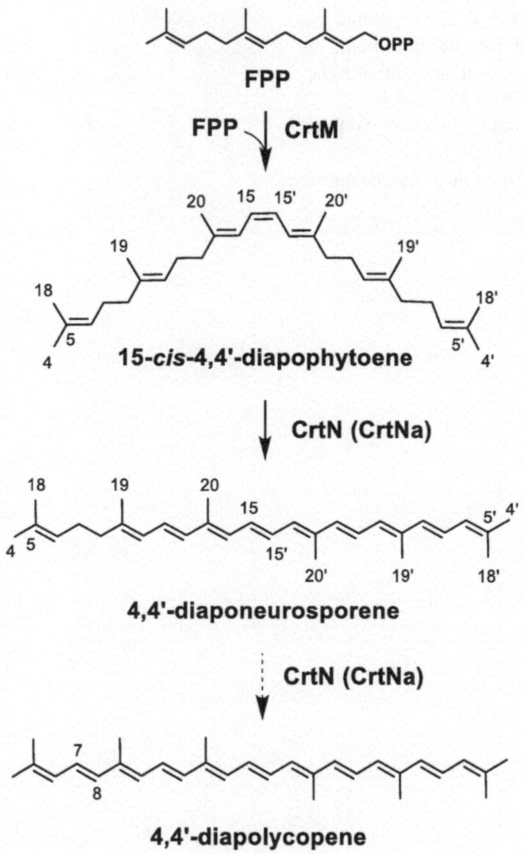

Fig. 4.2 Initial biosynthetic pathway of C_{30}-carotenoids from farnesyl diphosphate for production of 15-*cis*-4,4′-diapophytoene, 4,4′-diaponeurosporene, and 4,4′-diapolycopene. Two sorts of 15-*cis*-4,4′-diapophytoene desaturase CrtN (CrtNa) exist, i.e., one that synthesizes 4,4′-diaponeurosporene and can stop the desaturation reaction there, and the other that synthesizes 4,4′-diaponeurosporene and can take one more desaturation step to 4,4′-diapolycopene

way of 4,4′-diapophytofluene and 4,4′-diapo-ζ-carotene (Raisig, and Sandmann 1999). CrtN is also very likely to utilize FAD as a coenzyme, since this protein retains a classical FAD-, NAD(P)-binding domain at the N-terminus (Wieland et al. 1994) and since CrtI (phytoene desaturase) that is significantly homologous to CrtN (Raisig and Sandmann 2001) proved to require FAD (Fraser et al. 1992). Taking these findings into consideration we speculated the reaction mechanism of CrtN as follow: The reaction starts with radical generation at C-12 in 15-*cis*-4,4′-diapophytoene by the dehydrogenation with FAD, and the radical conjugates to the triene structure in the intermediate, as shown in Fig. 4.3. Then, the 15-*cis* olefin structure isomerizes to a more stable 15-*trans* olefin when further dehydrogenated to 4,4′-diapophytofluene with FADH, which is successively desaturated to 4,4′-diapo-ζ-carotene and 4,4′-diaponeurosporene with FAD. This reaction mechanism from 15-*cis*-4,4′-diapophytoene to 4,4′-diaponeurosporene with CrtN should be applied to CrtI as well, which catalyzes lycopene synthesis from 15-*cis*-phytoene by way of phytofluene, ζ-carotene and neurosporene in diverse C_{40}-carotenogenic bacteria. Since it is likely that the ^{1}H and ^{13}C NMR assignment data of

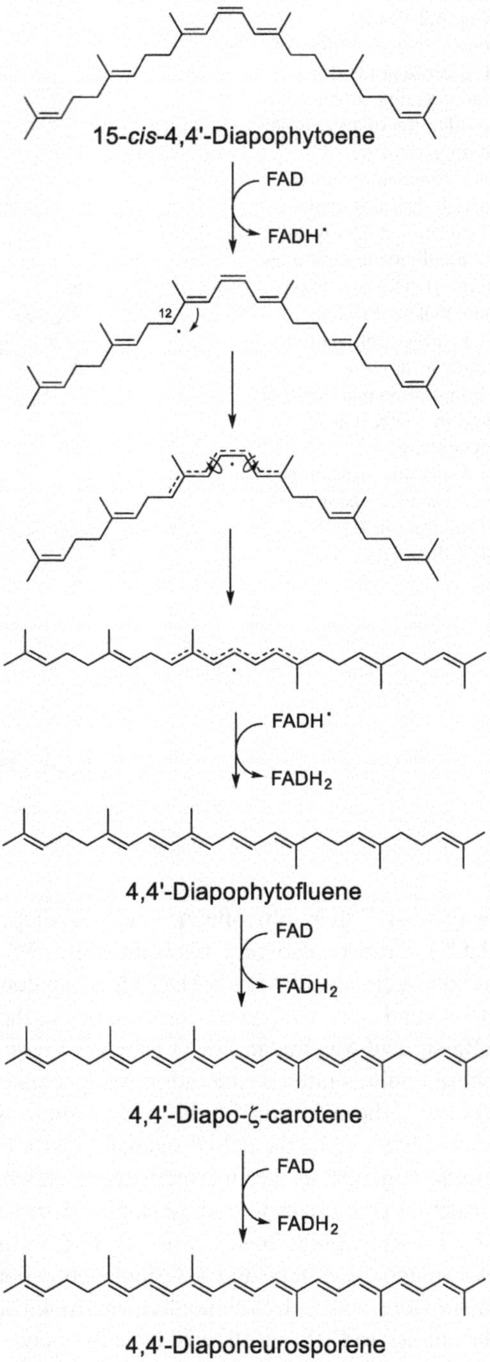

Fig. 4.3 Speculated biosynthetic route of 4,4′-diaponeurosporene from 15-*cis*-4,4′-diapophytoene, catalyzed by 15-*cis*-4,4′-diapophytoene desaturase CrtN (CrtNa)

15-*cis*-4,4′-diapophytoene and 4,4′-diaponeurosporene have not fully been available, we described them in Methods in Enzymology (Misawa et al. 2022) for your references.

5 Biosynthesis of C_{30}-Carotenoid Aglycones

The preceding section has shown detailed biosynthetic route from FPP up to 4,4′-diaponeurosporene (7,8-dihydro-4,4′-diapo-ψ,ψ-carotene) as the last common metabolite by CrtM and CrtN (CrtNa). This carotenoid 4,4′-diaponeurosporene can be further desaturated to 4,4′-diapolycopene (4,4′-diapo-ψ,ψ-carotene) according to the catalytic capability of CrtN originated in a bacterial host. Thus, 4,4′-diaponeurosporene and 4,4′-diapolycopene can be called the CrtMN products. In this subsection we explain modification reactions of the CrtMN products toward the biosynthesis of a diversity of C_{30}-carotenoids. Figure 4.1 in the Sect. 3 exhibited the structures of several glycosylated C_{30} carotenoids that were isolated as final products.

We found that the structures of their aglycones converge on three types, i.e., 4,4′-diaponeurosporen-4-oic acid (7′,8′-dihydro-4,4′-diapo-ψ,ψ-caroten-4-oic acid) as staphyloxanthin-type aglycone (Fig. 4.1a), 4,4′-diapolycopene-4,4′-dioic acid as dioic acid-type aglycone (Fig. 4.1b), and 5-hydroxy-5,6-dihydro-4,4′-diapolycopen-4′-oic acid (methyl ester) as high modification-type aglycone (Fig. 4.1c). This section shows biosynthetic routes of the three aglycones from the CrtMN products, followed by subsequent glycosylation and acylation reactions.

5.1 Staphyloxamthin-Type Aglycone

The biosynthesis of C_{30}-carotenoids has mostly been studied in *Staphylococcus aureus* (Marshall and Wilmoth 1981; Kim and Lee 2012; Pelz et al. 2005; Wieland et al. 1994). The final product in this bacterium is staphyloxanthin, as described in Fig. 4.1a. Staphyloxanthin is biosynthesized from 4,4′-diaponeurosporene by way of the synthesis of staphyloxanthin aglycone, i.e., 4,4′-diaponeurosporen-4-oic acid (7′,8′-dihydro-4,4′-diapo-ψ,ψ-caroten-4-oic acid). The typical biosynthetic pathway of this aglycone is shown in Fig. 4.4a. In the first place the terminal 4-methyl group of 4,4′-diaponeurosporene is oxidized to its aldehyde to synthesize 4,4′-diaponeurosporen-4-al (7′,8′-dihydro-4,4′-diapo-ψ,ψ-caroten-4-al). This reaction is catalyzed by 4,4′-diaponeurosporene 4-oxidase (simply referred to as diaponeurosporene oxidase), also referred to as 4,4′-diaponeurosporene/4,4′-diapolycopene terminal oxidase (aldehyde synthase). This enzyme was found to be encoded by a *crtN*-homologous gene named *crtP* (Pelz et al. 2005) or *crtNb* (Tao et al. 2005). The gene named *crtNb* was functionally identified using the corresponding genes from *Methylomonas* sp. strain 16a additionally. We want to

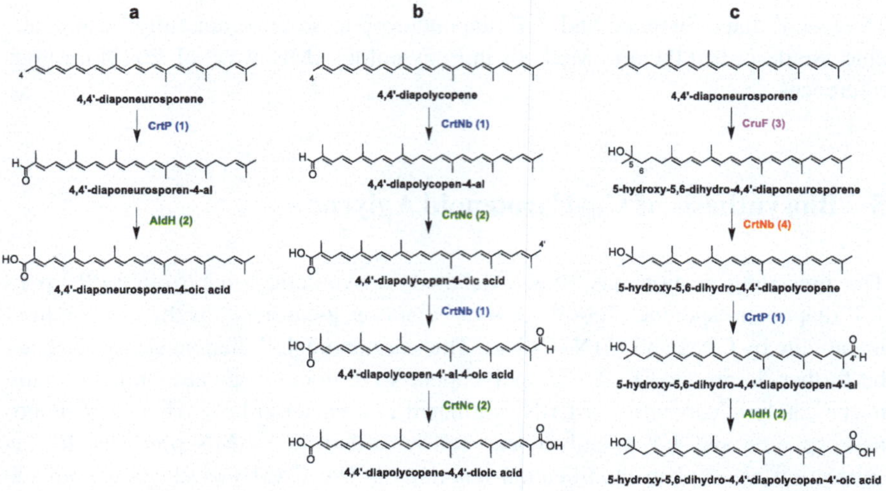

Fig. 4.4 Biosynthetic pathway of C_{30}-carotenoids from the CrtMN products for production of three types of aglycones. (**a**), biosynthesis of staphyloxanthin-type aglycone from 4,4′-diaponeurosporene; (**b**), biosynthesis of dioic acid-type aglycone from 4,4′-diapolycopene; (**c**), biosynthesis of high modification-type aglycone from 4,4′-diaponeurosporene. Numbers parenthesized in (1), (2), (3), and (4): the same number, the same enzyme regardless of difference in the gene-product name; different number, different enzyme even with the same gene-product name

assign "(1)" to this terminal oxidase enzyme (Fig. 4.4a) for easy-to-understand explanation in this review article, since several gene names exist for a single gene in the biosynthesis of C_{30}-carotenoids probably due to rapid progress in this field.

Subsequently, 4,4′-diaponeurosporen-4-al 4-dehydrogenase (simply called aldehyde dehydrogenase), also referred to as 4,4′-diaponeurosporen-4-al /4,4′-diapolycopen-4-al dehydrogenase, mediates the formation of the carboxylic acid from the aldehyde group of 4,4′-diaponeurosporen-4-al to synthesize 4,4′-diaponeurosporen-4-oic acid (staphyloxanthin aglycone). This enzyme was found to be encoded in another *crtN*-homolog, designated *aldH* (Kim and Lee 2012) or *crtNc* (Steiger et al. 2015). We want to assign "(2)" to this aldehyde dehydrogenase enzyme, called AldH or CrtNc (Fig. 4.4a).

In order to biosynthesize the final product staphyloxanthin from the aglycone two more enzymes are required. Aglycone 4,4′-diaponeurosporen-4-oic acid is esterified with D-glucose at the 4-position, and the added glucosyl group is further esterified with 12-methyltetradecanoic acid. These two reactions are catalyzed by glycosyltransferase (designated CrtQ or GT) and acyltransferase (designated CrtO or AT), respectively, which lead to the production of staphyloxanthin (Kim and Lee 2012) (Fig. 4.1a). Except for *Staphylococcus* genus, C_{30}-carotenogenic bacteria that can biosynthesize staphyloxanthin-type aglycone have not been found.

5.2 Dioic Acid-Type Aglycone

Considerable numbers of C_{30}-carotenogenic bacterial species are likely to generate dioic acid-type aglycone, which includes *Cytobacillus firmus* (Steiger et al. 2012a, 2015) and *Sporosarcina aquimarina* (Steiger et al. 2012b) of the phylum Firmicutes, in addition to *Methylomonas* sp. strain 16a (Tao et al. 2005) and *Methylobacterium extorquens* (Mo et al. 2023), which belong to the phylum Proteobacteria.

Staphyloxanthin-type (Fig. 4.4a) and high modification-type (Fig. 4.4c) aglycones are synthesized from 4,4′-diaponeurosporene, whereas dioic acid-type aglycone, i.e., 4,4′-diapolycopene-4,4′-dioic acid, is synthesized from 4,4′-diapolycopene, as shown in Fig. 4.4b. Thus, CrtN (CrtNa) originated in bacteria that can produce this aglycone, should generate 4,4′-diapolycopene through 4,4′-diaponeurosporene (Fig. 4.2). This substrate is converted to 4,4′-diapolycopen-4-al by terminal oxidase CrtNb (1), and further converted to 4,4′-diapolycopen-4-oic acid by aldehyde dehydrogenase (CrtNc) (2) (Fig. 4.4b) (Tao et al. 2005; Steiger et al. 2015). These two reactions are substantially the same to those of Fig. 4.4a. The 4′-methyl group of 4,4′-diapolycopen-4-oic acid is further converted to carboxylic acid via aldehyde in the same way by the action of terminal oxidase CrtNb (1) and aldehyde dehydrogenase (CrtNc) (2), to synthesize 4,4′-diapolycopene-4,4′-dioic acid (dioic acid-type aglycone) (Fig. 4.4b). Here we showed the metabolic route with the order of (1) (2) (1) (2), while the route with the order (1) (1) (2) (2) is also feasible.

Obtained dioic acid-type aglycone, 4,4′-diapolycopene-4,4′-dioic acid, is esterified with D-glucose at the 4-position, and the added glucosyl group is occasionally esterified with an aliphatic acid. These two reactions are catalyzed by glycosyltransferase (designated GT) and acyltransferase (designated AT), respectively.

5.3 High Modification-Type Aglycone

There are other bacterial species that produce glycosylated C_{30} carotenoids with highly modified aglycone and are capable of metabolizing 4,4′-diaponeurosporene (Fig. 4.4c). They are genus *Planococcus*, containing *Planococcus maritimus* strain iso-3 (Takemura et al. 2021; Hagiwara et al. 2023; Shindo et al. 2008b, 2014), *Planococcus* sp. CP5-4 (Moyo et al. 2022), and probably *Planococcus faecalis* AJ003^{T} (Kim et al. 2018; Lee et al. 2020), in addition to *Halobacillus halophilus* (Köcher et al. 2009; Osawa et al. 2010, 2014) and probably *Metabacillus indicus* (Steiger et al. 2012a, 2015).

Chemical structures of glycosylated C_{30}-carotenoids produced by *P. maritimus* strain iso-3 (Shindo et al. 2008b, 2014), *Planococcus* sp. CP5-4 (Moyo et al. 2022), and *H. halophilus* (Osawa et al. 2010, 2014) are clarified, as shown in Fig. 4.1. These aglycones were found to be the same 5-hydroxy-5,6-dihydro-4,4′-diapolycopen-4′-oic acid methyl ester, while

5-hydroxy-5,6-dihydro-4,4′-diapolycopen-4′-oic acid was also regarded as high modification-type aglycone. This aglycone is synthesized from 4,4′-diaponeurosporene with the action of 4 different enzymes (Fig. 4.4c). In the first place, 4,4′-diaponeurosporene is converted to 5-hydroxy-5,6-dihydro-4,4′-diaponeurosporene with 4,4′-diaponeurosporene 5,6-hydratase (simply referred to as diaponeurosporene hydratase) (Takemura et al. 2021). This enzyme was recently identified as the gene product 5,6-hydratase that was encoded by a gene (*orf7*) in the *Planococcus* iso-3 carotenogenic gene cluster (Fig. 4.5). This gene was named *cruF*, since the encoded 5,6-hydratase protein, referred to as CruF, was found to mediate the reaction similar to that of the C_{40}-caroteogenic CrtC and CruF proteins, and showed a slight sequence similarity to the latter (Sun et al. 2009). "(3)" was provided to this enzyme.

Next, 5-hydroxy-5,6-dihydro-4,4′-diaponeurosporene 7,8-desaturase (simply referred to as hydroxydiaponeurosporene desaturase) dehydrogenizes the 7–8 position of the substrate to synthesize 5-hydroxy-5,6-dihydro-4,4′-diapolycopene (Fig. 4.4c) (Takemura et al. 2021). This enzyme was also recently clarified as the gene product 7,8-desaturase that was encoded by a gene (*orf4*) in the carotenogenic gene cluster of *Planococcus maritimus* strain iso-3 (Fig. 4.5). This gene was designated *crtNb*, since the encoded 7,8-desaturase protein, referred to as CrtNb, was found to retain the highest sequence similarity to a gene product called CrtNb in *Halobacillus halophilus* (Kőcher et al. 2009). "(4)" was provided to this enzyme CrtNb.

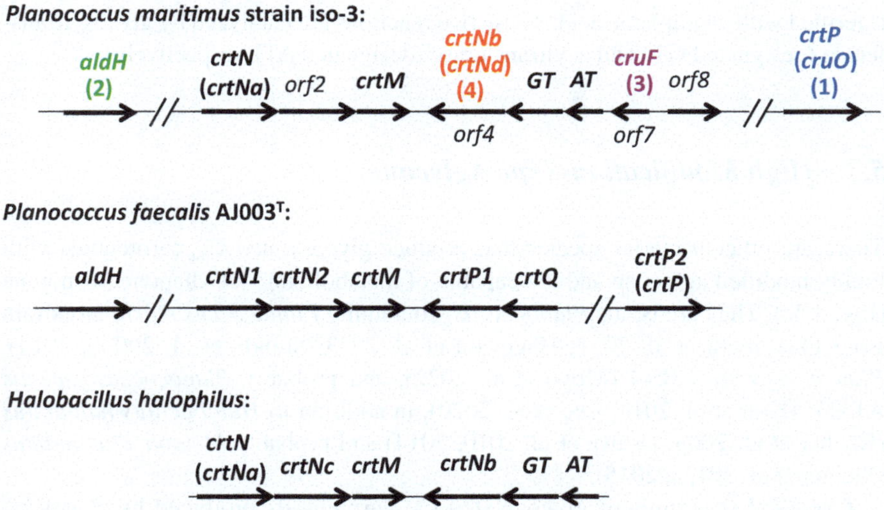

Fig. 4.5 Organization of the C_{30}-carotenoid biosynthesis genes in *Planococcus maritimus* strain iso-3 and other relevant bacterial strains. The *aldH* and *crtP* (renamed *cruO*) of *P. maritimus* iso-3 exist apart from the carotenogenic gene cluster. Direction and relative position of these two gene are described according to the corresponding genes of *Planococcus faecalis* AJ003^T (Lee et al. 2020)

Homologs of the *Planococcus* iso-3 *crtNb* gene were present in the carotenogenic gene clusters of other *Planococcus* strains such as *P. faecalis* AJ003^T (*crtP1*; Kim et al. 2018) and in the carotenogenic gene cluster of *H. halophilus* (Kőcher et al. 2009) at the same position (Fig. 4.5), whereas their functions had not been elucidated. The reaction with (4) was similar to that of the C_{40}-caroteogenic CrtD. Figure 4.6 shows the phylogenetic tree of *crtN* family. The *P. maritimus* iso-3 *crtNb*, *P. faecalis crtP1*, and *H. halophilus crtNb* genes for (4) formed one independent clade, which was a comparatively distant from the *crtNb* (*crtP*) clade containing genes for (1) and rather close to the *crtD* clade. We suggest that the gene name *crtNb* of *P. maritimus* iso-3 and *H. halophilus* should be changed to *crtNd*, since this *crtNb* is functionally different from a gene belonging to the *crtNb* (*crtP*) clade.

5-Hydroxy-5,6-dihydro-4,4′-diapolycopene is converted into aglycone 5-hydroxy-5,6-dihydro-4,4′-diapolycopen-4′-oic acid by a series of reaction with (1) and (2) (Fig. 5.4c). Here we showed the route with the order of (3) (4) (1) (2). Another route with the order of (3) (1) (4) (2), (3) (1) (2) (4), (1) (2) (3) (4), or (1) (3) (2) (4) may take place, if the substrate preference of (4) [and (3)] is broad. However, it was not observed that such a route occurred in our functional expression experiments with the corresponding genes using *E. coli*.

The high modification-type aglycone, 5-hydroxy-5,6-dihydro-4,4′-diapolycopen-4′-oic acid, is linked with D-glucose at the 4-position (Hagiwara et al. 2023), and the added glucosyl group is occasionally esterified with an aliphatic acid (Moyo et al. 2022). These two reactions are catalyzed by glycosyltransferase (GT)

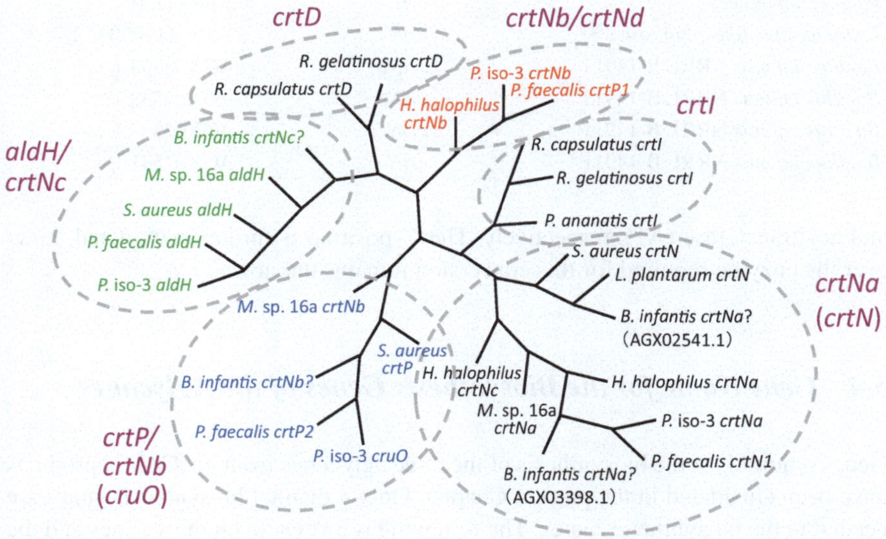

Fig. 4.6 Phylogenetic tree of *crtN* family. Homology search was performed using BLAST (http://blast.ncbi.nlm.nih.gov/Blast.cgi). Amino acid alignment of the encoded proteins was performed using MAFFT(http://www.mafft.ccbrc.jp/) and phylogenetic tree was constructed using the neighbor joining method, both as implemented on the GenomeNet (https://www.genome.jp/tools/ete/). Accession numbers of the individual sequences are listed in Table 4.1

Table 4.1 Examples of *crtN*-homologous genes and their accession numbers

Bacterial strain	Gene	Accession No.
Staphylococcus aureus	*crtP*	WP_102782088.1
Staphylococcus aureus	*crtN*	WP_057520765.1
Staphylococcus aureus	*aldH*	WP_001084326.1
Halobacillus halophilus DSM 2266	*crtNa*	ACM07424.1
Halobacillus halophilus DSM 2266	*crtNb*	ACM07427.1
Halobacillus halophilus DSM 2266	*crtNc*	ACM07425.1
Methylomonas sp. strain 16a	*crtNa*	AAX46183.1
Methylomonas sp. strain 16a	*crtNb*	AAX46185.1
Methylomonas sp. strain 16a	*aldH*	AAX46184.1
Planococcus maritimus strain iso-3	*crtNa*	LC620265
Planococcus maritimus strain iso-3	*crtP* (*cruO*)	LC722835.1
Planococcus maritimus strain iso-3	*crtNb* (*crtNd*)	LC620265
Planococcus maritimus strain iso-3	*aldH*	LC722836.1
Planococcus faecalis AJ003^T	*crtN1*	AQU80595.1
Planococcus faecalis AJ003^T	*crtP2*	AQU78488.1
Planococcus faecalis AJ003^T	*crtP1*	AQU80592.1
Planococcus faecalis AJ003^T	*aldH*	AQU79998.1
Rhodobacter capsulatus	*crtI*	CAA77540.1
Rhodobacter capsulatus	*crtD*	CAA77544.1
Rubivivax gelatinosus	*crtI*	BAA94063.1
Rubivivax gelatinosus	*crtD*	BAA94046.1
Pantoea ananatis	*crtI*	AER34890
Lactiplantibacillus plantarum	*crtN*	WP_011102097.1
Bacillus infantis NRRL B-14911	*crtNa*?	AGX03398.1
Bacillus infantis NRRL B-14911	*crtNb*	AGX02538.1
Bacillus infantis NRRL B-14911	*crtNc*	AGX02539.1
Bacillus infantis NRRL B-14911	*crtNa?*	AGX02541.1

and acyltransferase (AT), respectively. The 4′-position is further methylated, however the enzyme required for this conversion remains unknown.

5.4 Gene Name for the Biosynthesis Genes of the Aglycones

Genes required for the biosynthesis of the three aglycones from the CrtMN products have been elucidated in the present Chapter. Only 4 distinct biosynthesis genes are needed in the biosynthetic routes. The following is a keynote on the 4 genes and the corresponding gene products. (1) (2) (3) and (4) are the assigned numbers (Fig. 4.4):

(1) 4,4′-diaponeurosporene 4-oxidase (diaponeurosporene oxidase)
4,4′-diaponeurosporene/4,4′-diapolycopene terminal oxidase (aldehyde synthase).

The gene was named *crtP* (Pelz et al. 2005), *crtNb* (Tao et al. 2005), or *cruO* (Hagiwara et al. 2023).

crtP has extensively been used as the cyanobacterial phytoene desaturase gene. Thus, we thought that the same gene name should be confusing and proposed that it should be renamed, e.g., *cruO* (Hagiwara et al. 2023). However, this *crtP* for the C_{30}-carotenoid metabolism gains wider use especially in *Staphylococcus aureus*, and thus, this *crtP* is likely to be recognized as one of the appropriate gene names. *crtNb* also gains wider use, except for *S. aureus* (Fig. 4.6). Consequently, we suggest that the gene name (and the gene-product name) should be *crtP* (CrtP), *crtNb* (CrtNb) or *crtP*/*crtNb* (CrtP/CrtNb), or *cruO* (CruO).

(2) 4,4′-diaponeurosporen-4-al 4-dehydrogenase (aldehyde dehydrogenase)

4,4′-diaponeurosporen-4-al/4,4′-diapolycopen-4-al 4-dehydrogenase.

The gene was designated *aldH* (Kim and Lee 2012) or *crtNc* (Steiger et al. 2015). Similarly, we suggest that the gene name (and the gene-product name) should be *aldH* (AldH), *crtNc* (CrtNc), or *aldH*/*crtNc* (AldH/CrtNc).

(3) 4,4′-diaponeurosporene 5,6-hydratase (diaponeurosporene hydratase)

The gene was named *cruF* (Takemura et al. 2021). The gene-product name is CruF.

(4) 5-hydroxy-5,6-dihydro-4,4′-diaponeurosporene 7,8-desaturase (hydroxydiaponeurosporene desaturase)

The gene was designated *crtNb* (Takemura et al. 2021).

This *crtNb* gene for (4) is functionally different from the *crtNb* gene for (1). Such a same gene name for two distinct genes is confusing and may lead to misunderstanding. Thus, we suggest that the gene name (and the gene-product name) should be renamed *crtNd* (CrtNd), or described as *crtNb*/*crtNd* (CrtNb/CrtNd).

6 Biological Functions of C_{30}-Carotenoids

It has been considered that the conjugation length of a carotenoid is correlated with the strength of activity quenching singlet oxygen ($^1O_2$). C_{30} carotenoids generally can retain long conjugated bonds, e.g. as for 4,4′-diapolycopene, number of the double bonds with conjugation function is 11, which is comparable with 11 of lycopene. Siziya et al. (2022a) exhibited anti-oxidative activity of various C_{30} carotenoids. Using the methylene blue procedure we determined $^1O_2$-quenching potential for C_{30} carotenoids that *P. maritimus* iso-3 can produce or *E. coli*, carrying combinations of the *P. maritimus* iso-3 carotenogenic genes, can produce (Takemura et al. 2021; Hagiwara et al. 2023; Shindo et al. 2014). The result is shown in Table 4.2. The examined C_{30} carotenoids showed moderate or potent antioxidative activity, except for 4,4′-diapophytoene. The more modification of C_{30}-carotenoid metabolites proceeded, the more potent antioxidative activities tended to be gained. Especially, the final product, methyl-5-glucosyl-5,6-dihydro-4,4′-diapolycopenoate, showed the most potent antioxidative activity (IC_{50} 5.1 μM), comparable with that of astaxanthin (IC_{50} 3.7 μM).

Table 4.2 Singlet oxygen ($^1O_2$)-quenching activity of C_{30}-carotenoids (intermediates) relevant to *Planococcus maritimus* strain iso-3

Carotenoid (Structure)	$^1O_2$-quenching IC_{50} (μM)
15-*cis*-4,4′-Diapophytoene (Fig. 4.2)	>100
4,4′-Diaponeurosporene (Fig. 4.2)	45
5-hydroxy-5,6-dihydro-4,4′-Diaponeurosporene (Fig. 4.4c)	56
5-hydroxy-5,6-dihydro-4,4′-diapolycopene (Fig. 4.4c)	30
5-hydroxy-5,6-dihydro-4,4′-Diapolycopen-4′-al (Fig. 4.4c)	9.3
5-hydroxy-5,6-dihydro-4,4′-Diapolycopen-4′-oic acid (Fig. 4.4c; high modification-type aglycone)	7.8
5-glucosyl-5,6-dihydro-4,4′-Diapolycopene	30
5-glucosyl-5,6-dihydro-4,4′-Diapolycopen-4′-al	8.3
5-glucosyl-5,6-dihydro-4,4′-Diapolycopen-4′-oic acid	12
methyl-5-glucosyl-5,6-dihydro-4,4′-Diapolycopenoate (Fig. 4.1c; the final metabolite)	5.1
Astaxanthin	3.7

6.1 *Case of* Staphylococcus aureus

Human- and animal-pathogenic *Staphylococcus aureus* causes a variety of suppurative diseases, or produces a thermo-stable enterotoxin to cause food poisoning. As the final product, this bacterium produces staphyloxanthin, as exhibited in Fig. 4.1a. This carotenoid was reported to confer oxidative stress tolerance on *S. aureus* (Clauditz et al. 2006). Recently, staphyloxanthin was shown to be one of the most prominent virulence factors of *S. aureus*, and multiple strategies were provided to attenuate this virulence factor (Elmesseri et al. 2022). Discovery of potential staphyloxanthin inhibitor candidates, is a next challenge. Thus, its biosynthetic pathway shown here should provide suggestive knowledge.

6.2 *Case of* Planococcus *Species*

Some species of *Planococcus* have been reported to degrade solvents such as aromatic hydrocarbons and alkanes (Engelhardt et al. 2001; Li et al. 2006) and to acquire tolerance to industrial waste-water brine effluent (Moyo et al. 2022). A yellow marine bacterium *Planococcus maritimus* strain iso-3 was isolated from an intertidal sediment that looked industrially polluted, from the Clyde estuary in the UK (Shindo et al. 2008b). Its yellow pigment was identified as methyl 5-glucosyl-5,6-dihydro-4,4′-diapolycopenoate, which showed considerably potent antioxidative activity, as shown (Shindo et al. 2014). The iso-3 strain was found to exert tolerance to 0.3% *p*-xylene, 0.3% *p*-toluic acid, 1% benzene, and 1% toluene, when examined by shaking culture with Marine Broth (Difco 2216) including each solvent (JBA report 2004).

The genus *Planococcus* retains two more carotenogenic genes *cruF* and *crtNb/crtNd,* compared to the carotenogenic genes of bacteria with the other types of aglycone. It is likely that this existence of the two additional genes made carotenoids of *Planococcus* highly modified ones with stronger antioxidative activity. It is also speculated that dipolar structure of the final C_{30}-carotenoids in *Planococcus* exists through the phospholipid bilayers of the cell membrane with intercalating the hydrophobic part in the middle lipid layer of the membrane, and there exerts tolerance against various stress such as solvents.

7 Conclusions

This chapter revealed biosynthetic pathway from FPP to the final C_{30}-carotenoid metabolites by way of the three aglycones in a comprehensive manner. Biological functions of such C_{30}-carotenoids are lastly discussed and await future studies.

Acknowledgements The authors are grateful to Prof. Gerhard Sandmann for valuable information about C_{30} carotenoids. NM thanks to "Development of a Technological Infrastructure for Industrial Bioprocess" project in 2000-2004, which was entrusted by NEDO (New Energy and Industrial Technology Development Organization), and Marine Biotechnology Institute, since they offered circumstances of the research for his group of the day to start study on C_{30} carotenoids with *P. maritimus* strain iso-3.

Authors' Contributions N. Misawa conceptualized the content and was a main writer in this paper. M. Takemura and K. Shindo also wrote the manuscript. W. Matsumoto did additional experiments to support the paper. P.D. Fraser provided *P. maritimus* strain iso-3, and improved the manuscript.

References

Britton G, Liaaen-Jensen S, Pfander H (eds) (2004) Carotenoid handbook. Birkhäuser Verlag, Basel, Boston, Berlin

Clauditz A, Resch A, Wieland KP, Peschel A, Götz F (2006) Staphyloxanthin plays a role in the fitness of *Staphylococcus aureus* and its ability to cope with oxidative stress. Infect Immun 74:4950–4953

Elmesseri RA, Saleh SE, Elsherif HM, Yahia IS, Aboshanab KM (2022) Staphyloxanthin as a potential novel target for deciphering promising anti-*Staphylococcus aureus* agents. Antibiotics 11(3):298. https://doi.org/10.3390/antibiotics11030298

Engelhardt MA, Daly K, Swannell RPJ, Head IM (2001) Isolation and characterization of a novel hydrocarbon-degrading, gram-positive bacterium, isolated from intertidal beach sediment, and description of *Planococcus alkanoclasticus* sp. nov. J Appl Microbiol 90:237–247

Fraser PD, Misawa N, Linden H, Yamano S, Kobayashi K, Sandmann G (1992) Expression in *Escherichia coli*, purification and reactivation of the recombinant *Erwinia uredovora* phytoene desaturase. J Biol Chem 267:19891–19895

Garrido-Fernández J, Maldonado-Barragán A, Caballero-Guerrero B, Hornero-Méndez D, Ruiz-Barba JL (2010) Carotenoid production in *Lactobacillus plantarum*. Int J Food Microbiol 140:34–39

Hagi T, Kobayashi M, Nomura M (2018) Whole-transcriptome analysis of oxidative stress response genes in carotenoid-producing *Enterococcus gilvus*. Biosci Biotechnol Biochem 82:1053–1057

Hagiwara M, Maehara C, Takemura M, Misawa N, Shindo K (2023) Production of highly modified C_{30}-carotenoids with singlet oxygen-quenching activities, 5-glucosyl-5, 6-dihydro-4,4'-diapolycopen-4'-oic acid, and its three intermediates using genes from *Planococcus maritimus* strain iso-3. Syn Biol Engineer 1:10002

Japan Bioindustry Association (JBA) (2004) Report on "Development of a Technological Infrastructure for Industrial Bioprocess" project, entrusted by NEDO (New Energy and Industrial Technology Development Organization) (Japanese)

Kim JW, Choi BH, Kim JH, Kang H-J, Ryu H, Lee PC (2018) Complete genome sequence of *Planococcus faecalis* AJ003T, the type species of the genus *Planococcus* and a microbial C30 carotenoid producer. J Biotechnol 266:72–76

Kim SH, Lee PC (2012) Functional expression and extension of Staphylococcal staphyloxanthin biosynthesis pathway in *Escherichia coli*. J Biol Chem 287:21575–21583

Köcher S, Breitenbach J, Müller V, Sandmann G (2009) Structure, function and biosynthesis of carotenoids in the moderately halophilic bacterium *Halobacillus halophilus*. Arch Microbiol 191:95–104

Krinsky NI (2001) Carotenoids as antioxidants. Nutrition 17:815–817

Lee JH, Kim JW, Lee PC (2020) Genome mining reveals two missing CrtP and AldH enzymes in the C30 carotenoid biosynthesis pathway in *Planococcus faecalis* AJ003^{T}. Molecules 25:5892–5900

Li H, Liu YH, Luo N, Zhang XY, Luan TG, Hu JM, Wang ZY, Wu PC, Chen MJ, Lu JQ (2006) Biodegradation of benzene and its derivatives by a psychrotolerant and moderately haloalkaliphilic *Planococcus* sp. strain ZD22. Res Microbiol 157:629–636

Maoka T (2020) Carotenoids as natural functional pigments. J Nat Med 74:1–16. https://doi.org/10.1007/s11418-019-01364-x

Marshall JH, Wilmoth GJ (1981) Proposed pathway of triterpenoid carotenoid biosynthesis in *Staphylococcus aureus*: evidence from a study of mutants. J Bacteriol 147:914–919

Matsumoto W, Takemura M, Nanaura H, Ami Y, Maoka T, Shindo K, Kurihara S, Misawa N (2024) Carotenoid productivity in human intestinal bacteria *Eubacterium limosum and Leuconostoc mesenteroides*with functional analysis of their carotenoid biosynthesis genes. Engineering Microbiology 4, 100147

Misawa N (ed) (2021) Carotenoids: biosynthetic and biofunctional approaches, Advances in experimental medicine and biology, vol 1261. Springer, Singapore

Misawa N, Maoka T, Takemura M (2022) Carotenoids: carotenoid and apocarotenoid analysis — use of *E. coli* to produce carotenoid standards. Methods Enzymol 670:87–137

Mo XH, Sun YM, Bi YX, Zhao Y, Yu GH, Tan LL, Yang S (2023) Characterization of C_{30} carotenoid and identification of its biosynthetic gene cluster in *Methylobacterium extorquens* AM1. Synthetic and Systems Biotechnology 8(3):527–535. https://doi.org/10.1016/j.synbio.2023.08.002

Moyo AC, Dufossé L, Giuffrida D, Joaquim Van Zyl L, Trindade M (2022) Structure and biosynthesis of carotenoids produced by a novel *Planococcus* sp. isolated from South Africa. Microb Cell Factories 21:43. https://doi.org/10.1186/s12934-022-01752-1

Osawa A, Ishii Y, Sasamura N, Morita M, Köcher S, Müller V, Sandmann G, Shindo K (2010) Hydroxy-3,4-dehydro-apo-8′-lycopene and methyl hydroxy-3,4-dehydro-apo-8′-lycopenoate, novel C_{30}-carotenoids produced by a mutant of marine bacterium *Halobacillus halophilus*. J Antibiot 63:291–295

Osawa A, Ishii Y, Sasamura N, Morita M, Köcher S, Müller V, Sandmann G, Shindo K (2014) 5-Hydroxy-5,6-dihydro-apo-4,4′-lycopene and methyl 5-hydroxy-5,6-dihydro-apo-4,4′-lycopenoate, novel C_{30}-carotenoids produced by a mutant of marine bacterium *Halobacillus halophilus*. J Antibiot, Corrigendum 67:733–735

Pelz A, Wieland K-P, Pitzbach K, Hentschel P, Alberl K, Götz F (2005) Structure and biosynthesis of staphyloxanthin from *Staphylococcus aureus*. J Biol Chem 16:32493–32948

Raisig A, Sandmann G (1999) 4, 4′-diapophytoene desaturase: catalytic properties of an enzyme from the C30 carotenoid pathway of *Staphylococcus aureus*. J Bacteriol 181:6184–6187

Raisig A, Sandmann G (2001) Functional properties of diapophytoene and related desaturases of C30 and C40 carotenoid biosynthetic pathways. Biochim Biophys Acta 1533:164–170

Sandmann G, Misawa N (2021) Carotenoid production in *Escherichia coli*: case of acyclic carotenoids. In: Misawa N (ed) Carotenoids: biosynthetic and biofunctional approaches. Advances in experimental medicine and biology, vol 1261. Springer, Singapore. https://doi.org/10.1007/978-981-15-7360-6_17

Shindo K, Asagi E, Sano A, Hotta E, Minemura N, Mikami K, Tamesada E, Misawa N, Maoka T (2008a) Diapolycopenedioic acid xylosyl esters A, B, and C, novel antioxidative glyco-C_{30}-carotenoic acids produced by a new marine bacterium *Rubritalea squalenifaciens*. J Antibiot 61:185–191

Shindo K, Endo M, Miyake Y, Wakasugi K, Morritt D, Bramley PM, Fraser PD, Kasai H, Misawa N (2008b) Methyl glucosyl-3,4-dehydro-apo-8′-lycopenoate, a novel antioxidative glyco-C_{30}-carotenoic acid produced by a marine bacterium *Planococcus maritimus*. J Antibiot 61:729–735

Shindo K, Endo M, Miyake Y, Wakasugi K, Morritt D, Bramley PM, Fraser PD, Kasai H, Misawa N (2014) Methyl 5-glucosyl-5,6-dihydro-apo-4,4′-lycopenoate, a novel antioxidative glyco-C_{30}-carotenoic acid produced by a marine bacterium *Planococcus maritimus*. J Antibiot, Corrigendum 67:731–732

Siziya IN, Hwang CY, Seo M-J (2022a) Antioxidant potential and capacity of microorganism-sourced C_{30} carotenoids—a review. Antioxidants 11:1963. https://doi.org/10.3390/antiox11101963

Siziya IN, Yoon DJ, Kim M, Seo M-J (2022b) Enhanced production of C_{30} carotenoid 4,4′-diaponeurosporene by optimizing culture conditions of *Lactiplantibacillus plantarum* subsp. *plantarum* KCCP11226T. J Microbiol Biotechnol 32(7):892–901. https://doi.org/10.4014/jmb.2204.04035. Epub 2022 May 31. PMID: 35637169; PMCID: PMC9628921

Stahl W, Sies H (2003) Antioxidant activity of carotenoids. Mol Asp Med 24:345–351. ISSN 0098-2997. https://doi.org/10.1016/S0098-2997(03)00030-X

Steiger S, Perez-Fons L, Cutting SM, Fraser PD, Sandmann G (2015) Annotation and functional assignment of the genes for the C30 carotenoid pathways from the genomes of two bacteria: *Bacillus indicus* and *Bacillus firmus*. Microbiology 161:194–202

Steiger S, Perez-Fons L, Fraser PD, Sandmann G (2012a) Biosynthesis of a novel C30 carotenoid in *Bacillus firmus* isolates. J Appl Microbiol 113:888–895

Steiger S, Perez-Fons L, Fraser PD, Sandmann G (2012b) The biosynthetic pathway to a novel derivative of 4,4′-diapolycopene-4,4′-oate in a red strain of *Sporosarcina aquimarina*. Arch Micribiol 194:779–784

Sun Z, Shen S, Wang C, Wang H, Hu Y, Jiao J, Ma T, Tian B, Hua Y (2009) A novel carotenoid 1,2-hydratase (CruF) from two species of the non-photosynthetic bacterium *Deinococcus*. Microbiology 155:2775–2783

Taylor RF (1984) Bacterial triterpenoids. Microbiol Rev 48(3):181–198

Taylor RF, Davies BH (1974) Triterpenoid carotenoids and related lipids. Triterpenoid monohydroxy- and monoglucosyloxy-carotenoids from *Streptococcus faecium* UNH 564P. Biochem J 139(3):761–769. https://doi.org/10.1042/bj1390761. PMID: 4212487; PMCID: PMC1166340

Takemura M, Takagi C, Aikawa M, Araki K, Choi SK, Itaya M, Shindo K, Misawa N (2021) Heterologous production of novel and rare C_{30}-carotenoids using *Planococcus* carotenoid biosynthesis genes. Microb Cell Factories 20(1):1–12. https://doi.org/10.1186/s12934-021-01683-3

Tao L, Schenzle A, Odom JM, Cheng Q (2005) Novel carotenoid oxidase involved in biosynthesis of 4,4′-diapophytoene dialdehyde. Appl Envion Microbiol 71:3294–3301

Wieland B, Feil C, Gloria-Maercker E, Thumm G, Lechner M, Bravo J-M, Poralla K, Gotz F (1994) Genetic and biochemical analyses of the biosynthesis of the yellow carotenoid 4,4′-diaponeurosporene of *Staphylococcus aureus*. J Bacteriol 176:7719–7726

Xue L, Yang Chen YY, Yan Z, Lu W, Wan D, Zhu H (2019) Staphyloxanthin: a potential target for antivirulence therapy. Infect Drug Resist 12:2151–2160. https://doi.org/10.2147/IDR.S193649

Chapter 5
Overview of Recent Trends on Non-conventional Approaches to Extract Carotenoids

Marielba de los Angeles Rodriguez Salazar, Jhonathan Vinícius Menezes Silva, Ramon Sousa Barros Ferreira, and Renan Campos Chisté

1 Introduction

Carotenoids are remarkable compounds widely found in nature that provide very attractive colors and key biological functions to plants. These certainly impact our daily diet, which is frequently composed of fruits and vegetables. Furthermore, carotenoids are also found in animals and microorganisms, and relevant information concerning understanding their physiological role in such organisms is increasingly present in the literature (Swapnil et al. 2021; Crupi et al. 2023).

Over decades, food industries have used carotenoids as natural colorants to impart yellowish, orangish, or reddish colors to food products by recovering the color lost during processing or providing attractive colors to new products. In addition to the coloring properties, food products with carotenoid addition certainly contribute to healthier eating habits as carotenoids are bioactive compounds with essential physiological functions to human body, such as the provitamin A activity of β-carotene (and other few structures), the strengthening of the immune system

Marielba de los Angeles Rodriguez Salazar, Jhonathan Vinícius Menezes Silva and Ramon Sousa Barros Ferreira contributed equally with all other contributors.

M. de los Angeles Rodriguez Salazar
Graduate Program of Food Science and Technology (PPGCTA), Institute of Technology (ITEC), Federal University of Pará (UFPA), Belém, PA, Brazil

J. V. M. Silva · R. S. B. Ferreira
School of Food Engineering (FEA), University of Campinas (UNICAMP), Campinas, São Paulo, Brazil

R. C. Chisté (✉)
Faculty of Pharmacy, Universidade Federal de Minas Gerais (UFMG), Belo Horizonte, MG, Brazil
e-mail: rcchiste@ufmg.br

R. Campos Chisté et al. (eds.), *Carotenoids*,
https://doi.org/10.1007/978-3-031-75322-0_5

associated to some biological properties, such as the antioxidant, anti-inflammatory and antimicrobial activities, which decrease the risk of the development of chronic degenerative diseases (Terao 2023). Chapters 10, 11, and 12 of this book discuss the carotenoid composition in foods and their biological properties.

As discussed in Chaps. 1 and 2, carotenoids are isoprenoid compounds mainly composed of 40-carbon tetraterpenes formed by eight units of isoprenes. However, smaller structures, such as apocarotenoids, are also common in nature. Carotenoids are lipophilic compounds with an extensive system of conjugated double bonds in their chemical structure that generate electron resonance systems (π) moving throughout the entire polyene chain. Due to these structural characteristics, carotenoids are highly reactive molecules and mostly absorb electromagnetic radiation in the visible region (360–780 nm), which is responsible for the yellow, orange, and red colors, yet colorless carotenoids, such as phytoene and phytofluene, have been receiving increasing attention by the scientific community (Mapelli-Brahm and Meléndez-Martínez 2021). To achieve coloring purposes, both high-purity carotenoids (isolated or synthetic) and mixed carotenoids in extracts are eligible being β-carotene and astaxanthin the most used high-purity carotenoids in foods and feeds, respectively, but extracts of natural sources, such as annatto seeds (bixin), paprika (capsanthin and capsorubin), saffron (crocin), and tomatoes (lycopene) are historically used over the centuries.

However, when extracted from their source and applied to foods, carotenoids exhibit low stability and are susceptible to loss of color and biological activity. The same structural characteristics that confer interesting biological properties to carotenoids are also responsible for making these compounds highly susceptible to degradation by isomerization and oxidation, especially during extraction or when exposed to some food processing and storage conditions, such as the presence of oxygen and reactive oxygen species (ROS), acidic media, exposure to light and high temperatures, action of oxidative enzymes that are present in food, among other factors (Boon et al. 2010), as illustrated in Fig. 5.1.

Thus, given the susceptibility of carotenoids to degradation (Fig. 5.1), it is important to develop reliable and sustainable strategies to overcome most of the limitations imposed by the usual conditions of solid-liquid extractions to produce natural food additives/extracts. Given this framework, this Chapter was focused on discussing non-conventional approaches to the extraction of carotenoids, namely microwave-assisted extraction, pressurized liquid extraction, supercritical fluid extraction, use of green solvents (ionic liquids, deep eutectic solvents, bio-based solvents), pulsed electric field-assisted extraction, enzyme-assisted extraction, and ohmic heating extraction. These technological approaches have been presented in the literature as alternative processes to increase stability and carotenoid yields in the extracts to be further used as food ingredients by the food industry or as nutraceuticals by the pharmaceutical industries.

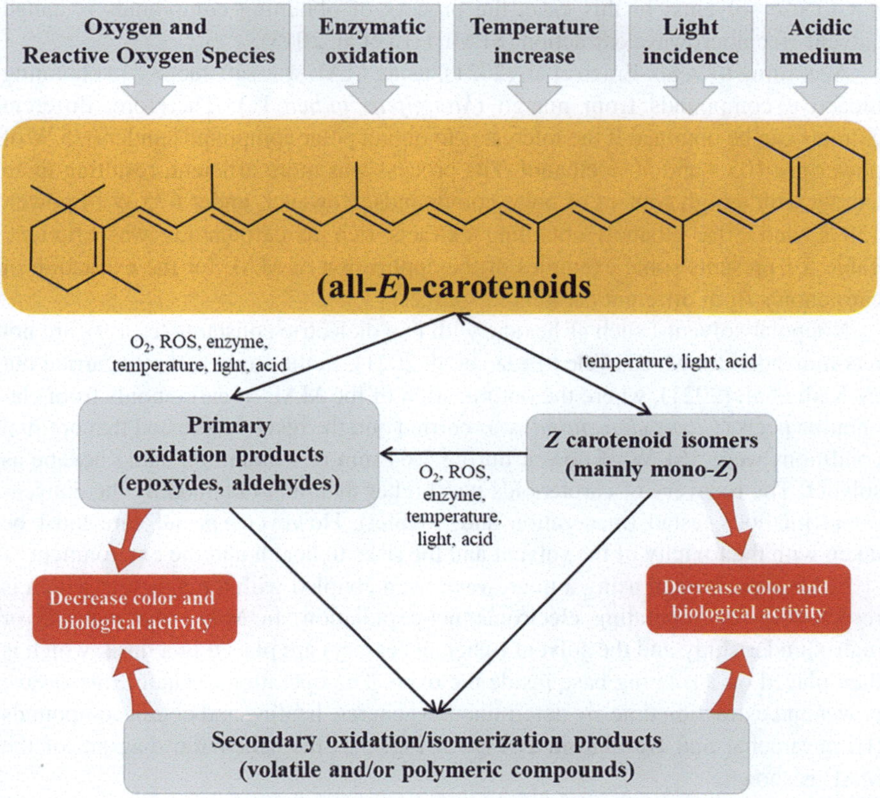

Fig. 5.1 General scheme of carotenoid degradation by oxidation and isomerization during carotenoid extraction, food processing, and storage conditions. (Adapted from Borsarelli and Mercadante (2009))

2 Microwave-Assisted Extraction (MAE)

MAE is a technology for obtaining bioactive compounds based on the interconnection of magnetic and electric fields of microwave energy with the material or matrix under study, resulting in heating. In the extraction process, a change from microwave energy to thermal energy is produced (López-Salazar et al. 2023). Thus, the MAE allows the transfer of mass and heat from the interior of the matrix under study to the exterior (Li et al. 2013).

Using solvents with greater dissipation and dielectric constants (ε_r) is advisable in MAE. Furthermore, solvent polarity is an important factor in the solubility and extraction of bioactive compounds. Polar solvents such as water (ε_r: 80.4) and ethanol (ε_r: 24.4) can easily absorb microwave energy (Nour et al. 2021). Also, due to the toxicity of some solvents and the negative effect on the environment and human health, in the MAE process, it is possible to use the water present in the matrix or not to use solvents; in this case, the process of obtaining compounds is called solvent-free microwave extraction (SFME) (Li et al. 2013).

According to Vélez-Erazo et al. (2021), using MAE was satisfactory in obtaining bioactive compounds from mango (*Mangifera indica* L.). Therefore, different extracts can be obtained if the interest is to obtain polar compounds under 675 W of power per 105 s and 50% ethanol. The process was more efficient, resulting in an extract with a high content of polar compounds. However, under 675 W of power, 150 s, and 80% ethanol, obtaining extracts rich in carotenoids was efficient. Table 5.1 presents some examples of the application of MAE for the extraction of carotenoids from different carotenoid sources.

Nonpolar solvents, such as hexane with low dielectric constants (ε_r: 1.9), are not recommended for MAE (Vélez-Erazo et al. 2021). In the study recently carried out by Kadi et al. (2021), where the optimization of the MAE of carotenoids from clementine peels (*Citrus clementina*) was carried out, the results indicated that optimal conditions were 561 W of power, during 7.64 min of irradiation, using hexane as solvent. The recovery of carotenoids was higher than that obtained by the conventional methods tested (maceration and Soxhlet). However, special care must be taken with the toxicity of the solvent and the risks to health and the environment.

MAE is performed using a microwave oven coupled with a magneton, which is responsible for generating electromagnetic radiation. In MAE, the material or matrix under study and the solvent (when necessary) are placed in a flask, which is then placed on a rotating base inside the oven. The operational data of microwave power and extraction time are determined to generate heating and obtain compounds (Hiranvarachat and Devahastin 2014). In Fig. 5.2, the schematic diagram of the MAE is shown.

For the MAE process to be efficient, the following parameters must be considered: type of solvent, volume, compounds of interest to be extracted, dielectric constants, exposure time, temperature, size and characteristics of the material or matrix under study, power, and type of equipment. These parameters are essential to avoid the degradation of compounds and thus obtain a higher extraction rate (Nour et al. 2021; López-Salazar et al. 2023).

In MAE, the separation of the substance is achieved by adjusting the parameters and heat generation caused by electromagnetic radiation. Although β-carotene can be extracted quickly using common equipment, thermal degradation and cis/trans isomerization of carotenoids can still occur (Huang et al. 2023). According to Rifqi et al. (2023), the extraction of β-carotene by MAE has a positive effect on the profile and quality of the final product as the method helps to reduce the amount of solvent used, reduce the extraction time and increase the yield of compounds in the extraction. Therefore, it can be considered that MAE is a viable and acceptable green technology for obtaining carotenoids.

As carotenoids are lipophilic and have good solubility in oil, MAE using vegetable oils as a green solvent can be a good alternative. In the study by Elik et al. (2020), MAE optimization was carried out to obtain β-carotene from carrot waste (*Daucus carota* L) using linseed oil as a solvent. The β-carotene content in flaxseed oil after MAE was improved, and it was observed that microwave power and

Table 5.1 Application of microwave-assisted extraction (MAE) of carotenoids from different plant sources

Plant sources	Part used	Extraction condition	Solvent	Carotenoid determination	Carotenoid contents	References
Tomato (*Lycopersicon esculentum*)	Shell and seeds Dried in a vacuum oven at 50 °C for 5 h) and ground	300 W power, time: 60 s	95% EtOH	UV-Vis spectrophotometry, solvent: acetone/hexane. λ: 453, 505, 663 and 645 nm	Lycopene 477.5 μg/g β-carotene 112.4 μg/g.	Ershidat (2023)
Carrot (*Daucus carota* L)	Juice residues. Freeze-dried (1% moisture) and ground (powder <0.85 mm)	165 W power, time: 9.39 min and <110 °C; oil/residue ratio: 8.06:1 g/g	Linseed oil	Total carotenoid contents by UV-Vis spectrophotometry, λ: 450 nm. Quantification of β-carotene by HPLC/UV-Vis, mobile phase: methanol/acetonitrile (90:10 v/v), C18 column.	Total carotenoid recovery: 77.87%	Elik et al. (2020)
Pumpkin (*Cucurbita maxima*)	Peel and pulp Freeze-dried and grinding	130 W power, time: 30 min, 45 °C, sample/oil (1:10)	Corn oil	Total carotenoid contents by UV-Vis spectrophotometry, solvent: cyclohexane λ: 470 nm	Bark: 34.94 μg/g extracts Pulp 31.067 μg/g extracts	Sharma and Bhat (2021)
Passion fruit (*Passiflora edulis*),	Bark Freeze-dried and ground (powder <0.250 mm)	200 W power, time: 25 min and < 110 °C;	Olive oil	Total carotenoid contents by UV-Vis spectrophotometry, λ: 450 nm	Total carotenoid recovery: 86.9% in oil	Chutia and Mahanta (2021)
Gac or Gak (*Momordica cochinchinensis*)	Bark Dried to 4% moisture and milled and sieved (250–500 μm particle)	360 W power, time: 4 min and 60 °C, solvent/sample (80:1 mL/g)	Ethyl acetate	Total carotenoid contents by UV-Vis spectrophotometry, λ: 450 nm	236 mg/100 g (dry basis)	Chuyen et al. (2018)

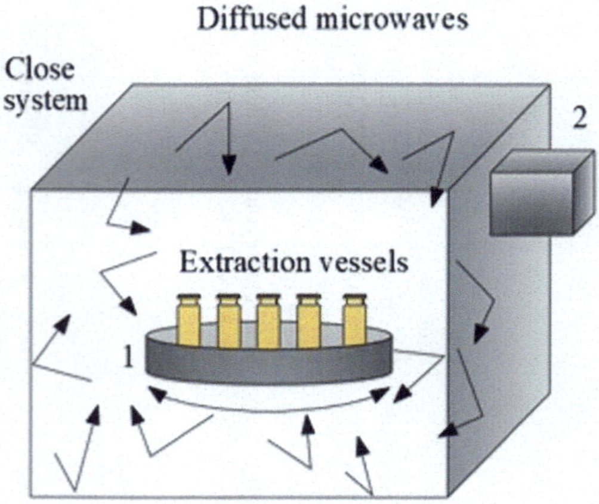

Fig. 5.2 Scheme of MAE systems using diffused microwaves. *1*- Turntable; *2*- Magnetron

extraction time affected extraction. Sharma and Bhat (2021) carried out a study on obtaining carotenoids from the peel and pulp of pumpkin (*Cucurbita maxima*) with MAE using corn oil as a solvent and conventional extraction with hexane/isopropyl alcohol (60:40 v/v), they found. It was found that carotenoid extraction and the oxidative stability of the extracts were improved with MAE than with conventional extraction.

Hiranvarachat and Devahastin (2014) carried out work on improving MAE via intermittent radiation to avoid excessive thermal degradation of β-carotene from carrot peel. The results obtained showed that the use of appropriate microwave powers and solvent/sample ratios (180 W/75 mL: 2 g and 300 W/150 mL: 2 g) resulted in greater amounts of extractable β-carotene compared to the use of continuous MAE.

When compared to a green extraction technology such as PLE, MAE demonstrates greater efficiency in extraction time and viability in terms of lycopene recovery. In a study carried out by Chada et al. (2022), tomato MAE extract (*Solanum lycopersicum* L.) dried in a spouted bed showed a higher lycopene content (59.66 μg lycopene/g extract), which represented a recovery rate of 66.93%, in just 3 min of extraction. However, the extract obtained with PLE presented 20.09 μg of lycopene/g extract.

The MAE process has advantages such as reduced extraction time, less or no use of solvent, reduced waste production, good selectivity, excellent heat transfer, and lower energy consumption (Chutia and Mahanta 2021; Sharma and Bhat 2021). However, MAE applies to thermally stable compounds due to the high temperatures involved in the process. Several studies confirm that extraction technology can be used to obtain carotenoids, as it allows the use of green solvents, and even if high temperatures are used, as short extraction times are used, it is possible to avoid the degradation of the compounds. In MAE, process conditions can vary from 60 to 110 °C to obtain carotenoids, with 130 to 360 W of power and 1–30 min of extraction.

3 Pressurized Liquid Extraction (PLE)

PLE is a technology for extracting bioactive compounds using solvents at high pressures and temperatures but below their critical conditions to be maintained in a liquid state during extraction (Barp et al. 2023). In PLE, high temperatures break interactions between the matrix and compounds of interest. Also, the high pressure helps to keep the solvents in a liquid state and reduces surface tension, causing an increase in the transfer of compounds extracted by the solvent (Mustafa and Turner 2011; Trentini et al. 2017).

Green technologies to obtain bioactive compounds seek the reduction and/or application of green solvents. Ethanol, methanol, and water are the main solvents used in PLE (Kultys and Kurek 2022). In the study carried out by Vardanega et al. (2023), in order to choose the solvent for the PLE of carotenoids from granadilla (*Passiflora ligularis* Juss) bark, conventional extractions were carried out using different solvents. The extract obtained with ethanol (0.40 mg/g) had the highest carotenoid content, followed by the extract obtained with ethyl acetate (0.25 mg/g) and n-hexane (0.11 mg/g). Ethanol was chosen for PLE based on these results. Table 5.2 presents studies where PLE was applied to extract carotenoids.

Garcia-Mendoza et al. (2015) evaluated mango peel *(Mangifera indica* L.) to obtain bioactive extracts using SFE with CO_2 as the solvent and PLE using ethanol. Both extractions were carried out at 40 °C and 300 bar. The best results were obtained in SFE and the authors described that the carotenoid content is dependent on the solvent, being high for supercritical CO_2, with carotenoid contents of 5.60 mg β- carotene/g ext., compared to value obtained by PLE (0.70 mg β- carotene/g ext). In this study, the polarity of CO_2 and the low temperatures used were taken advantage of to ensure better extraction of carotenoids.

However, Achour et al. (2023) described the efficiency of PLE for the recovery of carotenoids from the green and orange biomasses of the microalgae *Dunaliella salina*. The highest levels of carotenoids were found in the orange biomass extract using n-hexane, with the β-carotene isomers being the main carotenoids. The authors proved that the use of this technology is promising in terms of carotenoid recovery. However, when there is an option to use a green solvent, toxic solvents are not recommended, as these types of solvents require a removal step and limit their applications.

The PLE process follows the following steps: the first is packaging the matrix under study in the extraction cell (static extraction, closed system), where the solvent is fed into the extraction cell at a determined pressure and temperature. Dynamic extraction (open system), where the solvent percolates through the matrix, is where it is possible to collect the extract or oil during system depressurization. In the process, an inert gas is used as a purging or entrainment solvent (Cardenas-Toro et al. 2015; Perez-Vazquez et al. 2023). Figure 5.3 shows the schematic diagram of PLE.

As in SFE, this extraction technology requires the matrix under study to undergo pre-treatment. It is dried to remove as much water as possible and ground to reduce

Table 5.2 Application of Pressurized Liquid Extraction (PLE) in the extraction of carotenoids from different plant sources

Plant source	Part used	Extraction condition	Solvent	Determination method	Carotenoids	References
Granadilla or sweet passion fruit (*Passiflora ligularis Juss*)	Bark Dried in a vacuum oven at 30 °C for 24 h, ground and sieved	60 °C/100 bar, flow rate: 1.58 g/min, time: 30 min	Ethanol	HPLC, C18 column, mobile phase: water/acetone, flow: 1.0 mL/min, and λ: 450 nm.	β-carotene: 0.49 mg/g Lutein 0.050 mg/g	Vardanega et al. (2023)
Tomato (*Solanum lycopersicum* L.)	Shell and seed dried in a spouted bed at 70 °C for 1.5 h, ground and sieved (particle size 0.48 mm)	90 °C/100 bar, flow rate: 2 mL/min	Ethanol /ethyl acetate (1:1, v/v)	HPLC/UV-Vis, C18 column, mobile phase: methanol/ acetonitrile, flow: 0.8 mL/min and λ: 470 nm	Lycopene: 20.09 µg/g β-carotene: 46.51%	Chada et al. (2022)
Macauba (*Acrocomia aculeata*)	Pulp Dry in an oven at 60 °C for 8 h, ground and sieved (particle size 0.5 mm)	60 °C/100 bar flow rate: 3 mL/min time: 62 min	Ethanol and isopropanol	β-carotene content by UV-Vis spectrophotometry, solvent: hexane and λ: 450 nm	β-carotene: 232.44 mg/100 g	Trentini et al. (2017)
Palm oil (*Elaeis guineensis*)	Palm fiber Dried in an oven at 30 °C for 24 h, ground and sieved	55 °C/80 bar flow rate: 2.4 g/min time: 62 min	Ethanol	UHPLC-DAD, C18 column, mobile phase: water/acetonitrile, flow: 0.6 mL/min and λ: 450 nm	β-carotene α-carotene	Cardenas-Toro et al. (2015)
Mango (*Mangifera indica* L.)	Bark Freeze-drying for 48 h, ground and sieved (particle size 0.29 mm)	40 °C/300 bar flow rate: 1.1 mL/min time: 5 h	Ethanol	Total carotenoid contents by UV-Vis spectrophotometry, solvent: ethanol and λ: 450 nm	0.35 mg β-carotene/g extract	Garcia-Mendoza et al. (2015)

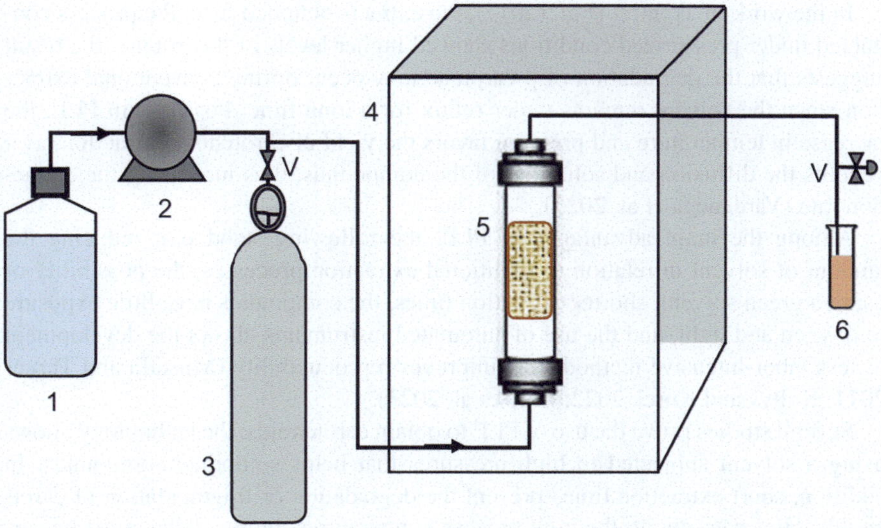

Fig. 5.3 Schematic diagram of PLE system. *1*- Solvent; *2*- Pump; *3*- N2 cylinder; *4*- Oven; *5*- Extractor; *6*- Collection vial; *V*- Flow control valves

particle size and increase the contact surface between the sample and the solvent, thus increasing the mass transfer rate and, consequently, the extraction yield (Chada et al. 2022; Perez-Vazquez et al. 2023).

PLE depends on several parameters to obtain compounds, such as solvent, temperature, pressure, time, and number of extraction cycles. It is important to note that PLE conditions allow the extraction of target analytes quickly, using less solvent and maintaining a high yield (Mustafa and Turner 2011).

Carotenoids are highly sensitive compounds. Thus, the fragmentation or isomerization of β-carotene into compounds of lower molecular mass can occur during exposure to high temperatures, as well as light and acid (Rifqi et al. 2023). The study by Mustafa et al. (2012) carried out PLE of carotenoids from carrot (*Daucus carota*) byproducts, using ethanol as a solvent. The extraction procedure was optimized by varying the extraction time and temperature. The results showed that extraction time and temperatures affect the yield of carotenoids; at 60 °C, the β-carotene content was higher (22.9 mg/100 g) than when 180 °C was used (2. 9 mg/100 g). Analysis of β-carotene in the extraction cycles showed that around 80% was recovered after 20 min.

Also, the effects of PLE with dimethyl sulfoxide and water as solvents on the recovery of carotenoids and other compounds from the diet of microalgae *Spirulina, Chlorella, and Phaeodactylum tricornutum* was evaluated. The results of the study showed that water extracted more proteins, while dimethyl sulfoxide extracted more carotenoids and chlorophylls. Therefore, PLE combined with a suitable extraction solvent can reduce the dependence on high-temperature extraction conditions, which can generate carotenoid degradation (Wang et al. 2023).

In the work of Trentini et al. (2017), the extracts obtained from the process conducted under pressurized conditions showed higher levels of β-carotene; the result suggests that the degradation of β-carotene may occur during conventional extraction since the solvent remains under reflux for a long time. Possibly, in PLE, the increase in temperature and pressure favors the yield of carotenoid extraction, as it benefits the diffusion and solubility of the compounds, thus increasing the extraction rate (Vardanega et al. 2023).

Among the main advantages of PLE, the following stand out: reducing the amount of solvent in relation to traditional extraction processes, the possibility of using a green solvent, shorter extraction times, the compounds have little exposure to oxygen and light, and the use of automated instruments allows the development of less labor-intensive methods and improves reproducibility (Mustafa and Turner 2011; Kultys and Kurek 2022; Barp et al. 2023).

Several studies prove the use of PLE to obtain carotenoids; the technology allows using a solvent subjected to high pressures that helps solubilize compounds. In addition, short extraction times prevent the degradation or fragmentation of carotenoids. However, due to the high process temperatures, more studies must be carried out to improve the PLE of thermosensitive compounds.

4 Supercritical Fluid Extraction (SFE)

SFE is an extraction technology based on using a solvent or pure substance maintained at a temperature and pressure above its critical point (supercritical phase). A pure substance is a supercritical fluid (SF) when, under critical conditions, there is no change in its physical properties (Brunner 2005; Soquetta et al. 2018). SF has a density and diffusivity similar to liquid, presenting high-density values and giving it high solubility power. Furthermore, SF has low viscosity, being similar to that of gases, facilitating the penetration of the solvent into the solid matrix (Larocca et al. 2023;

Carbon dioxide (CO2) is the solvent most used as SF, as it presents mild and easy-to-reach cytic conditions (Tc = 31 °C and Pc = 73 bar), it is generally recognized as safe (GRAS), it has a nonpolar character, it is inert and has low cost (Kultys and Kurek 2022; Viñas-Ospino et al. 2023a). Supercritical CO2 protects bioactive compounds during extraction by providing a non-oxidizing environment, preventing the formation of oxidation products (Vafaei et al. 2022). Therefore, this solvent has excellent applications in extraction processes for thermosensitive and nonpolar compounds such as carotenoids.

The food, cosmetic, and pharmaceutical industries are interested in using carotenoids for various applications obtained through sustainable technology (Mezzomo and Ferreira 2016; de Andrade Lima et al. 2019; Popescu et al. 2022).

Ethanol and water are considered the most suitable cosolvents for obtaining bioactive compounds, as they have GRAS status, do not present toxicity, when present in the extract or oil, can be evaporated, and have good miscibility in CO_2 (Mezzomo and Ferreira 2016; Viñas-Ospino et al. 2023a). Table 5.3 presents studies where SFE was successfully applied to obtain carotenoids from different plant matrices.

According to Pires et al. (2021a, b), in the evaluation of bioactive compounds obtained by SFE from murici pulp (*Byrsonima crassifolia*), extraction with supercritical CO_2 facilitated the obtaining of more nonpolar compounds (oil) and defatted murici pulp. In a subsequent step, SFE with CO_2 + 10% ethanol cosolvent made it possible to obtain a mixture of compounds (lutein and phenolic compounds) with good antioxidant capacity.

SFE can be divided into two stages. The first is extraction, where the ternary system is formed by the plant matrix + extract + CO_2, which is a static extraction mode (closed system). Then, the CO_2 flows through the matrix, solubilizing the components, this being the dynamic extraction mode (open system). The extracted components and CO_2 mixture leaves the extractor and passes to the separator. The second stage is separation, in which the interaction of extract or oil + CO_2 is considered, the solvent is evaporated, and the extract or oil is precipitated in the collecting flask (Bello et al. 2023; Larocca et al. 2023). Figure 5.4 shows the schematic diagram of the SFE.

The matrices under study before SFE require pre-treatment to rupture the cell walls and increase the extraction rate of the compounds of interest. This pretreatment includes drying to obtain less than 10% water content and grinding or reducing particle size to increase the contact surface between the matrix and the solvent (Patel et al. 2019).

Pressure and temperature are the main SFE parameters that directly influence obtaining carotenoids (Vafaei et al. 2022). SFE is based on using solvents at high pressures; therefore, strict operational care must be taken in the processes (de Andrade Lima et al. 2019). In the SFE of β-carotene from bitter melon (*Momordica charantia* L.), the results showed that pressure significantly affected extraction efficiency among the extraction parameters (Patel et al. 2019).

In SFE, very high pressure and temperature conditions could cause the degradation and isomerization of carotenoids. According to the study by Popescu et al. (2022), an oil rich in carotenoids (203.59 mg/100 g extract) and an oleoresin rich in lycopene (1172.32 mg/100 g extract) were obtained using SFE under conditions of 70 °C and 450 bar, using camelina oil as cosolvent. However, Lima et al. (de Andrade Lima et al. 2019) suggest that SFE under conditions of 59 °C, 350 bar, and 15.5% ethanol cosolvent are optimal extraction conditions to extract α-carotene and β-carotene from the pulp and peels of different vegetable matrices. In most matrices, the carotenoid recovery value was greater than 95%.

Obtaining and identifying carotenoids and apocarotenoids directly with SFE and SF chromatography with mass spectrometry detection is a technique that provides a valuable tool that reduces compound degradation, analysis execution time, and sample contamination risks. (Giuffrida et al. 2020). Murador et al. (2019a, b) evaluated the composition of carotenoids in non-saponified extracts of orange peel (*Citrus sinensis* L.) using SFE followed by chromatography with chemical ionization and mass spectrometry detection (SFC-APCI/MS). The method was successfully applied and allowed the identification of 9 carotenoids, 6 carotenoid esters, 19 apocarotenoids,

Table 5.3 Application of supercritical fluid extraction (SFE) in the extraction of carotenoids from different plant sources

Plant source	Part used	Extraction condition	Solvent	Determination method	Carotenoid	References
Banana (*Musa* spp)	Bark Dried in an oven at 50 °C for 72 h and ground (powder: 0.5 mm)	70 °C/250 bar, flow rate: 9.8 g/min, time 150 min	CO_{2+} 5% ethanol	HPLC-DAD, C18 column, mobile phase: methanol/trichloromethane (98:2, v/v), flow 1.0 mL/min and λ: 454 nm	β-carotene in the extract: 1.14 mg/g	Bello et al. (2023)
Tucumã-do-Pará (*Astrocaryum vulgare* Mart.)	Pulp Freeze-dried and sieved (powder: 1.22 mm)	40 °C/400 bar flow rate: 5.31 g/min, dynamic time: 30 min	CO_2	Total carotenoid contents by UV-Vis spectrophotometry, solvent: petroleum ether λ: 450 nm	1351.8 μg carotenoids/g	Menezes et al. (2022)
Murici (*Byrsonima crassifolia*)	Pulp Freeze-dried at −40 °C for 48 h	70 °C/220 bar, dynamic time: 1 h	CO_{2+} 10% ethanol	Total carotenoid contents by UV-Vis spectrophotometry, solvent: ethanol λ: 445 nm	Lutein content: 224.77 μg/g	Pires et al. (2021a, b)
Grapefruit (*Citrus paradisi*)	Endocarp/soft part Freeze-dried at −52 °C for 96 h, ground and sieved	70 °C/305 bar, flow rate: 35 g/min, dynamic time: 135 min	CO_{2+} 5% ethanol	Supercritical fluid chromatography, BEH 2-EP column, mobile phase: CO_2/methanol (85:15, v/v), and λ: 452 nm	Lycopene recovery: 93%	Priyadarsani et al. (2021)
Peach (*Prunus persica*)	Pulp and peel Freeze-dried at −20 °C for 72 h, ground and sieved	59 °C/350 bar, flow rate: 15 g/min, time: 30 min	CO_{2+} 15.5% ethanol	HPLC-DAD, C30 column, mobile phase: methanol/MTBE/water, flow: 1.0 mL/min and λ:450 nm	Total carotenoid recovery for β-carotene: pulp: 99.8% and peel: 94.2%	de Andrade Lima et al. (2019)

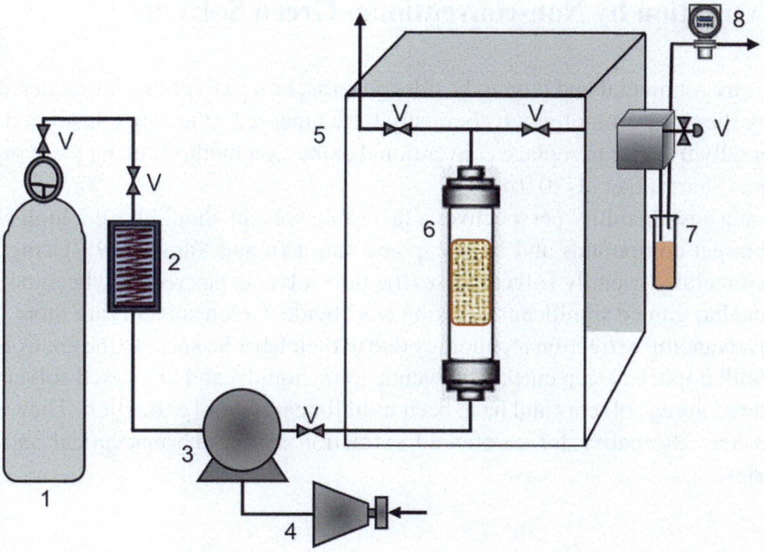

Fig. 5.4 Schematic diagram of SFE system. *1*- CO_2 cylinder; *2*- Cooling bath; *3*- CO_2 pump; *4*- Compressor; *5*- Oven; *6*- Extractor vessel; *7*- Separation vessel; *8*- Flowmeter; *V*- Flow control valves

and 8 apo-esters. The authors suggest excluding the saponification process, as it can lead to the modification of carotenoids, such as isomerization.

In addition, SFE allows the scale-up of extraction processes. Recently, Menezes et al. 2022) carried out work to obtain tucumã-do-Pará (*Astrocaryum vulgare* Mart) oil rich in carotenoids using SFE and applied a new scale-up equation. The process developed was efficient. The study reinforced the strategic potential of using tucumã-do-Pará as a new and promising source of vegetable oil rich in carotenoids on a large scale.

Supercritical technology has certain advantages, such as the extraction solvent can be easily removed from the solute by reducing pressure and temperature and can be reused. It requires the use of little or no organic solvents, requiring less energy and enabling rapid extraction; extraction times usually are short and range from 2 to 3 h, due to the high solubility power; and negligible presence of light and oxygen during extraction (Brunner 2005; Priyadarsani et al. 2021; Kultys and Kurek 2022).

Therefore, several studies highlight the use of SFE to obtain carotenoids from plant matrices, mainly because it is a clean and sustainable technology that uses nonpolar solvents and avoids using organic solvents. In addition, using low or mild temperatures during the process prevents the degradation of compounds. In SFE, conditions can be 40 and 70 °C/300 and 450 bar to extract carotenoids, with the aim of improving the extraction process without sudden changes in temperature and using high pressures.

5 Extraction by Non-conventional Green Solvents

Due to environmental and human health concerns, new extraction techniques stimulated by the expansion of green chemistry have emerged. They are considered environmentally friendly to replace conventional extraction methods using toxic organic solvents (Sharmila et al. 2016).

From a sustainability perspective, a favorable solvent should have a high capacity for target compounds and be cheap and safe (Yu and Yang 2019). Using more environmentally friendly solvents as extraction solvents has recently become a priority and has gained significant attention worldwide. Green solvents are more favorable in advancing extraction technology due to their harmlessness to the environment and health impacts. Deep eutectic solvents, ionic liquids, and bio-based solvents are considered green solvents and have been used for carotenoid extraction. They would open a new alternative for carotenoid extraction in the pharmaceutical and food industries.

5.1 *Ionic Liquids Used for Carotenoid Extraction*

Ionic liquids are described in the literature as molten salts, which means that they are liquids in a thermodynamic definition, but also liquid crystals with a melting temperature below 100 °C. Its properties of interest are self-assembly capacity, low vapor pressure, flammability, thermal stability, and wide liquid phase range (Toledo Hijo et al. 2016). In the last two decades, literature has revealed several applications of ionic liquids in the pharmaceutical and biotechnology industry, among several other interests in the area of food engineering and chemistry, such as bioseparation processes for bioactive compounds, most recently for the recovery of carotenoids, which include, lycopene, α-carotene, β-carotene, lutein, zeaxanthin, and β-cryptoxanthin (Cheng et al. 2020; De Souza Mesquita et al. 2020; Farias et al. 2023; Mayumi Ueda et al. 2023; Murador et al. 2019a, b; Mussagy et al. 2019, 2021; Rajabi et al. 2017; Viñas-Ospino et al. 2023a).

Obtaining ionic liquids in several combinations allows for the production of solvents with specific characteristics for each type of molecule. Table 5.4 summarizes some of the ionic liquids applied in recent years to recover different carotenoids.

In ionic liquids based on choline bicarbonate, it was observed that the recovery of the carotenoids astaxanthin and β-carotene is significantly influenced (p-value < 0.05) by the nature of the anion of the ionic liquid: octanoic acid ([Oct^-]) > hexanoic acid ([Hex^-]) > butyric acid ([But^-]) > lactic acid ([Lac^-]), demonstrating that the hydrophobicity relative to IL is an important factor for the recovery of astaxanthin and β-carotene (Mussagy et al. 2022a). These authors also demonstrated that the recovery yield of both carotenoids was significantly ($p \leq 0.05$) influenced by the nature of the IL anion ([Oct^-] ≥ [Hex^-] ≥ [But^-] ≥ [Lac^-]), demonstrating that the relative hydrophobicity of IL is a crucial factor for the recovery

Table 5.4 Main ionic liquids used to extract carotenoids from different sources

Cation	Anion	Abbreviation	Carotenoid	Source	Main results	Reference
Tetramethylammonium ($N_{1111}{}^{+}$) and Trimethylethylammonium ($N_{1112}{}^{+}$)	Acetate (OAc^{-})	N_{1111} OAc and N_{1112} OAc	β-carotene	Model solution β-carotene solubilized in hexane (100 ppm)	Maximum extraction efficiency of 63.09% and 37.89% was achieved for aqueous solutions of N_{1111} OAc and N_{1112} OAc, respectively	Rajabi et al. (2017)
Propylamine (Pro^{+})	Octanoic acid (Oct^{-})	[Pro][Oct]:EtoH (50%, v/v)	β-carotene and astaxanthin	*Phaffia rhodozyma*	Four extraction cycles were required to recover 99% of both carotenoids. In the first cycle, recovery yields of β -carotene (86.3%) and astaxanthin (76.6%) were achieved.	Silva et al. (2023a, b)
1-butyl-3-methylimidazolium [C_4mim^{+}]	Tetrafluoroborate ($BF4^{-}$)	[C_4mim][BF_4])	β-carotene, lycopene andγ-carotene	Peach palm waste	Extraction yield of 172 μg of carotenoids/g dried biomass: 35% (all-*E*)-β-carotene, 18% (all-*E*)-γ-carotene, 9% (all-*E*)-lycopene, and 38% other minor compounds (Z-isomers)	De Souza Mesquita et al. (2019); Santamarina et al. (2022)

(continued)

Table 5.4 (continued)

Cation	Anion	Abbreviation	Carotenoid	Source	Main results	Reference
3-Diethylamino-propylammonium ($DEAPA^+$)	Hexanoate (Hex^-)	[DEAPA] [Hex]	β -carotene, torularhodin, and torulene	*Rhodotorula glutinis* CCT-2186	[Hex]⁻based PILs were the most efficient in the recovery of carotenoids with extraction rates six-fold higher than those of the common volatile organic solvent (DMSO). [DEAPA][Hex] extracted 206.65 μg/mL β- carotene, 112.82 μg/mL torularhodin, and 17.21 μg/mL torulene.	Mussagy et al. (2019)
1-butyl-3-methylimidazolium ([$BMIM^+$])	Chloride ([Cl^-])	[BMIM][Cl]	(all-*E*)-Lutein, (all-*E*)-β-cryptoxanthin, and (all-*E*)-β-carotene	Orange peel	[BMIM][Cl] was the most effective ionic liquid, leading to a total carotenoid content of 32.08 μg/g, while 7.88 μg/g was obtained with acetone.	Murador et al. (2019a, b)
Choline bicarbonate [Ch^+]	Octanoic acid (Oct^-)	[Ch][Oct]	β-carotene and astaxanthin	*Phaffia rhodozyma*	Cholinium-based ionic liquids was efficient for the recovery of β-carotene and astaxanthin (≈ 40% w/w) in one-step extraction;	Mussagy et al. (2022a)
1-butyl-3-methylimidazolium $[BMIM]^+$	Tetrafluoroborate $[BF4]^-$	[BMIM] [BF4]	(all-*E*)-Lutein, (all-*E*)-zeaxanthin, (all-*E*)-β-cryptoxanthin, and (all-*E*)-β-carotene	Pequi (*Caryocar brasilense*)	[BMIM][BF4]/ethanol (1:1, v/v) demonstrated the highest-extracted yields of carotenoids (107.90 μg/g), but also 1.69 times higher in antioxidant activity when compared with carotenoids extracted with acetone.	Ramos-Souza et al. (2023)

of astaxanthin and β-carotene. A relationship between the length of the anionic alkyl chain of IL and the recovery of β-carotene and astaxanthin was demonstrated by correlating the recovery yields with the carbon number of the corresponding IL anion, i.e., $n_{[Oct]} = 8$, $n_{[Hex]} = 6$, $n_{[But]} = 4$ and $n_{[Lac]} = 3$. Therefore, there is a linear correlation between the carbon number of the IL anion and the relative strength for recovery of astaxanthin and β-carotene. As these carotenoids are very hydrophobic, increasing the hydrophobic character of the IL increases the recovery of carotenoids.

Protic ionic liquids were used to recover β-carotene, torularhodin, and torulene from the biomass of the yeast *Rhodotorula glutinis*, using aqueous solutions of ammonium-based ILs, and it was observed that the extraction suitability was also affected by the nature of the alkyl chain of the anion. Hexanoic acid $[Hex]^-$ > butyric acid $[But]^-$ > propanoic acid $[Pro]^-$ > and acetic acid $[Ac]^-$ leading to higher recovery yields of carotenoids (Mussagy et al. 2019).

De Oliveira et al. (2021) also studied the extraction of intracellular dyes from *T. amestolkiae fungi* using imidazolium-based ILs. They demonstrated that increasing the length of the cationic alkyl chain (from 1 to 4) allows the highest recovery yields of red, orange, and yellow dyes. However, above four alkyl chains (from 6 to 10), no linear dependence was observed with increasing alkyl chain length, since from 6 to 10, the concentration of red dyes remains practically constant (De Oliveira et al. 2021).

Ionic liquids have shown promise for recovering carotenoids from different matrices, providing superior results when compared to conventional solvents such as acetone, methanol, ethanol, and DMSO (De Oliveira et al. 2021; Mussagy et al. 2019, 2022a).

Although IL solutions are effective for the recovery of carotenoids, integrating the process for recovery of carotenes and their removal from the extracting solvent (ionic liquid) is critical. Fortunately, there are well-established protocols for the selective separation of molecules from aqueous solutions of ILs, for example, cost-effective processes that include precipitation and reverse extraction approaches by adding antisolvents or strong salting agents (De Souza Mesquita et al. 2020; Mussagy et al. 2019, 2022a).

Comparing the extraction results of ILs with acetone, ILs even show 100% higher recovery results than acetone. On the other hand, acetone has a protective effect for both carotenoids at low temperatures. At the same time, aqueous solutions of LI [Ch][Oct] (50% by weight) showed significant losses (of around 20–30%) of β- carotene and astaxanthin (Mussagy et al. 2022a). In general, aqueous solutions of [Ch][Oct] have environments with the lowest stabilizing ability, leading to losses of β-carotene and astaxanthin of up to 80% and 70%, respectively (maximum losses at 65 °C). The same authors suggest that temperature has a marked influence on the stability of both carotenoids and revealed that, over 14 days, β-carotene and astaxanthin were much more stable in acetone than in aqueous [Ch] solution. [Oct], emphasizing a favorable molecular environment in the presence of the organic solvent or some unfavorable interactions between carotenoids, water, and anion/cation of ILs. The aqueous solution of LI based on $[Oct]^-$ is a liquid-crystalline salt consisting of cations ([Ch]$^{+)}$ and anions ([Oct]$^{-}$) solvated in water molecules with some nonpolar microdomains in which the β molecules -Carotene and astaxanthin are solubilized during extraction. However, even with the presence of more hydrophobic regions, a highly oxidizing environment caused by the high water content (50% by weight) in the solution is generated with the increase in temperature (from 4 to 65 °C), which favors easy oxidation of β-carotene and astaxanthin molecules (Mussagy et al. 2022a).

Murador et al. (2019a, b) evaluated the stability of carotenoid extracts obtained with LI [BMIM][Cl] + ethanol (1:2, v/v) and acetone in aqueous and oily media. In both systems (aqueous and oily), the carotenoids present in IL extracts were more stable than those obtained with acetone. The ionic liquids provided greater thermal stability to the carotenoids at 60–90 °C. The authors attribute the improvement in stability to micellar formation with stable non-aqueous microemulsions containing double-chain anionic active surfaces.

Martins and De Rosso (2016) evaluated the thermal stability (40–80 °C) and the stability in the presence and absence of light in aqueous and oily media of tomato carotenoid extracts obtained with IL and acetone. It was observed that in all experiments, the extracts obtained with LI [BMIM][Cl] + ethanol (1:1, v/v) were more stable than those obtained with acetone. The medium also influenced stability, with carotenoid extracts evaluated in an oily medium being more stable. Regarding light, after 12 days of exposure, trans-lycopene in tomato products was reduced by approximately 28%.

Assessing the thermal stability of carotenoids from different sources is essential to determining their properties and consequently applying these pigments, especially the more stable ones, for scale-up processes and industrial food applications since many of them involve thermal heating. These results are also important to better understand the structure-stability relationships of the carotenoid/extraction medium and solvent.

5.2 Deep Eutectic Solvents Employed for Carotenoid Extraction

The term *Deep Eutectic Solvents* is used to describe mixtures of quaternary ammonium salts with carboxylic acids, polyols, amino acids, and/or sugars that have melting points much lower than those of their pure components. These components of carboxylic acids, polyols, amino acids, and/or sugars are known as hydrogen bond donors (HBD), *and* quaternary ammonium salts are hydrogen bond acceptors (HBA). (Hydrogen-Bond Acceptor) (Abbott et al. 2003; Martins et al. 2019).

From a more thermodynamic point of view, in the opinion of Martins et al. (2019), a Deep eutectic solvent is a mixture of two or more pure compounds where the eutectic point temperature is quite different from the ideal eutectic point and below the operating temperature, and which presents significant negative deviations from ideality. Otherwise, only the term "eutectic solvent" could describe mixtures that do not meet these criteria.

Based on solid-liquid equilibrium (SLE) diagrams, the temperature drop in a deep eutectic solvent can be defined as the difference ΔT_2 between the ideal eutectic point ($T_{E,ideal}$) and real (T_E) and not as the difference ΔT_1 between the linear combination of the melting points of the pure components and the real eutectic point (Smith et al. 2014).

DES have been reported to be easy to prepare and can be obtained from a solid mixture of two or three components under stirring and heating until the formation of a homogeneous liquid using temperatures ranging from 70 to 100 °C (Bajkacz and Adamek 2017; Cui et al. 2015; Shang et al. 2019; Xia et al. 2021; Xu et al. 2019).

Currently, DES has been used to extract various bioactive compounds from different plant matrices and biomasses (Da Costa et al. 2020; Ferreira et al. 2023; Li et al. 2019; de Almeida Pontes et al. 2021; Toledo Hijo et al. 2022). From literature reports, DES have many advantages in the context of extraction processes, and these advantages not only include their non-toxicity, but also their low cost and easy preparation. Because they can be obtained from components that are generally recognized as safe (GRAS), one of the most promising benefits is that recovery of target compounds may not be necessary, and this slows down processes, saving time and energy (Viñas-Ospino et al. 2023c).

There are two groups of DES; the first is based on hydrophilic compounds and has been well studied for the extraction of more polar compounds such as polyphenols, flavonoids, anthocyanins, and proteins (Ferreira et al. 2023; de Almeida Pontes et al. 2021, 2022; Toledo Hijo et al. 2022). The second group is based on hydrophobic compounds recently developed for the extraction of nonpolar compounds such as carotenoids (Dharani et al. 2022; Li et al. 2019; Mayumi Ueda et al. 2023; Pitacco et al. 2022; Tang et al. 2023; Viñas-Ospino et al. 2023b). This group of hydrophobic DES is more recent and is generally obtained by combining terpenes such as thymol, menthol, and camphor with carboxylic acids such as pyruvic acid, acetic acid, l-lactic acid and lauric, caprylic, capric, myristic, palmitic and stearic acid (Martins et al. 2018) (Table 5.5).

The contents of astaxanthin, astaxanthin ester, and lutein were analyzed by Li et al. (2022) in DL-menthol:acetic acid (DES1), DL-menthol:propionic acid (DES2), DL-menthol:valeric acid (DES3) and DL-menthol:octanoic acid (DES4) used in the preparation of microemulsions ME1, ME2, ME3 and ME4, respectively. The stability of carotenoids was evaluated during storage and investigated at 4 °C and 25 °C. During storage, different loss levels were observed for all carotenoids, and the concentrations of astaxanthin, astaxanthin ester, and lutein in DES-based MEs decreased with storage time. As reported by the authors, the percentages of degradation of astaxanthin ester stored at 4 °C for 6 days in ME1, ME2, ME3, and ME4 were only 23.37%, 17.13%, 9.29%, and 8.29%, respectively. When the storage temperature was increased to 25 °C, the degradation percentages increased to 59.73%, 56.40%, 53.02%, and 42.85%, respectively.

Table 5.5 Main Deep Eutectic Solvents (DES) used to extract carotenoids from different sources

DES	Abbreviation HBA:HBD	Carotenoid	Source	Main results	Reference
1-nonanoic acid (C9), n-decanoic acid (C10), and undecanoic acid (C11)	C9:C10:C11 (2:1:1)	β-Carotene and lycopene	Watermelon juice, grape fruit juice, tomato juice, and guava juice	Revealed the highest extraction efficiencies compared with other single fatty acid and binary fatty acid DESs	Li et al. (2019)
Choline chloride ([Ch]Cl) and glycerol (gly)	[Ch]Cl:Gly (1:2)	Total carotenoids	*Rhodotorula mucilaginosa* cells	The [Ch]Gly treatment presented the greatest total carotenoids accumulation in DES fluid phase (0.110 mg/mL), which was about two-fold higher than other treatments.	da Costa et al. (2020)
DL-menthol:octanoic acid	Me: OA (1:2)	Astaxanthin and lutein	Solubility study	Compared with ethanol, methanol, and acetone, all the DES studied enhanced the solubility of carotenoids due to the stronger hydrogen bonding and Van der Waals interactions.	Li et al. (2022)
Thymol:oleic acid, dl-menthol: oleic acid, and geraniol: oleic acid	Thy:OA (1:1) Me:OA (1:1), and Ger:OA (1:1)	Astaxanthin	*Haematococcus pluvialis*	All DES gave good recovery percentages without any thermal, mechanical or chemical pre- treatment; the extraction of dried biomass gave an astaxanthin recovery of about 60% in 6 h, independently from the DES used.	Pitacco et al. (2022)
Menthol:Camphor	Me:cam (1:1)	β -carotene and β-cryptoxanthin	Orange peels	High efficient extraction yields when using hydrophobic DES such as Menthol: Camphor (163.5 mg/100 g), and the stability of carotenoids in this solvent was 56% for total carotenoids after 60 days at 4 °C.	Viñas-Ospino et al. (2023c)
DL-Menthol: camphor DL-Menthol: Eucalyptol Lauric acid: octanoic acid	Me:cam (1:1); Me: Me (1:1); C12: C8 (1:3);	Total carotenoids	Orange peel	High total carotenoid contents were obtained with hydrophobic DESs using DL-Menthol: camphor (163.5 mg/100 g_{fw}), DL-Menthol: Eucalyptol (168.7 mg/100 g_{fw}), and lauric acid: octanoic acid (153.1 mg /100 g_{fw})	Viñas-Ospino et al. (2023b)

The results showed that DL-menthol:octanoic acid microemulsions preserved the studied carotenoids. Furthermore, astaxanthin ester showed greater stability at 4 °C compared to astaxanthin and lutein.

Viñas-Ospino et al. (2023b) evaluated the stability of total carotenoids in DES extracts during storage for 60 days at 25 °C with exposure to light and in the dark at 25 °C, 4 °C and −20 °C. A hexane extraction was used as a control. It was observed that at room temperature, the degradation rate is higher than at low temperatures, where the most significant degradation of carotenoids occurred in extracts with hexane, C12:C8 (Lauric acid: octanoic acid), and LA:Glu (Lactic acid: Glucose) at 25 °C with light. This trend is also similar to room temperature in the dark, where greater degradation of carotenoids to hexane and C12:C8 occurred. The results also demonstrated that the lowest degradation rate occurred in the Pro:Ma (Proline: malic acid) and Me:Cam (DL-Menthol: Camphor) extracts; however, it is important to mention that even when degradation is lower in Pro:Ma, the concentration of carotenoids after 60 days of storage was still higher in Me:Cam. At 4 °C, carotenoids showed greater stability in all DES extracts evaluated, and Me:Cam maintained 56.0% carotenoids after 60 days. Finally, the results at −20 °C also showed that the carotenoids in Me:Cam were more stable, with 64.1% carotenoids after 60 days. The results revealed that carotenoids from orange peels are more stable when incorporated into DESs than other traditional solvents, such as hexane. These data suggested that DES Me:Cam is the most promising carotenoid extractant and stabilizer among the tested DES, and this characteristic may be useful for the development of food formulations.

Other previous research has also demonstrated that DES improve the stability of bioactive compounds (Panić et al. 2019; Stupar et al. 2021; Viñas-Ospino et al. 2023b, c)due to hydrogen bond interactions between the solvent and target compounds. This interaction decreases the flow of solute molecules. It limits their contact with oxygen at the interface between DESs and air, thus minimizing oxidative degradation, which is the most common type of degradation (Cvjetko Bubalo et al. 2018; Viñas-Ospino et al. 2023b).

5.3 Bio-based Solvents Employed for Carotenoid Extraction

Bio-based solvents are produced from renewable biomass sources, such as energy crops, forest products, aquatic biomass, and waste. Some bio-based solvents are alcohols such as bio-butanol, bio-2-octanol, bio-1,3-propanediol, and bio-1,3-butanediol and glycerol; esters such as lactate and ethyl acetate; terpenes such as α-pinene, D-Limonene, p-cymene and p-cresol; carboxylic acids such as acetic, formic, levulinic lactic, ascorbic, among others (Silva et al. 2023b; Torres-Valenzuela et al. 2019). Viscosities are generally low, which makes them easy to handle in extraction processes. Despite their great potential, few studies still apply these solvents to extract bioactive compounds, especially carotenoids, making them potential solvents to be used together with DES and ILs to replace volatile and toxic organic solvents of petroleum origin.

The bio-based solvents ethyl acetate (EtOAc), ethyl lactate (EtOLac), limonene (Lim), 2-methyl tetrahydrofuran (2-MeTHF), and cyclopentylmethyl ether (CPME) were applied for extraction of β-carotene and α-carotene of dried carrot waste (Mussagy et al. 2023)and astaxanthin β-carotene from *Phaffia rhodozyma biomass* (Mussagy et al. 2022b).

For the extraction of carotenoids from dried carrot residue, the results of this investigation suggest that EtOAc:2-MeTHF biobased solvent mixtures have significant potential as a carotenoid extractor, producing high recovery rates (60.01 ± 4.27 μg/g of β-carotene and 46.62 ± 3.32 μg/g of α-carotene) under optimized operating parameters. The investigation also revealed that the nature and type of solvent, storage temperature, and the presence or absence of light influenced the stability of carotenoids in all experimental tests. Half-life results revealed that carotenoids can be stored without degradation using EtOAc, 2-MeTHF, or even EtOAc:2-MeTHF at room temperature and in the absence of light for approximately 1 week (Mussagy et al. 2023).

Other researchers have also investigated the extraction of carotenoids with oleic acid, limonene, CPME, and ethyl lactate and observed recovery yields superior to conventional solvents hexane and acetone (Bandić et al. 2023; Scurria et al. 2020; Szabo et al. 2022; Vo et al. 2023).

The absence of studies employing bio-based hydrophobic solvents to extract carotenoids raises the need for future investigations related to the recovery and stability of these compounds in solvents, highlighting them as possible substitutes for conventional organic solvents.

6 Pulsed Electric Field-Assisted Extraction of Carotenoids

The pulsed electric field technique is a non-thermal technology that uses low temperatures and can be strategically used to treat samples rich in thermosensitive compounds, such as carotenoids (Bocker and Silva 2022). The mechanism of action of this technology is based on the phenomenon of electroporation, where high-intensity charges are applied to the plant matrix, capable of increasing the permeability of cell membranes, promoting the transport of material present in the intracellular environment, and facilitating the extraction of compounds of interest., such as the release of carotenoids that are located in chromoplasts (Schweiggert et al. 2012; Saini and Keum 2018). A systematic scheme of pulsed electric fields is presented in Fig. 5.5.

The extraction of carotenoids in samples treated by the pulsed electric field (PEF) is a favorable method for obtaining higher yields of carotenoids (Luengo et al. 2014; Parniakov et al. 2015). Furthermore, PEF treatment did not demonstrate the ability to change the qualitative profile of carotenoids in a comparison of treated and untreated samples (Andreou et al. 2020; Martínez et al. 2020). This probably

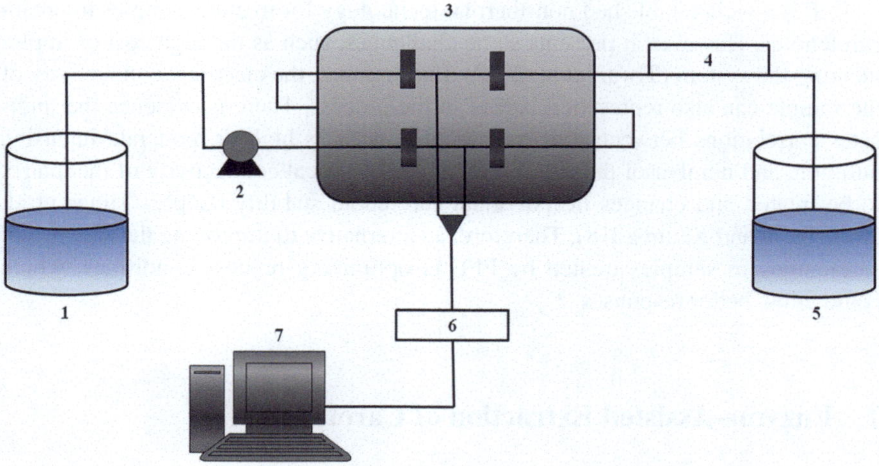

Fig. 5.5 Schematic diagram of the pulsed electric field system. *1*- Untreated matrix; *2*- Pump; *3*- Treatment chamber; *4*- Cooling Coil; *5*- Matrix treated; *6*- High-voltage pulse generator; *7*- Computer control system. (Adapted from Bocker and Silva (2022))

occurs due to the absence of high temperatures during processing, which can induce the *E-Z* isomerization process and cause loss of biological activity, improving the stability of the compounds (Zhang et al. 2017).

Pulsed electric field treatment has been strongly associated with improved carotenoid extraction. González-Casado et al. (2018) reported that the total carotenoid content in tomato fruits treated with PEF, with electric field strength varying between 0 and 200 kV m − 1, increased by ≈ 50%. Vallverdú-Queralt et al. (2013), when evaluating the impact of PFE treatment on tomato fruits for juice production, observed that the application of medium-intensity PFE (1 kV/cm using 16 monopolar pulses of 4 μs) improved the carotenoid content. Furthermore, the application of high-intensity PFE (35 kV/cm for 1500 μs with 4 μs square wave bipolar pulses) in the juice produced with treated tomatoes showed a synergistic effect, promoting an increase of up to 19% in the concentration of all-*E* -Lycopene. The combined treatment between medium and high-intensity PFE also promoted ≈ 35 and 50% increases in lutein and β-carotene content, respectively.

The application of PEF has also been strategically used to value co-products generated from tomato processing. Pataro et al. (2018), when evaluating the effect of treatment with PFE on previously treated tomato peels, revealed that application at 0.25, 0.50, and 0.75 kV/cm promoted an increase in the content of (all-E)-lycopene without causing reactions of isomerization. In another study, Pataro et al. (2020) carried out a moderate-intensity PEF treatment (5 kV/cm) on tomato co-products. In addition, they observed that the previously used treatment influenced the increase in lycopene content but without inducing the formation of cis isomers in the chromatographic profile and without altering carotenoid selectivity.

PEF is a well-established non-thermal technology for treating samples to obtain carotenoids. However, it presents some challenges, such as the high cost of implementing the system (Tiwari et al. 2022). Furthermore, the electrical conductivity of the sample can also represent a barrier in the process. There is evidence that indicates correlations between process variables, such as high or moderate intensity, duration, and number of pulses, as well as the electrical conductivity of the matrix to be treated and changes in individual carotenoid stability (López-Gámez et al. 2021; Saini and Keum 2018). Therefore, an alternative to improving the stability of carotenoids in samples treated by PFE is optimizing process conditions, which could allow better responses.

7 Enzyme-Assisted Extraction of Carotenoids

Enzyme-assisted extraction (EAE) is an efficient technology for extracting compounds of interest from different matrices and is usually used as a pre-treatment to obtain value-added compounds (Nadar et al. 2018). The principle of this technique is based on the action of enzymes such as pectinases, cellulases, and hemicellulase acting on the hydrolysis of polysaccharides found in the structure of the cell wall, allowing the release of intracellular compounds, such as carotenoids (Danalache et al. 2018; Ali et al. 2022). In addition to improving extraction yields, the application of EAE is also considered a green and environmentally friendly technology, which does not require the application of organic solvents that are toxic to the environment and human health (Nadar et al. 2018; Marathe et al. 2017).

The choice of enzyme used in enzymatic extraction procedures is based on the complexity of the matrix to be studied. Some of the enzymes used to extract compounds of interest are endo-1,4-β-D-glucanase (EC 3.2.1.4), endo-β-1,4- xylanase (EC 3.2.1.8), pectinases (EC 3.2.1.15) and β-D xylosidase (EC 3.2. 1.3) (López-Martínez et al. 2023). The synergistic action of enzymes is also an alternative to improve cell rupture in plant matrices. The joint application of polygalacturonase, pectin lyase, cellulose, and xylanase, strategically designed, proved to be efficient in the breakdown of polysaccharides in the cell wall of tomatoes, promoting greater release of carotenoids (Lombardelli et al. 2020).

In a study on the optimization of the enzymatic extraction of astaxanthin in algae of the species *Haematococcus pluvialis,* Zhao et al. (2019) observed that the use of pectinase promoted an increase of $\approx 10\%$ in the yield of this carotenoid compared to the use of cellulase. On addition, the application of the pectinase enzyme reduced the extraction time of astaxanthin, probably due to the efficiency of the enzyme in rupture of plant tissue. The enzyme pectinase, as well as viscozyme L, also proved to be more efficient in extracting carotenoids from red capsicum (*Capsicum annuum*) and apricot juice (Nath et al. 2016; Bashir et al. 2021). Strati et al. (2015) also observed that enzymatic hydrolysis using pectinase in tomato residues improved the total carotenoid contents and individual lycopene content by up to 6 times.

In EAE processes, optimizing the interaction conditions between the enzyme and the raw material is essential to guarantee better yields and stability of carotenoids. In a study that evaluated the efficiency of EAE in the extraction and stability of carotenoids, Jalali-Jivan et al. (2021) observed that the use of pectinase enzyme (Pectofruit) in the treatment of carrot pomace proved to be more efficient in treatments that considered 30 and 60 min compared to the use of cellulase (Celluclast BG and Celluclast 1.5 L), while at 90 min, the enzymes pectinase and cellulase (Celluclast BG) showed equal responses. On the other hand, after 120 minutes of enzymatic liquefaction, it was noted that cellulase enzymes showed a better response compared to pectinase, indicating that the relationship between treatment and enzyme can affect the stability of carotenoids.

Associated with enzymatic action can also be combined with extraction techniques such as ultrasound, microwaves, and supercritical fluid extraction to obtain better responses (Encalada et al. 2019; Marathe et al. 2017). Lenucci et al. (2015) reported that tomato purees enzymatically pre-treated with glycosidase presented a higher concentration of carotenoids extracted by supercritical Co2; in addition, enzymatic hydrolysis was not shown to be capable of inducing the formation of cis-lycopene isomers. Enzymatic treatment associated with extraction with eutectic liquids also proved to be more efficient in extracting carotenoids from sunflower waste samples pre-treated with the enzyme viscozyme (Ricarte et al. 2020).

Enzyme-assisted extraction has some limitations, the main one being the high cost of enzymes, which can represent a barrier in terms of industrial scale. On the other hand, efficiency, short extraction time, and the absence of application of toxic organic solvents are important factors when considering this technique for carotenoid extraction (Cheng et al. 2020). In this sense, approaches such as these contribute to the concept of green chemistry and, consequently, can enhance the use of natural sources, such as fruits and co-products, for the enzymatic extraction of carotenoids intended for application as natural pigments in the development of foods and beverages.

8 Ohmic Heating Extraction Carotenoids

Ohmic heating (OH) is a technology that presents a process based on thermal heating caused by the action of electrical currents conducted on a material. Therefore, this thermal sample treatment technology presents a mechanism of action based on Joule's law, where heat is generated in a matrix due to the material's resistance to an applied electric current (Jan et al. 2023). In this sense, different paths can be traced for the application of OH in foods, such as pasteurization, sterilization, and even extraction of bioactive compounds, such as carotenoids (Rocha et al. 2018; Schottroff et al. 2019; Balthazar et al. 2022).

Food treatment with OH technology is advantageous concerning different attributes, such as preserving quality characteristics and bioactive compounds (Rodríguez et al. 2021; Pires et al. 2021a, b; Coelho et al. 2023). Furthermore, in

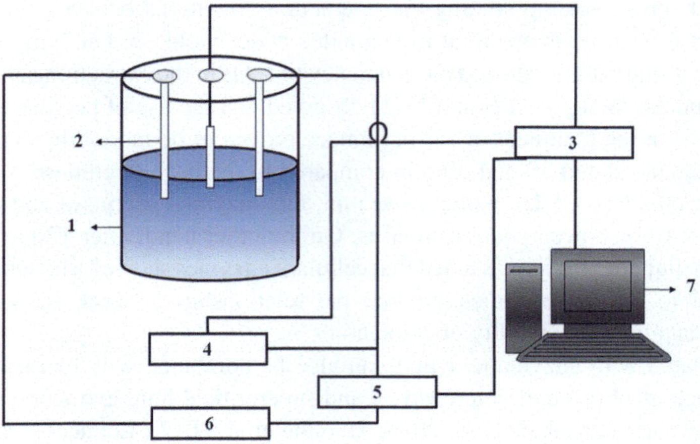

Fig. 5.6 Schematic diagram of the ohmic heating extraction system. *1*- Ohmic cell; *2*- Electrode; *3*- Data acquisition system; *4*- Voltmeter; *5*- Ammeter; *6*- AC power supply; *7*- Computer. (Adapted from Aamir and Jittanit et al. (2017))

terms of extraction, OH has stood out as an excellent technique for obtaining better yields of compounds. This efficiency can be explained by the synergistic effect caused by the action of electrical currents on the electropermeabilization of the cell wall associated with the softening of the tissue caused by increasing temperature (Mannozzi et al. 2018; Aamir and Jittanit 2017). A representative schematic of the ohmic heating system can be found in Fig. 5.6.

The application of ohmic heating is often reported to recover value-added compounds from co-products. Recent research shows that OH can have a positive effect on the extraction of carotenoids in tomato residues, associating better responses with a rapid increase in temperature, which contributes to the generation of damage to the cell wall, resulting in increased permeabilization and, consequently, greater release of these compounds. (Coelho et al. 2019). OH is also considered a more stable thermal technology for treating samples that are sources of heat-sensitive compounds, such as carotenoids, being able to reduce losses of epoxyxanthophyll and hydroxyxanthophyll carotenoids (30 and 20%) compared to treatment with conventional heating (70 and 40%) (Achir et al. 2016).

Aamir and Jittanit (2017) observed that ohmic heating treatment for oil extraction in gac aril samples improved individual β-carotene concentration by 93% compared to oil samples extracted with conventional heating. Furthermore, the authors observed, through scanning electron micrographs on the residues, that the improvement in response occurred due to the OH generating a collapse in the cell wall caused by disturbing effects on the surface of the structure, allowing an increase in the contact area between the intracellular constituents and the solvent used.

The efficient extraction of carotenoids through the application of OH technology is a promising method for obtaining better yields of this pigment. However, as it is a thermal approach, evaluating the impact of process conditions is essential to minimize losses resulting from the low stability of carotenoids at high temperatures. Pereira et al. (2021) when comparing the obtaining of extracts from seaweed of the species *Gracilaria vermiculophylla* by extraction with conventional heating and ohmic heating, it was observed that the treatment carried out with frequency and electric field adjusted to 25 kHz and 2–8 V/cm at 82 °C, improved carotenoid extraction by 40% after 1 h of treatment. However, after 2 h of treatment, a counter effect was observed, attributed to thermal degradation.

Giuliangeli et al. (2022), when evaluating the retention of carotenoids in guava pulp, it was observed that ohmic heating treatment (60–80 °C at 21.2 V/m and 60 Hz) was more efficient compared to treatment with conventional heating at different temperatures (60, 70 and 80 °C), on the other hand, OH treatment induced losses in carotenoid concentration. Jaeschke et al. (2016) observed that the non-thermal effects (electrical effects) associated with ohmic heating did not influence the stability of carotenoids in acerola pulp; on the other hand, the presence of oxygen proved to be a limiting factor, presenting a degradation rate of 21.4% in oxygen-rich samples, 3.8% in O_2- poor samples and 1.3% in unmodified samples. Mannozzi et al. (2018) also attributed the degradation of carotenoids in carrot juice treated with OH to the presence of oxygen.

Carotenoids extracted using OH are a good alternative for obtaining this pigment, as it is an efficient technology that obtains better yields and does not require the use of toxic solvents, which may pose fewer environmental and health-related risks. On the other hand, studies in the literature indicate the need to optimize extraction conditions to obtain better yields and greater stability of carotenoids extracted by ohmic heating.

References

Aamir M, Jittanit W (2017) Ohmic heating treatment for Gac aryl oil extraction: effects on extraction efficiency, physical properties and some bioactive compounds. Innovative Food Sci Emerg Technol 41:224–234. https://doi.org/10.1016/j.ifset.2017.03.013

Abbott AP, Capper G, Davies DL, Rasheed RK, Tambyrajah V (2003) Novel solvent properties of choline chloride/urea mixtures. Chem Commun 70–71. https://doi.org/10.1039/b210714g

Achir N, Dhuique-Mayer C, Hadjal T, Madani K, Pain JP, Dornier M (2016) Pasteurization of citrus juices with ohmic heating to preserve the carotenoid profile. Innovative Food Sci Emerg Technol 33:397–404. https://doi.org/10.1016/j.ifset.2015.11.002

Achour HY, Llamero CB, Saadi SA, Bouras N, Zitouni A, Señoráns J (2023) Pressurized liquid extraction for the recovery of carotenoids and functional compounds from green and Orange Dunaliella salina biomasses. Period Polytech Chem Eng 67(2):278–286. https://doi.org/10.3311/PPch.21386

Ali A, Riaz S, Sameen A, Naumovski N, Iqbal MW, Rehman A, Mehany T, Zeng XA, Manzoor MF (2022) The disposition of bioactive compounds from fruit waste, their extraction, and analysis using novel technologies: a review. PRO 10(10):2014. https://doi.org/10.3390/pr10102014

de Almeida Pontes PV, Ayumi Shiwaku I, Maximo GJ, Caldas Batista EA (2021) Choline chloride-based deep eutectic solvents as potential solvent for extraction of phenolic compounds from olive leaves: extraction optimization and solvent characterization. Food Chem 352:129346. https://doi.org/10.1016/j.foodchem.2021.129346

de Almeida Pontes PV, Czaikoski A, Almeida NA, Fraga S, de Oliveira Rocha L, Cunha RL, Maximo GJ, EAC B (2022) Extraction optimization, biological activities, and application in O/W emulsion of deep eutectic solvents-based phenolic extracts from olive pomace. Food Res Int 161:111753. https://doi.org/10.1016/J.FOODRES.2022.111753

de Andrade Lima M, Kestekoglou I, Charalampopoulos D, Chatzifragkou A (2019) Supercritical fluid extraction of carotenoids from vegetable waste matrices. Molecules 24(3). https://doi.org/10.3390/molecules24030466

Andreou V, Dimopoulos G, Dermesonlouoglou E, Taoukis P (2020) Application of pulsed electric fields to improve product yield and waste valorization in industrial tomato processing. J Food Eng 270:109778. https://doi.org/10.1016/j.jfoodeng.2019.109778

Bajkacz S, Adamek J (2017) Evaluation of new natural deep eutectic solvents for the extraction of isoflavones from soy products. Talanta 168:329–335. https://doi.org/10.1016/j.talanta.2017.02.065

Balthazar CF, Cabral L, Guimarães JT, Noronha MF, Cappato LP, Cruz AG, Sant'Ana AS (2022) Conventional and ohmic heating pasteurization of fresh and thawed sheep milk: energy consumption and assessment of bacterial microbiota during refrigerated storage. Innovative Food Sci Emerg Technol 76:102947. https://doi.org/10.1016/j.ifset.2022.102947

Bandić LM, Kahlina KV, Jurić S (2023) Optimization of solid–liquid extraction technique for carotenoids from Mandarin Peels using green solvents. J Anal Chem 78:1480–1486. https://doi.org/10.1134/S1061934823110023

Barp L, Višnjevec AM, Moret S (2023) Pressurized liquid extraction: a powerful tool to implement extraction and purification of food contaminants. Food Secur 12(10):1–25. https://doi.org/10.3390/foods12102017

Bashir O, Hussain SZ, Gani G, Jan N, Rather AH, Reshi M, Amin T (2021) Evaluating the physicochemical and antioxidant characteristics of apricot juice prepared through pectinase enzyme-assisted extraction from Halman variety. Food Measure 15:2645–2658. https://doi.org/10.1007/s11694-021-00833-w

Bello U, Amran NA, Samsuri S, Ruslan MSH (2023) Kinetics, thermodynamic studies, and parametric effects of supercritical CO2 extraction of Banana Peel wastes. Sustain Chem Pharm 31(November 2022):100912. https://doi.org/10.1016/j.scp.2022.100912

Bocker R, Silva EK (2022) Pulsed electric field assisted extraction of natural food pigments and colorings from plant matrices. Food Chem X 21(15):100398. https://doi.org/10.1016/j.fochx.2022.100398

Boon CS, McClements DJ, Weiss J, Decker EA (2010) Factors influencing the chemical stability of carotenoids in foods. Crit Rev Food Sci Nutr 50(6):515–532. https://doi.org/10.1080/10408390802565889

Borsarelli CD, Mercadante AZ (2009) Thermal and photochemical degradation of carotenoids. CRC Press, Boca Raton. https://doi.org/10.1201/9781420052312

Brunner G (2005) Supercritical fluids: technology and application to food processing. J Food Eng 67(1–2):21–33. https://doi.org/10.1016/j.jfoodeng.2004.05.060

Cardenas-Toro FP, Alcázar-Alay SC, Coutinho JP, Godoy HT, Forster-Carneiro T, Meireles MAA (2015) Pressurized liquid extraction and low-pressure solvent extraction of carotenoids from pressed palm fiber: experimental and economical evaluation. Food Bioprod Process 94:90–100. https://doi.org/10.1016/j.fbp.2015.01.006

Chada PSN, Santos PH, Rodrigues LGG, Goulart GAS, Azevedo dos Santos JD, Maraschin M, Lanza M (2022) Non-conventional techniques for the extraction of antioxidant compounds and lycopene from industrial tomato pomace (Solanum lycopersicum L.) using spouted bed drying as a pre-treatment. Food Chem X 13:100237. https://doi.org/10.1016/j.fochx.2022.100237

Cheng SH, Khoo HE, Kong KW, Prasad KN, Galanakis CM (2020) Extraction of carotenoids and applications. In: Carotenoids: properties, processing and applications. Academic Press, London, pp 259–288. https://doi.org/10.1016/B978-0-12-817067-0.00008-7

Chutia H, Mahanta CL (2021) Green ultrasound and microwave extraction of carotenoids from passion fruit peel using vegetable oils as a solvent: optimization, comparison, kinetics, and thermodynamic studies. Innov Food Sci Emerg Technol 67:102547. https://doi.org/10.1016/j.ifset.2020.102547

Chuyen HV, Nguyen MH, Roach PD, Golding JB, Parks SE (2018) Microwave-assisted extraction and ultrasound-assisted extraction for recovering carotenoids from Gac peel and their effects on antioxidant capacity of the extracts. Food Sci Nutr 6(1):189–196. https://doi.org/10.1002/fsn3.546

Coelho M, Pereira R, Rodrigues AS, Teixeira JA, Pintado ME (2019) Extraction of tomato by-products' bioactive compounds using ohmic technology. Food Bioprod Process 117:329–339. https://doi.org/10.1016/j.fbp.2019.08.005

Coelho MC, Ghalamara S, Campos D, Ribeiro TB, Pereira R, Rodrigues AS, Teixeira JA, Pintado M (2023) Tomato processing by-products valorization through ohmic heating approach. Food Secur 12(4):818. https://doi.org/10.3390/foods12040818

Crupi P, Faienza MF, Naeem MY, Corbo F, Clodoveo ML, Muraglia M (2023) Overview of the potential beneficial effects of carotenoids on consumer health and well-being. Antioxidants 12:1069. https://doi.org/10.3390/antiox12051069

Cui Q, Peng X, Yao XH et al (2015) Deep eutectic solvent-based microwave-assisted extraction of genistin, genistein and apigenin from pigeon pea roots. Sep Purif Technol 150:63–72. https://doi.org/10.1016/j.seppur.2015.06.026

Cvjetko Bubalo M, Vidović S, Radojčić Redovniković I, Jokić S (2018) New perspective in extraction of plant biologically active compounds by green solvents. Food Bioprod Process 109:52–73. https://doi.org/10.1016/J.FBP.2018.03.001

Da Costa WA, de Araújo Padilha CE, de Medeiros Dantas JM, Silva MMCL, de Oliveira Rocha HA, da Silva FLH, Ferrari M, dos Santos ES (2020) Fractional recovery of oleaginous bioactive produced by Rhodotorula mucilaginosa CCT3892 using deep eutectic solvents. Bioresour Technol Rep 12:100561. https://doi.org/10.1016/j.biteb.2020.100561

Danalache F, Mata P, Alves VD, Moldão-Martins M (2018) Chapter 10 – enzyme-assisted extraction of fruit juices. In: Fruit juices: extraction, composition, quality and analysis. Academic, London, pp 183–200. ISBN 9780128022306. https://doi.org/10.1016/B978-0-12-802230-6.00010-2

De Oliveira F, Hirai PR, Teixeira MFS, Pereira JFB, Santos-Ebinuma VC (2021) Talaromyces amestolkiae cell disruption and colorant extraction using imidazolium-based ionic liquids. Sep Purif Technol 257:117759. https://doi.org/10.1016/J.SEPPUR.2020.117759

De Souza Mesquita LM, Ventura SPM, Braga ARC, Pisani LP, Dias ACRV, De Rosso VV (2019) Ionic liquid-high performance extractive approach to recover carotenoids from Bactris gasipaes fruits. Green Chem 21:2380–2391. https://doi.org/10.1039/C8GC03283A

De Souza Mesquita LM, Martins M, Maricato É, Nunes C, Quinteiro PSGN, Dias ACRV, Coutinho JAP, Pisani LP, De Rosso VV, Ventura SPM (2020) Ionic liquid -mediated recovery of carotenoids from the Bactris gasipaes fruit waste and their application in food-packaging chitosan films. ACS Sustain Chem Eng 8:4085–4095. https://doi.org/10.1021/acssuschemeng.9b06424

Dharani S, Ramalingam P, Priya, Basavaraju S, Saraswathy N (2022) Extraction of lutein from targetes erecta using deep eutectic solvents. AIP Conf Proc 2446:020019. https://doi.org/10.1063/5.0110158/2823700

Elik A, Yanık DK, Göğüş F (2020) Microwave-assisted extraction of carotenoids from carrot juice processing waste using flaxseed oil as a solvent. LWT – food. Sci Technol 123:109100. https://doi.org/10.1016/j.lwt.2020.109100

Encalada AMI, Pérez CD, Flores SK, Rossetti L, Fissore EN, Rojas AM (2019) Antioxidant pectin enriched fractions obtained from discarded carrots (Daucus carota L.) by ultrasound-enzyme assisted extraction. Food Chem 289:453–460. https://doi.org/10.1016/j.foodchem.2019.03.078

Ershidat OTM (2023) Evaluation of cold extracted bioactive compounds from tomato pomace and its use in food fortification. Int J Food Prop 26(1):2137–2145. https://doi.org/10.1080/10942912.2023.2212141

Farias FO, Toledo Hijo AAC, Mussagy CU, Ferreira RSB, Santos-Ebinuma VC, Batista EAC, Meirelles AJA (2023) Exploring the potential of naturally-derived ionic liquids for sustainable vegetable oil deacidification. Food Bioprod Process 141:140–149. https://doi.org/10.1016/J.FBP.2023.08.005

Ferreira RSB, Farias FO, de Araujo EJS, Martínez J, Batista EAC (2023) Deep eutectic solvents as an alternative for extraction of flavonoids from soybean (Glycine max (L) Merrill) and okara: an experimental and computational approach based on COSMO-SAC model. Food Res Int 173:113266. https://doi.org/10.1016/J.FOODRES.2023.113266

Garcia-Mendoza MP, Paula JT, Paviani LC, Cabral FA, Martinez-Correa HA (2015) Extracts from mango Peel by-product obtained by supercritical CO2 and pressurized solvent processes. LWT Food Sci Technol 62(1):131–137. https://doi.org/10.1016/j.lwt.2015.01.026

Giuffrida D, Zoccali M, Mondello L (2020) Carotenoid and apocarotenoid analysis by SFE-SFC-QqQ/MS. In: Rodríguez-Concepción M, Welsch R (eds) Plant and food carotenoids: methods and protocols. Springer, New York, pp 103–116. https://doi.org/10.1007/978-1-4939-9952-1_16

Giuliangeli VC, Ströher GR, Shirai MA (2022) Comparison of energy consumption, color, ascorbic acid and carotenoid degradation in guava (Psidium guajava) pulp during conventional and ohmic heating. J Food Sci Technol 60:222–232. https://doi.org/10.1007/s13197-022-05607-w

González-Casado S, Martín-Belloso O, Elez-Martínez P, Soliva-Fortuny R (2018) Enhancing the carotenoid content of tomato fruit with pulsed electric field treatments: effects on respiratory activity and quality attributes. Postharvest Biol Technol 137:113–118. https://doi.org/10.1016/j.postharvbio.2017.11.017

Hiranvarachat B, Devahastin S (2014) Enhancement of microwave-assisted extraction via intermittent radiation: extraction of carotenoids from carrot peels. J Food Eng 126:17–26. https://doi.org/10.1016/j.jfoodeng.2013.10.024

Huang A, Liu H, Pan Y (2023) The antioxidant ability and extraction yield of Beta carotene. Highlights Sci Eng Technol 58:256–264. https://doi.org/10.54097/hset.v58i.10094

Jaeschke DP, Marczak LDF, Mercali GD (2016) Evaluation of non-thermal effects of electricity on ascorbic acid and carotenoid degradation in acerola pulp during ohmic heating. Food Chem 199:128–134. https://doi.org/10.1016/j.foodchem.2015.11.117

Jalali-Jivan M, Fathi-Achachlouei B, Ahmadi-Gavlighi H, Jafari SM (2021) Improving the extraction efficiency and stability of β-carotene from carrot by enzyme-assisted green nanoemulsification. Innovative Food Sci Emerg Technol 74:102836. https://doi.org/10.1016/j.ifset.2021.102836

Jan B, Aga MB, Dar AH, Majid I, Shams R, Rizvi QUEH, Khan SA (2023) Principles of ohmic heating for the food industry. In: Jafari SM (ed) Unit operations and processing equipment in the food industry, emerging thermal processes in the food industry. Woodhead Publishing, Cambridge, pp 229–243ISBN 9780128221075. https://doi.org/10.1016/B978-0-12-822107-5.00005-2

Kadi A, Boudries H, Bachir-bey M, Teffane M, Taibi A, Boulekbache-Makhlouf L (2021) Optimization of microwave-assisted extraction of carotenoids from Citrus clementina peels. Curr Bioact Compd 18(6). https://doi.org/10.2174/1573407218666211230152122

Kultys E, Kurek MA (2022) Green extraction of carotenoids from fruit and vegetable byproducts: a review. Molecules 14(2):738–754. https://doi.org/10.3390/molecules27020518

Larocca V, Martino M, Trupo M, Magarelli RA, Spagnoletta A, Ambrico A (2023) Evaluation of carbon dioxide supercritical fluid extraction (CO2-SFE) on carotenoids recovery from red yeast cells. Biomass Convers Biorefinery. https://doi.org/10.1007/s13399-023-04434-z

Lenucci MS, Caroli M, Marrese PP, Iurlaro A, Rescio L, Böhm V, Dalessandro G, Piro G (2015) Enzyme-aided extraction of lycopene from high-pigment tomato cultivars by supercritical carbon dioxide. Food Chem 170:193–202. https://doi.org/10.1016/j.foodchem.2014.08.081

Li Y, Fabiano-tixier AS, Vian MA, Chemat F (2013) Solvent-free microwave extraction of bioactive compounds provides a tool for green analytical chemistry. Trends Anal Chem 47:1–11. https://doi.org/10.1016/j.trac.2013.02.007

Li H, Zhao C, Tian H, Yang Y, Li W (2019) Liquid–Liquid microextraction based on acid–base-induced deep eutectic solvents for determination of β -carotene and lycopene in fruit juices. Food Anal Methods 12:2777–2784. https://doi.org/10.1007/S12161-019-01639-9

Li Y, Hu K, Huang C, Hu Y, Ji H, Liu S, Gao J (2022) Improvement of solubility, stability and antioxidant activity of carotenoids using deep eutectic solvent-based microemulsions. Colloids Surf B Biointerfaces 217:112591. https://doi.org/10.1016/J.COLSURFB.2022.112591

Lombardelli C, Liburdi K, Benucci I, Esti M (2020) Tailored and synergistic enzyme-assisted extraction of carotenoid-containing chromoplasts from tomatoes. Food Bioprod Process 121:43–53. https://doi.org/10.1016/j.fbp.2020.01.014

López-Gámez G, Elez-Martínez P, Martín-Belloso O, Soliva-Fortuny R (2021) Changes of carotenoid content in carrots after application of pulsed electric field treatments. LWT 147:111408. https://doi.org/10.1016/j.lwt.2021.111408

López-Martínez LX, Rojo-Gutiérrez E, Márquez-Molina O, Saenz-Hidalgo HK, Baeza-Jiménez R (2023) Chapter 15 – Bioactive extraction from tropical fruit residues by enzyme-assisted processes. In: Kuddus M, Ramteke P (eds) Value-addition in Agri-food industry waste through enzyme technology. Academic Press, London, pp 209–220. https://doi.org/10.1016/B978-0-323-89928-4.00022-5

López-Salazar H, Camacho-Díaz BH, Ocampo MLA, Jiménez-Aparicio AR (2023) Microwave-assisted extraction of functional compounds from plants: a review. Bioresources 18:6614–6638. https://doi.org/10.15376/biores.18.3.Lopez-Salazar

Luengo E, Condón-Abanto S, Álvarez I, Raso J (2014) Effect of pulsed electric field treatments on permeabilization and extraction of pigments from Chlorella vulgaris. J Membr Biol 247:1269–1277. https://doi.org/10.1007/s00232-014-9688-2

Mannozzi C, Fauster T, Haas K, Tylewicz U, Romani S, Rosa MD, Jaeger H (2018) Role of thermal and electrical field effects during the pre-treatment of fruit and vegetable mash by pulsed electric fields (PEF) and ohmic heating (OH). Innovative Food Sci Emerg Technol 48:131–137. https://doi.org/10.1016/j.ifset.2018.06.004

Mapelli-Brahm P, Meléndez-Martínez AJ (2021) The colored carotenoids phytoene and phytofluene: sources, consumption, bioavailability and health effects. Curr Opin Food Sci 41:201–209. https://doi.org/10.1016/j.cofs.2021.04.013

Marathe SJ, Jadhav SB, Bankar SB, Singhal RS (2017) Enzyme-assisted extraction of bioactives. In: Puri M (ed) Food bioactives. Springer, Cham. ISBN 978-3-319-51639-4. https://doi.org/10.1007/978-3-319-51639-4_8

Martínez JM, Schottroff F, Haas K, Fauster T, Sajfrtová M, Álvarez I, Raso J, Jaeger H (2020) Evaluation of pulsed electric fields technology for the improvement of subsequent carotenoid extraction from dried Rhodotorula glutinis yeast. Food Chem 323:126824. https://doi.org/10.1016/j.foodchem.2020.126824

Martins PLG, de Rosso VV (2016) Thermal and light stabilities and antioxidant activity of carotenoids from tomatoes extracted using an ultrasound-assisted completely solvent-free method. Food Res Int 82:156–164. https://doi.org/10.1016/J.FOODRES.2016.01.015

Martins MAR, Crespo EA, Pontes PVA, Silva LP, Bülow M, Maximo GJ, Batista EAC, Held C, Pinho SP, Coutinho JAP (2018) Tunable hydrophobic eutectic solvents based on terpenes and monocarboxylic acids. ACS Sustain Chem Eng 6:8836–8846. https://doi.org/10.1021/acssuschemeng.8b01203

Martins MAR, Pinho SP, Coutinho JAP (2019) Insights into the nature of eutectic and deep eutectic mixtures. J Solut Chem 48:962–982. https://doi.org/10.1007/s10953-018-0793-1

Mayumi Ueda K, Castro Leal F, Oliveira Farias F, Mussagy CU, Igarashi-Mafra L, Mafra MR (2023) Enhancing the carotenoid recovery from Eugenia uniflora l. pulp: a comparative study of supramolecular solvents, ionic liquids, and deep eutectic solvents. Sep Purif Technol 324:124632. https://doi.org/10.1016/J.SEPPUR.2023.124632

Menezes EGO, Barbosa JR, Pires FCS, Ferreira MCR, de Souza e Silva AP, Siqueira LMM, de Carvalho Junior RN (2022) Development of a new scale-up equation to obtain Tucumã-of-

Pará (Astrocaryum vulgare Mart.) oil rich in carotenoids using supercritical CO2 as solvent. J Supercrit Fluids 181:105481. https://doi.org/10.1016/j.supflu.2021.105481

Mezzomo N, Ferreira SRS (2016) Carotenoids functionality, sources, and processing by supercritical technology: a review. J Chem 2016:1–16. https://doi.org/10.1155/2016/3164312

Murador DC, Salafia F, Zoccali M, Martins PLG, Ferreira AG, Dugo P, Mondello L, de Rosso VV, Giuffrida D (2019a) Green extraction approaches for carotenoids and esters: characterization of native composition from orange peel. Antioxidants 8(12):613. https://doi.org/10.3390/antiox8120613

Murador DC, Braga ARC, Martins PLG, Mercadante AZ, de Rosso VV (2019b) Ionic liquid associated with ultrasonic-assisted extraction: a new approach to obtain carotenoids from orange peel. Food Res Int 126:108653. https://doi.org/10.1016/J.FOODRES.2019.108653

Mussagy CU, Santos-Ebinuma VC, Gonzalez-Miquel M, Coutinho JAP, Pereira JFB (2019) Protic ionic liquids as cell-disrupting agents for the recovery of intracellular carotenoids from yeast Rhodotorula glutinis CCT-2186. ACS Sustain Chem Eng 7:16765–16776. https://doi.org/10.1021/acssuschemeng.9b04247

Mussagy CU, Khan S, Kot AM (2021) Current developments on the application of microbial carotenoids as an alternative to synthetic pigments. Crit Rev Food Sci Nutr 62:6932–6946. https://doi.org/10.1080/10408398.2021.1908222

Mussagy CU, Farias FO, Bila NM, Giannini MJSM, Pereira JFB, Santos-Ebinuma VC, Pessoa A (2022a) Recovery of β -carotene and astaxanthin from Phaffia rhodozyma biomass using aqueous solutions of cholinium-based ionic liquids. Sep Purif Technol 290:120852. https://doi.org/10.1016/J.SEPPUR.2022.120852

Mussagy CU, Kurnia KA, Dias ACRV, Raghavan V, Santos-Ebinuma VC, Pessoa A (2022b) An eco-friendly approach for the recovery of astaxanthin and β -carotene from Phaffia rhodozyma biomass using bio-based solvents. Bioresour Technol 345:126555. https://doi.org/10.1016/J.BIORTECH.2021.126555

Mussagy CU, Oshiro A, Farias FO, Haddad FF, dos Santos JL, Scarim CB, Herculano RD, Pessoa A, Santos-Ebinuma VC (2023) Emerging role of bio-based solvents mixtures to increase the solubility and recovery of carotenoids from processed carrot wastes for potential skin care application. Ind Crop Prod 205:117436. https://doi.org/10.1016/J.INDCROP.2023.117436

Mustafa A, Turner C (2011) Pressurized liquid extraction as a green approach in food and herbal plants extraction: a review. Anal Chim Acta 703(1):8–18. https://doi.org/10.1016/j.aca.2011.07.018

Mustafa A, Trevino LM, Turner C (2012) Pressurized hot ethanol extraction of carotenoids from carrot by-products. Molecules 17(2):1809–1818. https://doi.org/10.3390/molecules17021809

Nadar SS, Rao P, Rathod VK (2018) Enzyme assisted extraction of biomolecules as an approach to novel extraction technology: A review. Food Res Int 108:309–330. https://doi.org/10.1016/j.foodres.2018.03.006

Nath P, Kaur C, Rudra SG, Varghese E (2016) Enzyme-assisted extraction of carotenoid-rich extract from red capsicum (Capsicum annuum). Agric Res 5:193–204. https://doi.org/10.1007/s40003-015-0201-7

Nour AH, Oluwaseun AR, Nour AH, Omer MS, Ahmed N (2021) Microwave-assisted extraction of bioactive compounds (review). In: Churyumov GI (ed) Microwave heating – electromagnetic fields causing thermal and non-thermal effects. IntechOpen, London. https://doi.org/10.5772/intechopen.96092

Panić M, Gunjević V, Cravotto G, Radojčić Redovniković I (2019) Enabling technologies for the extraction of grape-pomace anthocyanins using natural deep eutectic solvents in up-to-half-liter batches extraction of grape -pomace anthocyanins using NADES. Food Chem 300:125185. https://doi.org/10.1016/J.FOODCHEM.2019.125185

Parniakov O, Barba FJ, Grimi N, Marchal L, Jubeau S, Lebovka N, Vorobiev E (2015) Pulsed electric field assisted extraction of nutritionally valuable compounds from microalgae Nannochloropsis spp. using the binary mixture of organic solvents and water. Innovative Food Sci Emerg Technol 27:79–85. https://doi.org/10.1016/j.ifset.2014.11.002

Pataro G, Carullo D, Siddique MDAB, Falcone M, Donsì F, Ferrari G (2018) Improved extractability of carotenoids from tomato peels as side benefits of PEF treatment of tomato fruit for more energy-efficient steam-assisted peeling. J Food Eng 233:65–73. https://doi.org/10.1016/j.jfoodeng.2018.03.029

Pataro G, Carullo D, Falcone M, Ferrari G (2020) Recovery of lycopene from industrially derived tomato processing by-products by pulsed electric fields-assisted extraction. Innovative Food Sci Emerg Technol 63:102369. https://doi.org/10.1016/j.ifset.2020.102369

Patel AS, Kar A, Dash S, Dash SK (2019) Supercritical fluid extraction of β-carotene from ripe bitter melon pericarp. Sci Rep 9(1):1–10. https://doi.org/10.1038/s41598-019-55481-4

Pereira SG, Teixeira-Guedes C, Souza-Matos G, Maricato É, Nunes C, Coimbra MA, Teixeira JA, Pereira RN, Rocha CMR (2021) Influence of ohmic heating in the composition of extracts from Gracilaria vermiculophylla. Algal Res 58:102360. https://doi.org/10.1016/j.algal.2021.102360

Perez-Vazquez A, Carpena M, Barciela P, Cassani L, Simal-Gandara J, Prieto MA (2023) Pressurized liquid extraction for the recovery of bioactive compounds from seaweeds for food industry application: a review. Antioxidants 12(3):612. https://doi.org/10.3390/antiox12030612

Pires FCS, de Oliveira JC, Menezes EGO, Silva AP d SE, MCR F, LMM S, Almada-Vilhena AO, Pieczarka JC, Nagamachi CY, de Carvalho Junior RN (2021a) Bioactive compounds and evaluation of antioxidant, cytotoxic and cytoprotective effects of Murici pulp extracts (Byrsonima crassifolia) obtained by supercritical extraction in HepG2 cells treated with H2O2. Food Secur 10(4):737. https://doi.org/10.3390/foods10040737

Pires RPS, Guimarães JT, Barros CP, Balthazar CF, Chincha AIA, Freitas MQ, Duarte MCKH, Silva PHF, Pimentel TC, Abud YKD, Sant'Anna C, Sant'Ana AS, Silva MC, Nascimento JS, Cruz AG (2021b) Ohmic heating increases inactivation and morphological changes of salmonella sp. and the formation of bioactive compounds in infant formula. Food Microbiol 97:103737. https://doi.org/10.1016/j.fm.2021.103737

Pitacco W, Samorì C, Pezzolesi L, Gori V, Grillo A, Tiecco M, Vagnoni M, Galletti P (2022) Extraction of astaxanthin from Haematococcus pluvialis with hydrophobic deep eutectic solvents based on oleic acid. Food Chem 379:132156. https://doi.org/10.1016/J.FOODCHEM.2022.132156

Popescu M, Iancu P, Plesu V, Bildea CS (2022) Carotenoids recovery enhancement by supercritical CO2 extraction from tomato using seed oils as modifiers. PRO 10(12):2656. https://doi.org/10.3390/pr10122656

Priyadarsani S, Patel AS, Kar A, Dash S (2021) Process optimization for the supercritical extraction of lycopene from ripe grapefruit (Citrus paradisi) endocarp. Sci Rep 11(1):1–8. https://doi.org/10.1038/s41598-021-89772-6

Rajabi MS, Moniruzzaman M, Mahmood H, Sivapragasam M, Bustam MA (2017) Extraction of β -carotene from organic phase using ammonium based ionic liquids aqueous solution. J Mol Liq 227:15–20. https://doi.org/10.1016/J.MOLLIQ.2016.12.008

Ramos-Souza C, Nass P, Jacob-Lopes E, Zepka LQ, Braga ARC, De Rosso VV (2023) Changing despicable me: potential replacement of azo dye yellow tartrazine for Pequi carotenoids employing ionic liquids as high-performance extractors. Food Res Int 174:113593. https://doi.org/10.1016/J.FOODRES.2023.113593

Ricarte GN, Coelho MAZ, Marrucho IM, Ribeiro BD (2020) Enzyme-assisted extraction of carotenoids and phenolic compounds from sunflower waste using green solvents. 3 Biotech 10:405. https://doi.org/10.1007/s13205-020-02393-0

Rifqi M, Haziman ML, Islamawan PA, Hariadi H, Yusuf D (2023) Use of Beta-carotene pigment to improve food product chemical and sensory qualities: a review. J Funct Food Nutraceutical 4(2):67–78. https://doi.org/10.33555/jffn.v4i2.92

Rocha CMR, Genisheva Z, Ferreira-Santos P, Rodrigues R, Vicente AA, Teixeira JA, Pereira RN (2018) Electric field-based technologies for valorization of bioresources. Bioresour Technol 254:325–339. https://doi.org/10.1016/j.biortech.2018.01.068

Rodríguez LMN, Arias R, Soteras T, Sancho A, Pesquero N, Rossetti L, Tacca H, Aimaretti N, Cervantes MLR, Szerman N (2021) Comparison of the quality attributes of carrot juice pasteurized by ohmic heating and conventional heat treatment. LWT 145:111255. https://doi.org/10.1016/j.lwt.2021.111255

Saini RK, Keum YS (2018) Carotenoid extraction methods: a review of recent developments. Food Chem 240:90–103. https://doi.org/10.1016/j.foodchem.2017.07.099

Santamarina AB, de Souza Mesquita LM, Casagrande BP, Sertorio MN, Vitor de Souza D, Mennitti LV, Ribeiro DA, Estadella D, Ventura SPM, de Rosso VV, Pisani LP (2022) Supplementation of carotenoids from peach palm waste (Bactris gasipaes) obtained with an ionic liquid mediated process displays kidney anti-inflammatory and antioxidant outcomes. Food Chem X13:100245. https://doi.org/10.1016/J.FOCHX.2022.100245

Schottroff F, Pyatkovskyy T, Reineke K, Setlow P, Sastry SK, Jaeger H (2019) Mechanisms of enhanced bacterial endospore inactivation during sterilization by ohmic heating. Bioelectrochemistry 130:107338. https://doi.org/10.1016/j.bioelechem.2019.107338

Schweiggert RM, Mezger D, Schimpf F, Steingass CB, Carle R (2012) Influence of chromoplast morphology on carotenoid bioaccessibility of carrot, mango, papaya, and tomato. Food Chem 135(4):2736–2742. https://doi.org/10.1016/j.foodchem.2012.07.035

Scurria A, Fabiano Tixier AS, Lino C, Pagliaro M, D'Agostino F, Avellone G, Chemat F, Ciriminna R (2020) High yields of shrimp oil rich in Omega-3 and natural astaxanthin from shrimp waste. ACS Omega 5:17500–17505. https://doi.org/10.1021/acsomega.0c01978

Shang X, Dou Y, Zhang Y, Tan JN, Liu X, Zhang Z (2019) Tailor-made natural deep eutectic solvents for green extraction of isoflavones from chickpea (Cicer arietinum L.) sprouts. Industrial Crops and Products 140(15):111724. https://doi.org/10.1016/j.indcrop.2019.111724

Sharma M, Bhat R (2021) Extraction of carotenoids from pumpkin Peel and pulp: comparison between innovative green extraction technologies (ultrasonic and microwave-assisted extractions using corn oil). Food Secur 10(4):787. https://doi.org/10.3390/foods10040787

Sharmila G, Nikitha VS, Ilaiyarasi S, Dhivya K, Rajasekar V, Kumar NM, Muthukumaran K, Muthukumaran C (2016) Ultrasound assisted extraction of total phenolics from Cassia auriculata leaves and evaluation of its antioxidant activities. Ind Crop Prod 84:13–21. https://doi.org/10.1016/j.indcrop.2016.01.010

Silva PGP, Mussagy CU, de Azevedo Lima C, Santos-Ebinuma VC, de Medeiros Burkert JF, Santos LO (2023a) Sustainable approach to recover β -carotene and astaxanthin from Phaffia rhodozyma grown in a stirred-tank bioreactor under the influence of magnetic fields. Bioresour Technol 390:129906. https://doi.org/10.1016/J.BIORTECH.2023.129906

Silva SS, Abranches DO, Pinto AS, Soares BP, Passos H, Ferreira AM, Coutinho JAP (2023b) Solubility enhancement of hydrophobic compounds in aqueous solutions using biobased solvents as Hydrotropes. Ind Eng Chem Res 62:12021–12028. https://doi.org/10.1021/acs.iecr.3c01469

Smith EL, Abbott AP, Ryder KS (2014) Deep eutectic solvents (DESs) and their applications. Chem Rev. https://doi.org/10.1021/cr300162p

Soquetta MB, de Marsillac Terra L, Bastos CP (2018) Green technologies for the extraction of bioactive compounds in fruits and vegetables. CYTA – J Food 16(1):400–412. https://doi.org/10.1080/19476337.2017.1411978

Strati IF, Gogou E, Oreopoulou V (2015) Enzyme and high pressure assisted extraction of carotenoids from tomato waste. Food Bioprod Process 94:668–674. https://doi.org/10.1016/j.fbp.2014.09.012

Stupar A, Šeregelj V, Ribeiro BD, Pezo L, Cvetanović A, Mišan A, Marrucho I (2021) Recovery of β -carotene from pumpkin using switchable natural deep eutectic solvents. Ultrason Sonochem 76:105638. https://doi.org/10.1016/J.ULTSONCH.2021.105638

Swapnil P, Meena M, Singh SK, Dhuldhaj UP, Harish; Marwal, A. (2021) Vital roles of carotenoids in plants and humans to deteriorate stress with its structure, biosynthesis, metabolic engineering and functional aspects. Current Plant Biology 26:100203. https://doi.org/10.1016/j.cpb.2021.100203

Szabo K, Teleky BE, Ranga F, Roman I, Khaoula H, Boudaya E, Ltaief AB, Aouani W, Thiamrat M, Vodnar DC (2022) Carotenoid recovery from tomato processing by-pthrough green chemistry. Molecules 27:3771. https://doi.org/10.3390/MOLECULES27123771

Tang W, Wu Y, Wang M, Row KH, Qiu H, Zhou JL (2023) Emerging application of extraction phase of ionic and non-ionic deep eutectic solvents toward natural herbal medicine. TrAC Trends Anal Chem 165:117137. https://doi.org/10.1016/J.TRAC.2023.117137

Terao J (2023) Revisiting carotenoids as dietary antioxidants for human health and disease prevention. Food Funct 14:7799–7824. https://doi.org/10.1039/D3FO02330C

Tiwari S, Yawale P, Upadhyay N (2022) Carotenoids: extraction strategies and potential applications for valorization of under-utilized waste biomass. Food Biosci 48:101812. https://doi.org/10.1016/j.fbio.2022.101812

Toledo Hijo AAC, Maximo GJ, Costa MC, Batista EAC, Meirelles AJA (2016) Applications of ionic liquids in the food and bioproducts industries. ACS Sustain Chem Eng 4:5347–5369. https://doi.org/10.1021/acssuschemeng.6b00560

Toledo Hijo AAC, Alves C, Farias FO, Peixoto VS, Meirelles AJA, Santos GHF, Maximo GJ (2022) Ionic liquids and deep eutectic solvents as sustainable alternatives for efficient extraction of phenolic compounds from mate leaves. Food Res Int 157:111194. https://doi.org/10.1016/J.FOODRES.2022.111194

Torres-Valenzuela LS, Ballesteros-Gómez A, Rubio S (2019) Green solvents for the extraction of high added-value compounds from Agri-food waste. Food Engineer Rev 12:83–100. https://doi.org/10.1007/S12393-019-09206-Y

Trentini CP, da Silva SB, Rodrigues GM, dos Santos Garcia VA, Cardozo-Filho L, da Silva C (2017) Pressurized liquid extraction of Macauba pulp oil. Can J Chem Eng 95(8):1579–1584. https://doi.org/10.1002/cjce.22789

Vafaei N, Rempel CB, Scanlon MG, Jones PJH, Eskin MNA (2022) Application of supercritical fluid extraction (SFE) of tocopherols and carotenoids (hydrophobic antioxidants) compared to non-SFE methods. AppliedChem 2(2):68–92. https://doi.org/10.3390/appliedchem2020005

Vallverdú-Queralt A, Odriozola-Serrano I, Oms-Oliu G, Lamuela-Raventós RM, Elez-Martínez P, Martín-Belloso O (2013) Impact of high-intensity pulsed electric fields on carotenoids profile of tomato juice made of moderate-intensity pulsed electric field-treated tomatoes. Food Chem 141(3):3131–3138. https://doi.org/10.1016/j.foodchem.2013.05.150

Vardanega R, Fuentes FS, Palma J, Bugueño-Muñoz W, Cerezal-Mezquita P, Ruiz-Dominguéz MC (2023) Extraction of valuable compounds from granadilla (Passiflora ligularis Juss) Peel using pressurized fluids technologies. Sustain Chem Pharm 34(May):101135. https://doi.org/10.1016/j.scp.2023.101135

Vélez-Erazo EM, Pasquel-Reátegui JL, Dorronsoro-Guerrero OH, Martínez-Correa HA (2021) Phenolics and carotenoids recovery from Agroindustrial mango waste using microwave-assisted extraction: extraction and Modeling. J Food Process Eng 44(9). https://doi.org/10.1111/jfpe.13774

Viñas-Ospino A, López-Malo D, Esteve MJ, Frígola A, Blesa J (2023a) Green solvents: emerging alternatives for carotenoid extraction from fruit and vegetable by-products. Food Secur 12(4):1–19. https://doi.org/10.3390/foods12040863

Viñas-Ospino A, Panić M, Bagović M, Radošević K, Esteve MJ, Radojčić Redovniković I (2023b) Green approach to extract bioactive compounds from orange peel using hydrophilic and hydrophobic deep eutectic solvents. Sustain Chem Pharm 31:100942. https://doi.org/10.1016/J.SCP.2022.100942

Viñas-Ospino A, Panić M, Radojčić- Redovniković I, Blesa J, Esteve MJ (2023c) Using novel hydrophobic deep eutectic solvents to improve a sustainable carotenoid extraction from orange peels. Food Biosci 53:102570. https://doi.org/10.1016/J.FBIO.2023.102570

Vo TP, Tran HKL, Ta TMN, Nguyen HTV, Phan TH, Nguyen THP, Nguyen VK, Dang TCT, Nguyen LGK, Chung TQ, Nguyen DQ (2023) Extraction and emulsification of carotenoids from carrot pomaces using oleic acid. ACS Omega 8:39523–39534. https://doi.org/10.1021/acsomega.3c05301

Wang M, Morón-Ortiz Á, Zhou J, Benítez-González A, Mapelli-Brahm P, Meléndez-Martínez AJ, Barba FJ (2023) Effects of pressurized liquid extraction with dimethyl sulfoxide on the recovery of carotenoids and other dietary valuable compounds from the microalgae spirulina, Chlorella and Phaeodactylum tricornutum. Food Chem 405(2022):134885. https://doi.org/10.1016/j.foodchem.2022.134885

Xia GH, Li XH, Jiang Y-h (2021) Deep eutectic solvents as green media for flavonoids extraction from the rhizomes of Polygonatum odoratum. Alex Eng J 60:1991–2000. https://doi.org/10.1016/j.aej.2020.12.008

Xu M, Ran L, Chen N, Fan X, Ren D, Yi L (2019) Polarity-dependent extraction of flavonoids from citrus peel waste using a tailor-made deep eutectic solvent. Food Chem 297:124970. https://doi.org/10.1016/j.foodchem.2019.124970

Yu C, Yang C (2019) Effect of ultrasound on the extraction of bioactive aglycone isoflavones for the green valorization of black soybean residue (okara). J Food Process Preserv 43:e13944. https://doi.org/10.1111/jfpp.13944

Zhang ZH, Wang LH, Zeng XA, Han Z, Wang MS (2017) Effect of pulsed electric fields (PEFs) on the pigments extracted from spinach (Spinacia oleracea L.). Innovative Food Sci Emerg Technol 43:26–34. https://doi.org/10.1016/j.ifset.2017.06.014

Zhao XY, Zhang XW, Liu H, Zhu H, Zhu Y (2019) Enzyme-assisted extraction of astaxanthin from Haematococcus pluvialis and its stability and antioxidant activity. Food Sci Biotechnol 28:1637–1647. https://doi.org/10.1007/s10068-019-00608-6

Chapter 6
Encapsulation Techniques for the Preservation of Carotenoids' Bioactivity

Hamdy A. Shaaban, Mahassen M. Sleem, and Haitham H. Abdel-Gawad

1 Introduction

Carotenoids are a significant natural source of provitamin A and antioxidant components (Šeregelj et al. 2017), which are ideal associated with some health benefits, such as the inhibition of LDL oxidation, alleviation of oxidative stress, anti-inflammatory properties, and enhancement of immune response (Kiokias and Gordon 2004; Šeregelj et al. 2020). Carotenoids stand out due to their nonpolar structure and highly unsaturated molecules, which contribute to their low water solubility, limited chemical stability, and reduced bio-accessibility. Additionally, they are more susceptible to oxidation and isomerization during various treatments or storage, leading to the diminishing of their bioactive properties and organoleptic characteristics (Sy et al. 2015; Xavier and Mercadante 2019). Furthermore, carotenoids might undergo chemical transformations in the gastrointestinal tract (GIT), impacting their potential health benefits (Donhowe and Kong 2014). As a result, it is essential to develop carotenoid delivery systems to address these challenges (Donhowe and Kong 2014). Colloidal delivery systems represent one of the most practical approaches for incorporating hydrophobic bioactive compounds, such as carotenoids, into food products (Patel and Velikov 2011; McClements 2012).

Carotenoids consist of a hydrophobic carbon chain and are characterized as natural pigments produced by plants, algae, bacteria, and fungi, with no synthesis

H. A. Shaaban (✉)
Department of Chemistry of Flavor and Aroma, National Research Centre, Dokki, Giza, Egypt

M. M. Sleem
Food Technology Research Institute, Agriculture Research Centre, Giza, Egypt

H. H. Abdel-Gawad
Egyptian Drug Authority, EDA, Giza, Egypt

R. Campos Chisté et al. (eds.), *Carotenoids*,
https://doi.org/10.1007/978-3-031-75322-0_6

occurring in humans and animals. They are responsible for the yellow, orange, and red colorations observed in several vegetables (Nisar et al. 2015).

Carotenoids are comprised of two groups, namely carotenes and xanthophylls obtained from edible sources. The former are hydrocarbons (e.g., α-carotene, β-carotene, and lycopene), while xanthophylls have oxygen in their chain structures (e.g., astaxanthin, lutein, and zeaxanthin) (Havaux 2013).

These bioactive compounds exhibit various functionalities, including their involvement in the immune system, provitamin A activity in certain carotenoids (α-carotene, β-carotene, and cryptoxanthin), regulation of the cell cycle, apoptosis, cell differentiation, and other biological processes (Ribeiro et al. 2018). Consequently, they can serve as a preventive factor against conditions such as cardiovascular diseases, obesity, and inflammation. The primary effects are closely associated with the notable antioxidant properties exhibited by carotenoids (Ribeiro et al. 2018).

Carotenoids function as physicochemical suppressors of reactive oxygen species (ROS), employing an efficient system to capture and eliminate free radicals, promoting the protection of biological membranes against the effects of oxidation and lipid peroxidation. Thus, they reduce risks of morbidity and mortality, with antioxidant activity being related to the number of conjugated double bonds and functional groups such as ketone groups and cyclopentane rings in its chemical structure (Delgado-Vargas et al. 2000; Fiedor and Burda 2014).

The use of these bioactive compounds has been growing due to their health benefits. However, due to high physicochemical instability related to factors such as heat and the presence of light, oxygen, and constituents present in the food matrix, it is essential to employ techniques that guarantee protection, thereby ensuring the integrity and preservation of functional properties (Xianquan et al. 2005).

From this perspective, encapsulation techniques arise, which together with encapsulating agents provide greater stability and efficiency. Besides, improving bioavailability and/or bioactivity, releasing them in a controlled manner at the target site, and mainly prolonging and enhancing its antioxidant action (Steiner et al. 2018).

It's important to highlight that there is no universally applicable method, as encapsulation techniques take into account various factors, including the compound of interest and its mechanisms of action. Among these factors are the molecule nature and possible chemical interactions with the encapsulating agent, cost-benefit, time, and storage conditions. In addition, the applicability and functionality that you want to preserve and/or improve through encapsulation, among other aspects (Martínez-Alvarez et al. 2020).

From this perspective, the particle size obtained is directly related to its function. It can be subdivided into nanoparticles (NPs), with sizes smaller than 100 nm, and microparticles (MPs), ranging from 1 to 1000 μm (Dias et al. 2015; Melo et al. 2020).

Therefore, techniques for encapsulating carotenoids take into account the hydrophobic nature of these molecules, with options like spray-drying offering advantages such as cost-effectiveness and methodological simplicity, but has limitations related to production on an industrial scale; coacervation, with high encapsulation efficiency, however, does not guarantee protection in acidic environments and;

emulsification, which despite using a significant amount of surfactant, acts improving aspects such as solubility, contact surface, stability and absorption at low concentrations (Rehman et al. 2020). Moreover, emulsification serves to mitigate toxicity risks and enhance the efficacy of these bioactive compounds, making it a promising encapsulation technique to enhance the potential utilization of carotenoids (Kumar et al. 2019).

One notable advantage for the food industry in encapsulating carotenoids is the opportunity to substitute artificial food colorings, thereby transforming food products into a source of these bioactive compounds and, consequently, making them functional. This application extends to various food items such as yogurt, bread, muffins, ice cream, and others. Furthermore, the enhanced solubility results in improved bio-accessibility, increased bioavailability, facilitating controlled release, and enabling the use of lower concentrations of the encapsulated active ingredient, thereby enhancing functionality (Eun et al. 2020; Santos et al. 2021; Sridhar et al. 2021).

In the food industry, encapsulation technology is employed to create effective barriers, using both liquid and solid ingredients, against environmental factors like oxygen, light, and free radicals (Desai and Park 2005). Bioactive compounds can be improved by using encapsulation techniques, which entrap sensitive ingredients inside a coating material (Saenz et al. 2009). Additionally, encapsulating agents find utility in enhancing the stability of bioactive compounds. Various encapsulation materials, including starch, maltodextrin, corn syrup, inulin, and Arabic gum, have been employed by researchers (Saenz et al. 2009; Ahmed et al. 2010a, b; Kha et al. 2010) to enhance the stability of bioactive compounds during spray drying. Certain reviews delve into encapsulation techniques, such as the microencapsulation of oils (Bakry et al. 2016) and microencapsulation of vitamin A (Gonçalves et al. 2016), exist. Nevertheless, these reviews do not provide information on the choice of encapsulating materials and modeling techniques for the encapsulation of carotenoids. Therefore, the objective of this review is to illustrate the impact of encapsulating agents on carotenoids from different sources, as well as the degradation kinetics during processing and storage using spray dryers and freeze dryers.

2 General Characteristics of Carotenoids

Carotenoids are isoprenoid compounds. Plants have yielded over 600 distinct fat-soluble carotenes through isolation (Gul et al. 2016). The synthesis of carotenoids through chemosynthesis entails the tail-to-tail linkage of two C20 geranyl-geranyl diphosphate molecules, giving rise to a precursor C40 carbon skeleton that serves as the foundation for all subsequent individual variations (Dutta et al. 2005). In plants, the most frequently encountered carotenoids include β-carotene, lutein, zeaxanthin, neoxanthin, violaxanthin, and chlorophylls (Gul et al. 2016; Lakshminarayana et al. 2005). The all-*trans*-β-carotene (chemical formula: $C_{40}H_{56}$; molecular weight: 536.88 g/mol) (Fig. 6.1) and its *cis*-isomers are widely distributed in plant-based

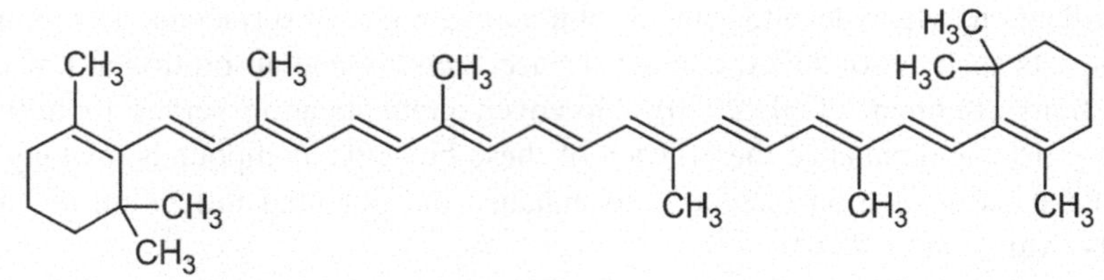

Fig. 6.1 Chemical structure of β-carotene

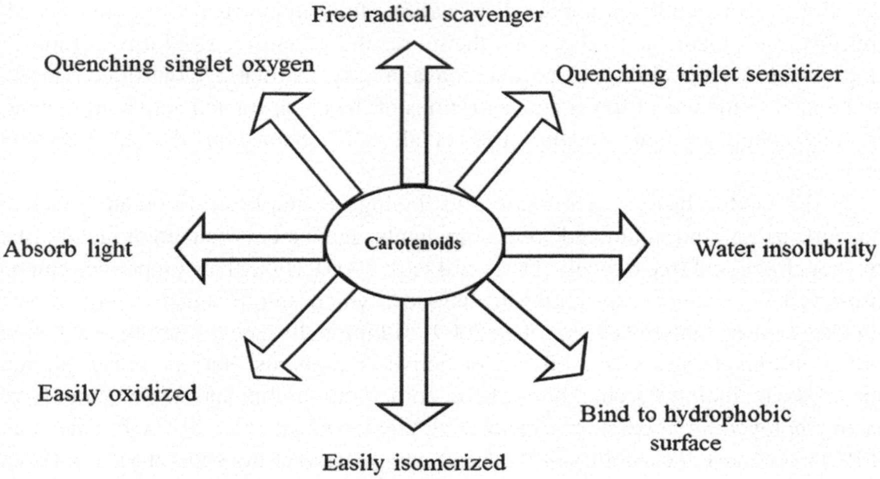

Fig. 6.2 Physical and chemical properties of carotenoids

foods and microorganisms (Khachik et al. 1995), and are generally the dominant precursors of vitamin A (Krinsky and Johnson 2005).

The physicochemical features including color, chemical reactivity, molecular shape, light absorbing intensity, and antioxidant properties of carotenoids depend on the length of the polyene chain (Fig. 6.2) (Britton 1995; Dutta et al. 2005). Carotenoids are classified into two categories: (i) α-, β-, γ-carotene and α-, β-cryptoxanthin, which can be metabolized by humans into retinol, and (ii) lutein, zeaxanthin, lycopene, which are unrelated to any vitamin A activity in humans (Bohn 2008).

A number of factors are responsible for the degradation of carotenoids. On exposure to heat, light, and oxygen during processing and storage, *trans*-β-carotene immediately undergoes thermal and chemical oxidation, isomerization, and photosensitization (Dutta et al. 2005; Gul et al. 2016). The instability of carotenoids is caused by their structural polyene chain. Carotenoid oxidation comprises the epoxidation process, resulting in the formation of apocarotenoids (Marty and Berset 1988), and low weight-molecular compounds responsible for the off-flavor in food (Falconer et al. 1964).

Throughout the digestion process, carotenoids also undergo various degradation pathways within the human body. Consequently, the preservation of carotenoids poses a significant challenge during both processing and storage. Employing microencapsulation techniques could augment the retention of carotenoids, safeguarding them against degradation caused by light, heat, and temperature fluctuations.

3 Chemistry of Carotenoids

Carotenoids are a group of pigments, mostly of plant origin, responsible for the yellow, orange and red colors in fruits and vegetables. All have antioxidant activity, and some are precursors of vitamin A. Moreover, carotenoids have a role in intercellular communication, immune system activation and disease prevention (Saini et al. 2015; Salter-Venzon et al. 2017), and hence they promote human health.

Carotenoids consist of eight repetitive isoprene units, featuring cyclic or linear structures at both ends of the carbon chains. This configuration gives rise to numerous *cis* and *trans* isomers, with the *trans* isomers being more prevalent in nature (Guerrero-Legarreta et al. 2013; Schawartz et al. 2017). Carotenoids are categorized into carotenes and xanthophylls. Carotenes, such as α-carotene, β-carotene, γ-carotene, and lycopene (Guerrero-Legarreta et al. 2013), are highly soluble in organic solvents and insoluble in polar solvents. Conversely, xanthophylls are soluble in medium polar solvents (such as alcohols) and organic solvents (such as ether and hexane). They represent oxygenated derivatives of carotenes, giving rise to alcohols, aldehydes, ketones, and acids. Fucoxanthin, lutein, and violaxanthin are examples of xanthophylls (Guerrero-Legarreta et al. 2013; Schawartz et al. 2017).

Plant tissues, specifically in plastids, serve as storage sites for carotenoids, including chromoplasts (colored plastids), amyloplasts (starch storage plastids), and elaioplasts (lipid storage plastids). Carotenoids are situated in the chromoplasts of fruits, flowers, and roots, while in grains and oil seeds, they are found in amyloplasts and elaioplasts, respectively (Howitt and Pogson 2006). In contrast, xanthophylls are freely found in green plant tissues, whereas in fruits and flowers they are found as esters of fatty acids (Perera and Yen 2007).

4 Natural Occurrence of Carotenoids

More than 700 carotenoids have been identified, yet only 40 are included in the human diet, with β-carotene, α-carotene, lycopene, β-cryptoxanthin, lutein, and zeaxanthin the most common (Olmedilla-Alonso 2017). Carotenoid intake comes primarily from foods of plant origin (fruits and vegetables). α- and β-Carotene are usually found in carrot, orange, pumpkin, pepper, sweet potato, mango, and vegetable leaves, varying in color from yellow, orange, and red. Lycopene provides a vibrant red hue to food and serves as the primary carotenoid in tomatoes. It is also found in watermelon, guava, and papaya. β-Cryptoxanthin is found in citrus fruit, melon, potato, guava and apple, giving them yellow and orange colors. The natural pairing of isomers, lutein and zeaxanthin, is observed in yellow corn and marigold flowers. Additionally, they are present in broccoli, pumpkin, peppers, vegetable leaves, seeds, and legumes. Paprika and peppers often contain the xanthophylls violaxanthin (yielding a yellow color), capsanthin, and capsorubin (imparting orange to red shades). Neoxanthin, distinguished by its yellow hue, is a naturally occurring element in vegetable leaves. Bixin, the principal component of annatto, imparts its distinctive red-brown color. The yellow coloration of saffron is attributed to crocin (Saini et al. 2015; Guerrero-Legarreta et al. 2013; Schawartz et al. 2017; Howitt and Pogson 2006).

Certain carotenoids are exclusive to algae and seafood. Astaxanthin, present in krill and microalgae *Haematococcus pluvialis*, gives shrimp, salmon, and flamingo feathers their pink-red color. This carotenoid is consumed by small crustaceans (shrimp and crawfish), fish (salmon), and birds (flamingo), respectively (Food-Info Foundation 2020; Liu et al. 2016). Fucoxanthin has a characteristic brown color and is only found in microalgae and the chloroplasts of brown algae (Kumar et al. 2013).

The most abundant carotenoids in the Western diet include β-carotene and α-carotene from carrot, pumpkin, apricot, pepper, mango, coriander and spinach (Saini et al. 2015; Olmedilla-Alonso 2017; Gul et al. 2015; Yoon et al. 2012); lutein from broccoli, pumpkin, spinach, corn, mango, papaya and lycopene from tomato, guava, papaya and watermelon (Saini et al. 2015; Olmedilla-Alonso 2017; Gul et al. 2015; Yoon et al. 2012; Bagetti et al. 2011; Chandrika et al. 2009).

Britton and Khachik (2009) proposed a ranking of fruits and vegetables based on their carotenoid levels, grouping the foods into the following categories: low level (0–1 μg/g), moderate level (1–5 μg/g), high level (5–20 μg/g), and very high level (>20 μg/g). The composition and quantity of carotenoids in food are influenced by numerous factors, encompassing those intrinsic to the plant (such as variety, genotype, and ripening stage) and external factors (including harvest season, growth conditions, post-harvest treatment, handling, storage conditions, plant diseases, and climatic conditions) (Saini et al. 2015; Olmedilla-Alonso 2017).

5 Stability of Carotenoids

Due to the presence of repetitive units of isoprene with double conjugated bonds, carotenoids exhibit high sensitivity to various environmental factors, including light, heat, elevated temperatures, acids, oxygen, metals, and free radicals. Consequently, these conditions induce structural alterations such as cycling, migration of double bonds, and the addition of oxygen molecules, leading to the creation of epoxy-carotenoids and apocarotenoids compounds devoid of biological activity (Butnariu 2016; Ribeiro et al. 2018; Thakur et al. 2017).

Light, high temperature, oil, and dehydration cause isomerization of carotenoids and formation of 15-*cis*-, 9-*cis*- and 13-*cis*-β-carotene and α-carotene isomers. The provitamin A potential of *cis*-isomers is lower than that of *trans*-isomers, consequently diminishing the nutraceutical properties of carotenoids (Butnariu 2016; Delgado-Vargas et al. 2000).

Several studies have documented the degradation of carotenoids in vegetable products subjected to thermal treatments. Addis et al. (2009) observed a decline in the carotenoid content of *Coccinia grandis* and *Trigonella foenum-graecum* leaves during drying (from 139.80 mg/100 g to 109.90 mg/100 g) and dehydration (from 139.80 mg/100 g to 63.20 mg/100 g and 116.60 mg/100 g to 96.20 mg/100 g, respectively). In contrast, Aschoff et al. (2015) found that pasteurization of orange juice reduced its carotenoid content (230.50 μg/100 g) compared to fresh oranges (328.70 μg/100 g). Furthermore, Odriozola-Serrano et al. (2009) observed a reduction in carotenoid levels in fresh tomato juice (14.10 mg/100 mL) following pasteurization and storage at 4 °C for 56 days (7.3 mg/100 mL). Despite the general trend of heat treatment leading to a decrease in carotenoid content, some carotenoids have been noted to increase in concentration during such treatment. This phenomenon is attributed to improved product stabilization, enzymatic denaturation, and moisture loss, which collectively alter the cell membrane and protein-carotenoid complexes, rendering carotenoids more accessible for extraction (Odriozola-Serrano et al. 2009). Plasma levels of β-carotene, α-carotene, β-cryptoxanthin, lutein, and zeinoxanthin were higher after the consumption of pasteurized orange juice instead of fresh oranges, increasing carotenoid bioavailability between 25% and 30% (Aschoff et al. 2015).

According to Addis et al. (2009), a brief blanching of green leaves results in the release of bound carotenoids, enhancing their extractability and consequently increasing their content compared to raw products. The depletion of carotenoids in fruit and vegetable items is primarily attributed to the oxidation of highly unsaturated carotenoid structures. Oxidation can be caused by auto-oxidation (spontaneous formation of free radicals in the presence of oxygen) or photo-oxidation (caused by oxygen in the presence of light) leading to carotenoid bleaching and resulting in colorless products. Some of these breakdown products are 5,6-epoxy-β-ionone, 5,6-epoxy-β-carotene, 5,8-epoxy-β-carotene, 5,6,5′6′-diepoxy-β-carotene, 2,6,6-trimethyl-cyclohexanone, dihydroactinidiolide, β-cyclocitral, 4-oxo-β-ionone, 2-methyl-2-hepten-6-one, β-ionone, pseudo-ionone, 6-methyl-3,5-heptadien-2-one, β-carotene 5,8-endoperoxide, geranial, neral, geronic acid, acycloretinal, aurochrome and mutatochrome (Odriozola-Serrano et al. 2009; Ribeiro et al. 2018; Delgado-Vargas et al. 2000; Burton et al. 2016; Carail and Caris-Veyrat 2006; Johnston et al. 2014).

6 Encapsulation Techniques for Carotenoid Protection

The utilization of carotenoids in the food industry has been constrained due to their susceptibility to chemical factors (oxygen, metals, free radicals) and environmental conditions (light and heat), as well as their insolubility in water, high melting points, and low bioavailability. The lipophilic nature of carotenoids restricts their direct integration into aqueous systems. Consequently, various carriers have been developed, including emulsions, nanoemulsions, liposomes, hydrogel particles (a network of hydrophilic polymer trapping solvent molecules), and solid-liquid particles (crystallized lipid dispersed in oil), to enhance their stability and applicability (Gul et al. 2015; Soukoulis and Bohn 2018). Among them, emulsions are widely used for carotenoid formulations.

Among these, emulsions stand out as widely employed for carotenoid formulations. Emulsions are colloidal dispersions of two immiscible liquids, with one liquid (dispersed phase) distributed as small droplets within the other (continuous phase) (McClements 2015). Emulsifying or stabilizing agents, which reduce surface tension between phases, are primarily used, facilitating the formation and maintaining the integrity of the emulsion (Walstra and van Vliet 2007). These stabilizing agents encompass small molecules like polyoxyethylene sorbitan fatty acid esters (Tweens) or larger molecules such as proteins (casein, whey protein, soybean protein, and bovine serum albumin), phospholipids (lecithin), and polysaccharides (gums, pectin, and modified starch) (Focsan et al. 2019).

Emulsions are categorized into Water-in-Oil (W/O) emulsions, where water is the dispersed phase within an oil continuous phase, and Oil-in-Water (O/W) emulsions, where oil is the dispersed phase within a water continuous phase (McClements 2015). O/W emulsions are typically prepared prior to carotenoid encapsulation to achieve high encapsulation efficiency and a narrow particle size distribution (Soukoulis and Bohn 2018; Janiszewska-Turak 2017).

In order to prevent the degradation of carotenoids and maintain their antioxidant activity, encapsulation techniques are employed, enhancing their stability and influencing their bioavailability and solubility (Focsan et al. 2019). Encapsulation techniques involve enveloping one or more delicate substances (such as pigments, antioxidants, essential oils, and drugs), referred to as the core material, and with another component serving as a protective barrier (referred to as the wall material, carrier, capsule, or membrane) (Santos and Meireles 2010; Gul et al. 2015).

6.1 *Microencapsulation*

Microencapsulation entails enveloping a liquid, solid, or gas with a surrounding material known as a carrier (Janiszewska-Turak 2017). Carrier agents are essential to safeguard the encapsulated substance against environmental factors, including moisture, heat, light, oxygen, pH fluctuations, and other compounds (Donhowe and

Kong 2014). Commonly used carriers encompass polysaccharides (including maltodextrin, starch, chitosan, inulin, sodium alginate, carrageenan, pectin, CMC, and citrus fibers), gums (such as Arabic gum, Mesquite gum, Guar gum, and locust bean gum), and proteins (including gluten, casein, gelatin, whey protein, soy protein, albumin, milk powder, and oligopeptides) (Janiszewska-Turak 2017; Arellano Corral 2011). The selection of an appropriate carrier is determined by its physical attributes, such as molecular weight, melting point, solubility, viscosity, diffusivity, film-forming capacity, and emulsifying properties (Arellano Corral 2011). Various microcapsules are shown in Fig. 6.3.

6.2 *Spray Drying*

Spray drying stands out as a widely employed microencapsulation technique, involving the creation of fine particles through the passage of a suspension through a sprayer, where compressed air at elevated temperatures is directed. The hot air flow dehydrates the particles, transforming them into powder, effectively encapsulating the substance of interest (Gul et al. 2015). Consequently, stable powders, falling within a size range of 1–1000 μm, are swiftly generated at a low cost and temperature. This feature allows for the encapsulation of thermolabile compounds while ensuring the stability of the encapsulated substance (Thakur et al. 2017; Mestry et al. 2011). The encapsulation process using spray drying encompasses various stages as shown in Fig. 6.4.

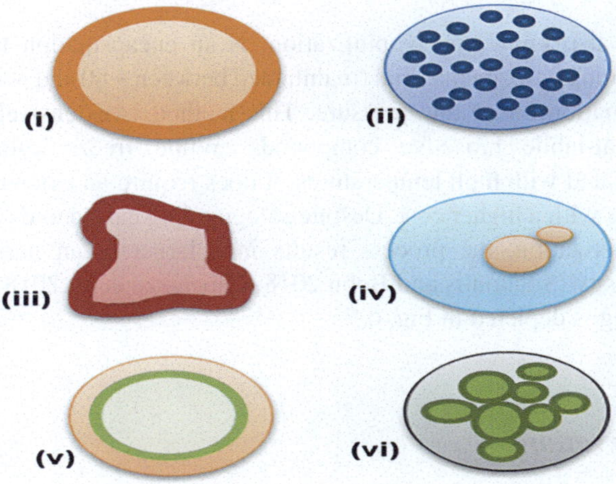

Fig. 6.3 Different types of microcapsules: (i) simple microcapsule, (ii) matrix (microsphere), (iii) irregular microcapsule, (iv) multicore microcapsule, (v) multiwall microcapsule, and (vi) assembly of a microcapsule

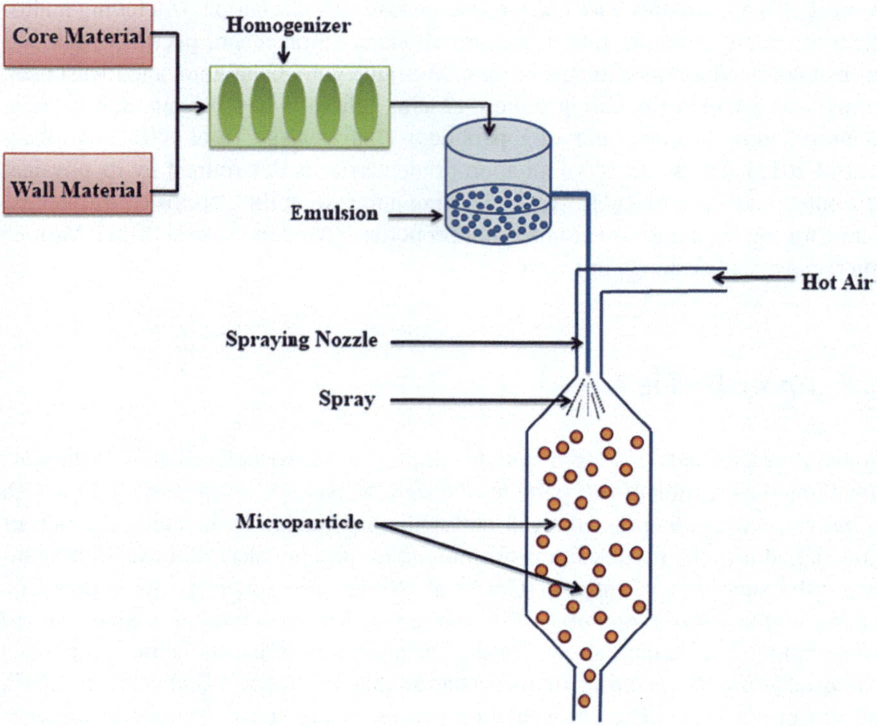

Fig. 6.4 Schematic representation of the microencapsulation process by spray drying

6.3 Freeze-Drying

Freeze-drying also known as lyophilization, is an encapsulation technique that entails dehydrating a frozen material (maintained between −80 and −40 °C) through vacuum sublimation under low pressure. This method is extensively utilized for preserving heat-labile bioactive compounds. While freeze-drying minimizes changes associated with high temperatures, it does require an extended processing time and comes with a higher cost. Despite safeguarding carotenoids against oxidation and isomerization, the process results in a larger mean particle size and increased porosity (Soukoulis and Bohn 2018; Capelezzo et al. 2018). The process of freeze-drying is depicted in Fig. 6.5.

6.4 Coacervation

Coacervation, a microencapsulation technique, involves the phase separation of a homogeneous polymer solution into a polymer-rich and a polymer-poor phase. This process comprises emulsification, followed by phase separation, cross-linking of

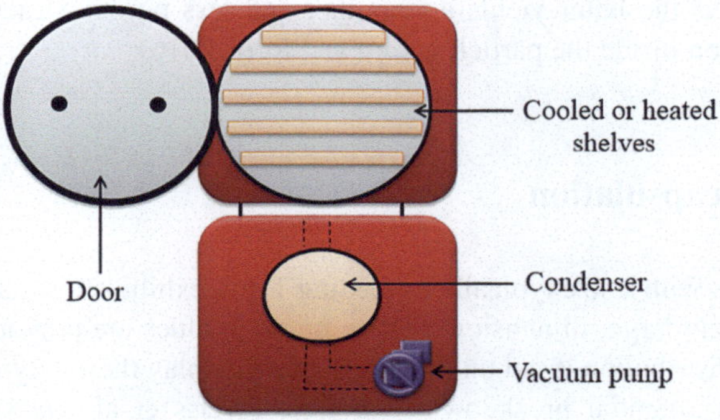

Fig. 6.5 Schematic diagram of a freeze dryer

coacervates, and lyophilization. The induction of phase separation is frequently accomplished by pH changes, disrupting the interaction between polymers (Thakur et al. 2017; Soukoulis and Bohn 2018).

An alternative microencapsulation method involves the creation of inclusion complexes with oligosaccharides and polysaccharides like cyclodextrins, glycyrrhizic acid, and arabinogalactan (Apanasenko et al. 2015). The inclusion complex possesses a hydrophobic center that forms non-covalent interactions with the hydrophobic structure of carotenoids, entrapping them within.

The utilization of inclusion complexes offers advantages such as enhanced chemical stability, protection of bioactive compounds from the environment, taste modification, and controlled release (Focsan et al. 2019; Capelezzo et al. 2018). This technique has been applied for the encapsulation of various carotenoids, including lycopene, β-carotene, lutein, canthaxanthin, zeaxanthin, and bixin (Nunes and Mercadante 2007).

The effectiveness of a microencapsulation process heavily relies on the chosen methodology, impacting key parameters such as moisture content, water activity, particle size, encapsulation efficiency, and the morphology of microcapsules (Gul et al. 2015).

For example, elevated temperatures accelerate the formation time of the membrane surrounding encapsulated compounds, thereby hindering their retention. Furthermore, high temperatures result in decreased carotenoid contents and increased isomerization (Janiszewska-Turak 2017). Juscamaita-Fabián et al. (2017) conducted a comparative study on the efficacy of Arabic gum (GA) and maltodextrin (MD) in forming carotenoid microcapsules from the petals of *Tropaeolum majus*. These authors achieved superior carotenoid retention with GA and a drying temperature of 130 °C, as higher temperatures led to greater carotenoid loss.

Conversely, encapsulation efficiency can be enhanced by increasing the ratio of core to wall material, allowing more carotenoids to be entrapped within the carrier (Soukoulis and Bohn 2018). Moreover, the degradation of carotenoids was reported as more pronounced in microcapsules obtained through freeze-drying than through spray drying, as the latter yields a smoother and less porous structure, reducing oxygen diffusion inside the particles (Lim and Roos 2016).

7 Nanoencapsulation

Microcapsules, with a size typically exceeding 1 μm, exhibit slower absorption due to their relatively larger dimensions. These microcapsules are polydisperse, meaning their size distribution is not uniform, and often display thermodynamic instability, leading to potential breakdown over time (Weiss et al. 2008). In contrast, nanoparticles (NPs) are thermodynamically stable or possess long-term kinetic stability, featuring a small and uniform size (<1000 nm) that ensures rapid release and absorption of bioactive compounds (Focsan et al. 2019).

Nanoparticles are characterized by minimal phase segregation, limited interaction between bioactive compounds and other food components, preservation of bioactive properties, enhanced absorption and bioavailability, and a reduced impact on sensory attributes (Sessa 2012).

Nanoencapsulation methods have been employed to safeguard vulnerable compounds from degradation, enhance their bioavailability, and maintain their biological activity. These techniques comprise both top-down and bottom-up approaches for nanoparticle (NP) production. The bottom-up technique involves the self-organization and self-assembly of molecules through processes such as nanoprecipitation, coacervation, and inclusion complexation. On the other hand, top-down techniques require specialized equipment to reduce particle size and produce NPs. These top-down methods encompass extrusion, homogenization, electrospinning/spraying, and emulsification-solvent evaporation processes (Rehman et al. 2020).

Nanocarriers such as nanoliposomes, nanoemulsions, and lipid-based nanoparticles represent examples of molecules employed for the protection of carotenoids (Dos Santos et al. 2018). Nanoemulsions consist of colloidal systems made of lipid droplets, smaller than100 nm in size, dispersed in a continuous aqueous phase (Soukoulis and Bohn 2018). They are used for the encapsulation of water-insoluble bioactive compounds. Nanoemulsions are colloidal systems composed of lipid droplets, each smaller than 100 nm in size, dispersed in a continuous aqueous phase (Soukoulis and Bohn 2018). These systems are utilized for encapsulating water-insoluble bioactive compounds. Nanoemulsions are recognized for their efficiency, cost-effectiveness, enhanced dispersibility, prolonged stability, and heightened bioavailability of active substances (Gul et al. 2015). While droplet formation is generally a straightforward process, further reduction into smaller droplets (50–200 nm) necessitates additional energy, often of a mechanical nature.

Nanoemulsion processes are commonly categorized into high-energy and low-energy techniques. High-energy techniques encompass high-pressure homogenization, sonication, and microfluidization, while low-energy techniques involve spontaneous emulsification and phase inversion temperature (PIT) (Dos Santos et al. 2018; Tasch Holkem et al. 2015).

High-energy techniques involve the utilization of specialized equipment to create nanoemulsions. High-pressure homogenization entails passing a solution through a narrow opening at elevated pressure (100–2000 bar) and high speed (1000 km/h). This process deforms and breaks droplets into smaller ones, approximately ~50 nm in size. Ultrasound technology involves the application of high-frequency ultrasonic waves that disrupt emulsion droplets. Microfluidization, akin to high-pressure homogenization, passes an emulsion through a chamber at a 180° angle at high speed, thereby reducing droplet size (Tasch Holkem et al. 2015).

Conversely, low-energy techniques do not rely on mechanical energy to break down droplets; instead, the chemical energy within their components is utilized. Spontaneous emulsification involves preparing an organic solution, using an oil- and water-miscible solvent, added to an aqueous solution containing a hydrophilic surfactant under constant stirring to form an oil-in-water emulsion. Subsequently, the organic phase is removed by evaporation under low pressure. In contrast, phase inversion temperature (PIT) does not necessitate organic solvents; instead, non-ionic surfactants, such as polyethyloxylates, are used, as they can alter their affinity between oil and water based on temperature. At elevated temperatures, polyethyloxylates become lipophilic, forming water-in-oil emulsions, whereas at lower temperatures, they become hydrophilic, forming oil-in-water emulsions. During cooling, interfacial tension is minimal, facilitating the creation of nanoemulsion. However, for this to occur, the surfactant concentration must surpass the critical micellar concentration (Tasch Holkem et al. 2015).

8 Supercritical Fluids

Supercritical fluids (SCF) have recently emerged as a viable alternative for encapsulating carotenoids. An SCF is a dense liquid possessing the physicochemical properties of a gas. The utilization of SCF for encapsulating bioactive and thermolabile compounds has gained traction in the food industry as a "green technology" owing to its non-toxic nature, easy removal without product damage, and the production of numerous microencapsulated and nanoencapsulated products (Soukoulis and Bohn 2018; Capelezzo et al. 2018; Soh and Lee 2019). Furthermore, this innovative approach addresses the drawbacks associated with conventional encapsulation techniques, including inadequate control over size and morphology, loss of thermolabile compounds, low encapsulation efficiency, and yield (Cocero et al. 2009).

Supercritical carbon dioxide (SC-CO_2) stands out as the most popular supercritical solvent in the food and pharmaceutical industry due to its low critical point ($T_c = 31.10$ °C, $P_c = 7.38$ MPa), safety, low viscosity and reactivity, ease of elimination, ability to inactivate microbes, relatively low cost, and improved solubility of certain lipophilic compounds (Janiszewska-Turak 2017). SC-CO_2 has notably enhanced the solubility of essential oils, bioactive compounds of plant origin, small molecular non-polar compounds, low molecular weight biopolymers, and low-polarity lipophilic compounds (Soh and Lee 2019). Janiszewska-Turak (2017) employed SC-CO_2 for the encapsulation of carotenoids such as astaxanthin, β-carotene, bixin, lutein, lycopene, and zeaxanthin. Supercritical encapsulation or micronization is categorized into the following types: supercritical anti-solvent or solution-enhanced dispersion by supercritical fluids (SAS/SEDS), rapid expansion of supercritical solutions (RESS), particle production from gas-saturated solutions (PGSS), and supercritical extraction from an emulsion (SFEE) (Soukoulis and Bohn 2018; Janiszewska-Turak 2017).

SAS/SEDS stands out as the most prevalent method for micronization and involves the use of an SCF, typically SC-CO_2, as an anti-solvent to decrease the solubility of the active substance and encapsulating agent in their solvent, leading to their precipitation into crystals of various morphologies (threads, sponges, leaves, needles, crystals, and spheres) (Janiszewska-Turak 2017). Nanoparticles are formed through the rapid diffusion of an organic solvent into the aqueous phase and the co-precipitation of active compounds and carriers at the solvent/anti-solvent interface. Subsequently, the solvent can be removed through evaporation (Soukoulis and Bohn 2018). The primary advantages of SAS/SEDS processes are associated with the reduction in particle size due to the swift precipitation of solutes, a higher solubility rate, nearly complete elimination of the solvent, and the ability to encapsulate thermolabile compounds (Silva and Meireles 2014).

Xia et al. (2017) and Zhao et al. (2017) have explored the use of supercritical fluids (SCF) as anti-solvents to develop lutein-loaded liposomes through solution-enhanced dispersion by supercritical fluids (SAS) using supercritical carbon dioxide (SC-CO_2) as an anti-solvent. Under specific conditions, such as pressure at 8 MPa, temperature at 35 °C, and flow rate at 1 mL/min, they achieved liposomes with particle sizes of 500 nm and 155 nm, and encapsulation efficiencies of 90% and 97%, respectively. Similarly, Machado Jr. et al. (2014) utilized supercritical fluid extraction of emulsion (SEDS) to encapsulate astaxanthin in poly(hydroxybutyrate-co-hydroxyvalerate) with SC-CO_2 as an anti-solvent. Through experimentation with various parameters, they identified that the highest encapsulation efficiency (48.25%) and smallest particle size (128 nm) were achieved at conditions of 35 °C, 100 bar, and 1 mL/min.

Rapid Expansion of Supercritical Solutions (RESS) encapsulation leverages the solvation capabilities of supercritical fluids (SCFs) by adjusting pressure and temperature. In this process, the bioactive compound and the encapsulating agent are dissolved in supercritical carbon dioxide (SC-CO_2) and then precipitated through rapid depressurization and nucleation upon passage through a nozzle (Soukoulis and Bohn 2018; Zhong and Dai 2011). RESS offers the advantage of single-step processing at low temperatures and can be applied to encapsulate a broad range of inorganic, organic, and polymeric materials (Soh and Lee 2019). However, a significant limitation is the low solubility of wall materials in SC-CO_2. Consequently, RESS is infrequently employed for carotenoid encapsulation using SCF, and only a few reports are available. One such report from a decade ago is by Quan et al. (2009), who utilized SC-CO_2 for astaxanthin encapsulation, yielding particles ranging from 0.3 to 0.8 μm. To address the solubility challenge, a liquid anti-solvent like ethanol or methanol can be used as a co-solvent for the polymer (Santos and Meireles 2010).

Particle Production from Gas-Saturated Solutions (PGSS) involves the use of SC-CO_2 as solute to saturate a solution containing the bioactive. As the gas-saturated solution passes through an atomizing nozzle at atmospheric pressure, gas vaporization cools the solution, inducing nucleation and precipitation of particles (Soukoulis and Bohn 2018; Janiszewska-Turak 2017). De Paz et al. (2012a, b) applied PGSS and SC-CO_2 for the microencapsulation of β-carotene, employing poly-(ε-caprolactones) as carriers to achieve micrometer-scale particles (111–652 μm). Alternatively, using soy lecithin for encapsulation with the PGSS drying technique resulted in smaller particles (10–500 μm), forming liposomes with a diameter of 0.9–6.1 μm upon rehydration.

Supercritical Fluid Extraction from an Emulsion (SFEE) involves the recrystallization of the active compound, incorporated in an oil-in-water (o/w) emulsion, utilizing the supercritical anti-solvent process. This method combines the efficiency of using supercritical fluids as an anti-solvent for particle formation with the emulsification effect on micro and nanoparticle formation (Silva and Meireles 2014). SC-CO_2 is employed to extract the organic solvent from the emulsion, leading to the supersaturation of the bioactive compound and polymer in the aqueous phase, resulting in their precipitation (Soh and Lee 2019). The presence of surface-active compounds in the emulsion facilitates particle formation without agglomeration and restricts particle size due to high nucleation rates. Compatible surfactants for this technique include polysorbates (Tweens), sodium caseinate, whey proteins, β-lactoglobulin, modified starch, maltodextrin (MD), gum Arabic (GA), and pectin (Soukoulis and Bohn 2018).

SFEE has been effectively employed by Tirado et al. (2019) for the encapsulation of astaxanthin using ethyl cellulose, resulting in nanoparticles ranging in size from 242 to 363 nm. In another study, Santos et al. (2012) utilized SC-CO_2 to produce nanoemulsions containing lycopene and β-carotene, achieving particle sizes within the range of 344 to 366 nm. The stability of carotenoids in micronized particles is contingent upon various factors, including the composition and nature of carotenoids, the type and concentration of the lipid phase, surfactant, particle size, interfacial composition, pH, ionic strength, and environmental conditions (light, oxygen, and temperature) (Soukoulis and Bohn 2018).

The selection of an appropriate wall material is crucial for the supercritical encapsulation of bioactive compounds, as carriers must shield the core material from environmental damage, minimize the loss of volatile compounds, and enable the controlled release of the active substance (Silva and Meireles 2014). Consequently, polysaccharides (gums, starches, celluloses, inulin, and cyclodextrins), proteins (casein, gelatin, and soy protein), and polymers (polylactic acid, polyhydroxy butyric acid, and polyethylene glycol) have been utilized as encapsulating agents in micronization processes (Capelezzo et al. 2018; Cocero et al. 2009).

9 Effects of Encapsulation on Carotenoids' Bioavailability

The bioavailability of bioactive compounds refers to the proportion of a specific nutrient that undergoes digestion, absorption, and metabolism through normal pathways (Gul et al. 2016). Various factors, as indicated by Ribeiro et al. (2008) and Bohn (2008), play a role in either stimulating or limiting the bioavailability of carotenoids. Mechanically or enzymatically releasing carotenoids from the food matrix is necessary for their bioavailability. Due to limited water solubility, carotenoids must be incorporated into lipid droplets.

The low bioavailability of carotenoids from natural sources is attributed to their existence as crystals or their location within protein complexes that cannot be adequately released in the gastrointestinal tract during digestion (Williams et al. 1998). The use of coating material during carotenoid processing enhances stability and facilitates controlled release of these bioactive compounds in the human body (Gul et al. 2016).

Numerous studies have reported that microencapsulation significantly affects the bioavailability of carotenoids in both in vitro and in vivo studies (Roman et al. 2012; Han et al. 2008; Donhowe and Kong 2014; Aissa et al. 2012). Higher water dispensability of carotenoids leads to increased bioavailability (Thurmann et al. 2002). Microencapsulation of carotenoids promotes their bioavailability (Soukoulis and Bohn 2015) through various mechanisms:

(a) Increasing stability against isomerization and degradation due to different factors.
(b) Improving solubility, resulting in increased bioaccessibility.
(c) Assisting targeted release kinetics.

9.1 *Effects of Encapsulated Carotenoids on In Vitro Studies*

In certain studies, the in vitro intestinal digestion of β-carotene encapsulated with alginate and chitosan was explored (Han et al. 2008; Roman et al. 2012). These studies revealed that complete transfer of β-carotene into the micelle phase during intestinal digestion is not necessarily correlated. The authors observed that the total concentration of β-carotene in the micelle cell was negligible, possibly due to the inhibition of micelle formation by the soluble fiber alginate. These findings highlighted the need for additional investigations into micelle formation to gain a clearer understanding of the bioavailability of microencapsulated β-carotene.

Encapsulated β-carotene also exhibited high resistance to gastric pH (Han et al. 2008). The resilience of encapsulated beads during transit through the stomach may be attributed to the presence of an external layer of chitosan-alginate coacervate. The release behavior of encapsulated β-carotene demonstrated efficient controlled release, maintaining a high percentage of release. This could be attributed to minimal water absorption by the encapsulating polymers, leading to swelling and preventing hydrophobic interaction within the polymeric matrix. Microencapsulating agents play a protective role, shielding micronutrients from enzymatic and acidic environments during release conditions and contributing to a controlled release mechanism of encapsulated micronutrients throughout gastrointestinal transit (Tiena et al. 2003). Similar outcomes were observed in isabgol fiber (Phyllium husk)-encapsulated carotenoids from crude palm oil, demonstrating effective regulation of release (Sen Gupta and Ghosh 2015).

9.2 Effects of Encapsulated Carotenoids on In Vivo Studies

In vivo studies comparing gum arabic microencapsulated β-carotene, produced through spray drying, with pure β-carotene, demonstrated that microencapsulated β-carotene exhibited higher potential against genotoxicity. This superiority was attributed to the microencapsulation process, which enhances the bioavailability of carotenoids, maintaining the antioxidant properties of β-carotene (Aissa et al. 2012).

Moreover, investigations into the physicochemical properties and bioavailability of microencapsulated lutein revealed that the encapsulated form exhibited greater bioavailability compared to a commercial reference sample. The wall material for the preparation of lutein microencapsulated product consisted of a modified food starch matrix. The increased bioavailability was attributed to the high solubility of the wall materials, promoting the aqueous solubility of lutein. Dissolution tests further indicated a rapid dissolution rate for lutein microencapsulated products, suggesting that their bioavailability was not restricted by dissolution behavior. Similar observations have been reported, emphasizing that encapsulated carotenoids experience enhanced solubility and release during digestion, facilitating their delivery into cellular compartments and ultimately improving bioavailability (Zhang et al. 2015; Wang and Bohn 2012).

10 Advantages of Encapsulated Carotenoids in Food Systems

Determining the stability of encapsulated carotenoids within a food system is a crucial consideration, yet there is a scarcity of studies in this area. In a study by (Mahfoudhi and Hamdi 2015), synthetic *trans*-β-carotene encapsulated with almond gum and gum arabic was employed in cake preparation. The results revealed that cakes made with almond gum exhibited a significantly higher a value compared to those made with gum Arabic. This outcome could be attributed to the superior protection of β-carotene when encapsulated with almond, as opposed to gum arabic.

Another investigation conducted by (Kha et al. 2015) focused on the development of yogurt, pasteurized milk, and cake incorporating encapsulated gac oil using whey protein concentrate and gum arabic. The results indicated that the color of products incorporating encapsulated gac oil remained unchanged, although slight decreases in β-carotene and lycopene levels were observed during storage. The authors attributed these changes to autooxidation, photooxidation, and photoisomerization over the storage period.

Coronel-Aguilera and Martin-Gonzalez (2015) produced spray-dried β-carotene powder using maltodextrin and sodium caseinate, followed by fluidized bed coating. The encapsulated β-carotene was applied to a yogurt system and compared with commercial peach yogurt. The study demonstrated that the total color value of yogurt containing encapsulated β-carotene was comparable to the standard value and remained stable in acidic media during a four-week storage period at 4 °C.

In a study by Donhowe et al. (2014), pudding and yogurt were manufactured using encapsulated β-carotene with maltodextrin, resulting in a water-dispersible powder with chitosan-alginate beds. The food matrix significantly influenced the release of β-carotene during digestion and its incorporation into the micelle phase. Yogurt containing encapsulated β-carotene exhibited a lower release rate and lower content of β-carotene in the micelle phase compared to pudding made with encapsulated β-carotene. The observed difference may be attributed to the pH of the food matrix, coagulation of the food matrix, and the protein composition of the food matrix (Donhowe et al. 2014).

11 Concluding Remarks

Carotenoids boast numerous health and nutritional benefits associated with health promotion and a reduced risk of various chronic degenerative diseases. However, their chemical instability makes them susceptible to oxidation under various storage conditions, including exposure to light, high temperatures, oxygen, acidity, and metal ions. To address this challenge, various encapsulation techniques, including microencapsulation, nanoencapsulation, and supercritical fluid encapsulation, have been employed.

Encapsulation serves not only to enhance the stability of carotenoids and protect them from degradation but also to increase their solubility in aqueous systems and improve bioavailability. Regardless of the encapsulation technology employed, the choice of an appropriate wall material is crucial for the effective protection and safe delivery of carotenoids.

Microencapsulation emerges as the most prevalent method for carotenoid encapsulation due to its simplicity and the production of high-quality products. However, microcapsules tend to exhibit decreased stability over time. In contrast, nanoencapsulation technology offers the advantage of producing more stable products with excellent absorption and bioavailability. Various carriers, including nanoemulsions, nanoliposomes, solid lipid nanoparticles (SLN), and nanostructured lipid carriers (NLC), can be utilized for carotenoid encapsulation.

Alternative or "green" technologies, such as supercritical encapsulation, provide viable options for the micro and nanoencapsulation of thermolabile compounds like carotenoids. These technologies are suitable for application in the food industry without adversely affecting sensory attributes.

Additionally, we have discussed the impact of different encapsulating agents on carotenoids from diverse sources, as well as the degradation kinetics of encapsulated carotenoids during processing and storage. The majority of studies conducted have focused on producing encapsulated carotenoids using various encapsulating agents through spray and freeze drying, with maltodextrins being commonly employed as encapsulating agents. Among the drying methods, spray drying stands out as the most suitable for encapsulation processes. However, several studies are still necessary in near future to cover the following aspects:

- How to improve encapsulated carotenoids absorption efficiency;
- The absorption mechanisms of encapsulated carotenoids;
- Which encapsulating agent is more efficient in improving the bioavailability of encapsulated carotenoids during digestive process?
- The effects of encapsulated carotenoids on human health, especially in the treatment of chronic degenerative diseases.

References

Addis G, Baskaran R, Raju M, Ushadevi A, Asfaw Z, Woldu Z, Baskaran V (2009) Effect of blanching and drying process on Carotenoids composition of underutilized Ethiopian (Coccinia grandis L. Voigt) and Indian (Trigonella foenum-graecum L.) Green Leafy vegetables. J Food Process Preserv 33:744–762

Ahmed M, Akter MS, Lee J-C, Eun J-B (2010a) Encapsulation by spray drying of bioactive components, physicochemical and morphological properties from purple sweet potato. LWT Food Sci Technol 43(9):1307–1312. https://doi.org/10.1016/j.lwt.2010.05.014

Ahmed M, Akter MS, Eun JB (2010b) Impact of a-amylase and maltodextrin on physicochemical, functional and antioxidant capacity of spray-dried sweet potato flour. J Sci Food Agric 90:494–502

Aissa AF, Bianchi MLP, Ribeiro JC, Hernandes LC, de Faria AF, Mercadante AZ, Antunes LMG (2012) Comparative study of b-carotene and microencapsulated b-carotene: evaluation of their genotoxic and antigenotoxic effects. Food Chem Toxicol 50(5):1418–1424. https://doi.org/10.1016/j.fct.2012.02.030

Apanasenko IE, Selyutina OY, Polyakov NE, Suntsova LP, Meteleva ES, Dushkin AV, Vachali P, Bernstein PS (2015) Solubilization and stabilization of macular carotenoids by water soluble oligosaccharides and polysaccharides. Arch Biochem Biophys 572:58–65

Arellano Corral CA (2011) Extracción de Luteína a Partir de Flores de Tagete (Tageteerecta) y EstabilizaciónporMicroencapsulación. Ph.D. Thesis, Facultad de CienciasQuímicas y Farmacéuticas, Universidad de Chile, Santiago, Chile, 44275.54, 17–21. [CrossRef]

Aschoff JK, Rolke CL, Breusing N, Bosy-Westphal A, Högel J, Carle R, Schweiggert RM (2015) Bioavailability of _-Cryptoxanthin is greater from pasteurized orange juice than from fresh oranges—a randomized cross-over study. Mol Nutr Food Res 59:1896–1904

Bagetti M, Facco EMP, Piccolo J, Hirsch GE, Rodriguez-Amaya D, Kobori CN, Vizzotto M, Emanuelli T (2011) Physicochemical characterization and antioxidant capacity of Pitanga fruits (Eugenia uniflora L.). Food Sci Technol 31:147–154 [CrossRef]

Bakry AM, Abbas S, Ali B, Majeed H, Abouelwafa MY, Mousa A, Liang L (2016) Microencapsulation of oils: a comprehensive review of benefits, techniques, and applications. Compr Rev Food Sci Food Saf 15(1):143–182. https://doi.org/10.1111/1541-4337.12179

Bohn T (2008) Bioavailability of non-pro vitamin A carotenoids. Curr Nutr Food Sci 4(4):240–258. https://doi.org/10.2174/157340108786263685

Britton G (1995) UV/Visible spectroscopy. In: Britton G, Liaaen-Jensen S, Pfander H (eds) Caotenoids, spectroscopy. Springer, Basel, pp 13–62

Britton G, Khachik F (2009) Carotenoids in food. In Carotenoids: Vol 5: Nutrition and Health (pp. 45–66). Basel: Birkhäuser Basel.

Burton GW, Daroszewski J, Mogg TJ, Nikiforov GB, Nickerson JG (2016) Discovery and characterization of carotenoid-oxygen copolymers in fruits and vegetables with potential health benefits. J Agric Food Chem 64:3767–3777

Butnariu M (2016) Methods of analysis (extraction, separation, identification and quantification) of carotenoids from natural products. J Ecosyst Ecograph 6:1–9

Capelezzo AP, Mohr LC, Dalcanton F, de Mello JMM, Fiori MA (2018) _-Cyclodextrins as encapsulating agents of essential oils. In: Cyclodextrin. a versatile ingdredient. IntechOpen, London

Carail M, Caris-Veyrat C (2006) Carotenoid oxidation products: from villain to saviour? Pure Appl Chem 78:1493–1503

Chandrika UG, Fernando KSSP, Ranaweera KKDS (2009) Carotenoid content and in vitro bioaccessibility of lycopene from guava (Psidium guajava) and watermelon (Citrullus lanatus) by high-performance liquid chromatography diode array detection. Int J Food Sci Nutr 60:558–566

Cocero MJ, Martín Á, Mattea F, Varona S (2009) Encapsulation and co-precipitation processes with supercritical fluids: fundamentals and applications. J Supercrit Fluids 47:546–555

Coronel-Aguilera CP, Martin-Gonzalez MFS (2015) Encapsulated of spray dried b-carotene emulsion by fluidized bed coating technology. LWT Food Sci Technol 62(1):187–193. https://doi.org/10.1016/j.lwt.2014.12.036

De Paz E, Martín Á, Cocero MJ (2012a) Formulation of _-carotene with soybean Lecithin by PGSS (particles from gas saturated solutions)-drying. J Supercrit Fluids 72:125–133

De Paz E, Martín Á, Duarte CMM, Cocero MJ (2012b) Formulation of _-carotene with poly-(-Caprolactones) by PGSS process. Powder Technol 217:77–83

de Santos PDF, Rubio FTV, da Silva MP, Pinho LS, Favaro-Trindade CS (2021) Microencapsulation of carotenoid-rich materials: a review. Food Res Int 147(July):110571. https://doi.org/10.1016/j.foodres.2021.110571

Delgado-Vargas F, Jimenez AR, Paredes-Lopez O, Francis FJ (2000) Natural pigments: carotenoids, anthocyanins, and betalains - characteristics, biosynthesis, processing, and stability. Crit Rev Food Sci Nutr 40(3):173–289. https://doi.org/10.1080/10408690091189257

Desai KGH, Park HJ (2005) Encapsulation of vitamin C in tripolyphosphate cross-linked chitosan microspheres by spray drying. J Microencapsul 22(2):179–192. https://doi.org/10.1080/02652040400026533

Dias MI, Ferreira ICFR, Barreiro MF (2015) Microencapsulation of bioactives for food applications. Food Funct 6(4):1035–1052. https://doi.org/10.1039/c4fo01175a

Donhowe EG, Kong F (2014) Beta-carotene: digestion, microencapsulation, and in vitro bioavailability. Food Bioprocess Technol 7(2):338–354. https://doi.org/10.1007/s11947-013-1244-z

Donhowe EG, Flores FP, Kerr LW, Wicker L, Kong F (2014) Characterization and in vitro bioavailability of b-carotene: effects of microencapsulation method and food matrix. LWT Food Sci Technol 57(1):42–48. https://doi.org/10.1016/j.lwt.2013.12.037

Dos Santos PP, de Andrade LA, Flôres SH, de Rios AO (2018) Nanoencapsulation of carotenoids: a focus on different delivery systems and evaluation parameters. J Food Sci Technol 55:3851–3860

Dutta D, Raychaudhuri U, Chakarborty R (2005) Structure, health benefits, antioxidant property and processing and storage of carotenoids. Afr J Biotechnol 4:1510–1520

Eun JB, Maruf A, Das PR, Nam SH (2020) A review of encapsulation of carotenoids using spray drying and freeze drying. Crit Rev Food Sci Nutr 60(21):3547–3572. https://doi.org/10.1080/10408398.2019.1698511

Falconer ME, Fishwick MJ, Land DG, Sayer ER (1964) Carotene oxidation and off flavor development in dehydrated carrot. J Sci Food Agric 15(12):897–901. https://doi.org/10.1002/jsfa.2740151215

Fiedor J, Burda K (2014) Potential role of carotenoids as antioxidants in human health and disease. Nutrients 6(2):466–488. https://doi.org/10.3390/nu6020466

Focsan AL, Polyakov NE, Kispert LD (2019) Supramolecular carotenoid complexes of enhanced solubility and stability—the way of bioavailability improvement. Molecules 24:3947

Food-Info Foundation E-Numbers: E100–E200 Food Colours. Available online: http://www.food--info.net/uk/e/e100-200.htm. Accessed on 2 June 2020

Goncalves A, Estevinho BN, Rocha F (2016) Microencapsulation of vitamin A: a review. Trends Food Sci Technol 51:76–87. https://doi.org/10.1016/j.tifs.2016.03.001

Guerrero-Legarreta I, López-Hernández E, Armenta-López RE, García-Barrientos R (2013) Pigmentos. In: de Alimentos Q, Badui-Dergal S (eds) . Pearson, Mexico City, p 379

Gul K, Tak A, Singh AK, Singh P, Yousuf B, Wani AA (2015) Chemistry, encapsulation, and health benefits of _-carotene—a review. Cogent Food Agric 1:1018696

Gul K, Tak A, Singh AK, Singh P, Yousuf B, Wani AA (2016) Chemistry, encapsulation, and health benefits of b-carotene –a review. Cogent Food Agric 1:1–12

Han J, Guenier A, Salmieri S, Lacroix M (2008) Alginate and chitosan functionalization for micronutrient encapsulation. J Agric Food Chem 56(7):2528–2535. https://doi.org/10.1021/jf703739k

Havaux M (2013) Carotenoid oxidation products as stress signals in plants. Plant J 79(4):597–606. https://doi.org/10.1111/tpj.12386

Howitt CA, Pogson BJ (2006) Carotenoid accumulation and function in seeds and non-green tissues. Plant Cell Environ 29:435–445

Janiszewska-Turak E (2017) Carotenoids microencapsulation by spray drying method and supercritical micronization. Food Res Int 99:891–901

Johnston JB, Nickerson JG, Daroszewski J, Mogg TJ, Burton GW (2014) Biologically active polymers from spontaneous Carotenoid oxidation: a new frontier in carotenoid activity. PLoS One 9:e111346

Juscamaita Fabián L, Pérez Salvatierra T, Espinoza Silva C, Quispe Solano M, Hinostroza Quiñónez G, Flores Ramos O, Manyari Cervantes G (2017) Evaluación de la Estabilidad de Carotenoides y Actividad Antioxidante de la Flor de Mastuerzo(Tropaeolum majus L.) en la Microencapsulaciónpor spray-drying. Rev Soc Quím Perú 83:282–293

Kha TC, Nguyen MH, Roach PD (2010) Effects of spray drying conditions on the physicochemical and antioxidant properties of the Gac (Momordicacochinchinensis) fruit aril powder. J Food Eng 98(3):385–392. https://doi.org/10.1016/j.jfoodeng.2010.01.016

Kha TC, Nguyen MH, Roach PD, Stathopoulos CE (2015) A storage study of encapsulatedgac (Momordicacochinchinensis) oil powder and its fortification into foods. Food Bioprod Process 96:113–125. https://doi.org/10.1016/j.fbp.2015.07.009

Khachik F, Beecher GR, Smith JC Jr (1995) Lutein, lycopene, and their oxidative metabolites in chemoprevention of cancer. J Cell Biochem 22:236–246. https://doi.org/10.1002/jcb.240590830

Kiokias S, Gordon MH (2004) Antioxidant properties of carotenoids in vitro and in vivo. Food Rev Int 20:99–121. https://doi.org/10.1081/FRI-120037155

Krinsky NI, Johnson EJ (2005) Carotenoid actions and their relation to health and disease. Mol Asp Med 26(6):459–516. https://doi.org/10.1016/j.mam.2005.10.001

Kumar S, Hosokawa M, Miyashita K (2013) Fucoxanthin: a marine carotenoid exerting anticancer effects by affecting multiple mechanisms. Mar Drugs 11:5130–5147

Kumar M, Bishnoi RS, Shukla AK, Jain CP (2019) Techniques for formulation of nanoemulsion drug delivery system: a review. Prev Nutr Food Sci 24(3):225–234. https://doi.org/10.3746/pnf.2019.24.3.225

Lakshminarayana R, Raju M, Krishnakantha TP, Baskaran V (2005) Determination of major carotenoids in a few Indian leafy vegetables by high-performance liquid chromatography. Journal of Agricultural and Food Chemistry 53(8):2838–42.

Lim ASL, Roos YH (2016) Spray drying of high hydrophilic solids emulsions with layered interface and Trehalose-Maltodextrinas glass formers for carotenoids stabilization. J Food Eng 171:174–184

Liu X, Luo Q, Cao Y, Goulette T, Liu X, Xiao H (2016) Mechanism of different Stereoisomeric Astaxanthin in resistance to oxidative stress in Caenorhabditis elegans. J Food Sci 81:H2280–H2287

Machado FRS Jr, Reis DF, Boschetto DL, Burkert JFM, Ferreira SRS, Oliveira JV, Burkert CAV (2014) Encapsulation of Astaxanthin from Haematococcus Pluvialis in PHBV by means of SEDS technique using supercritical CO2. Ind Crop Prod 54:17–21

Mahfoudhi N, Hamdi S (2015) Kinetic degradation and storage stability of b-carotene encapsulated by freeze-drying using almond gum and gum Arabic as wall materials. J Food Proc Preserv 39(6):896–906. https://doi.org/10.1111/jfpp.12302

Martínez-Alvarez O, Calvo MM, Gomez-Estaca J (2020) Recent advances in astaxanthin micro/nanoencapsulation to improve its stability and functionality as a food ingredient. Mar Drugs 18(8):1–25. https://doi.org/10.3390/MD18080406

Marty C, Berset C (1988) Degradation products of trans b-carotene produced during extrusion cooking. J Food Sci 53(6):1880–1886. https://doi.org/10.1111/j.1365-2621.1988.tb07866.x

McClements DJ (2012) Advances in fabrication of emulsions with enhanced functionality using structural design principles. Curr Opin Colloid Interface Sci 17:235–245. https://doi.org/10.1016/j.cocis.2012.06.002

McClements DJ (2015) Food emulsions: principles, practices, and techniques, 3rd edn. CRC Press, Boca Raton. ISBN 978-0-429-15403-4

Melo RLF, da Souza ICC, Carvalho AJR, Bezerra EM, da Costa RF (2020) As nanopartículascomo ferramentas biol'ogicas: umarevis˜aoexplorat'oria. Res Soc Dev 9(7):e363974155. https://doi.org/10.33448/rsd-v9i7.4155

Mestry AP, Mujumdar AS, Thorat BN (2011) Optimization of spray drying of an innovative functional food: fermented mixed juice of carrot and watermelon. Dry Technol 29:1121–1131

Nisar N, Li L, Lu S, Khin NC, Pogson BJ (2015) Carotenoid metabolism in plants. Mol Plant 8(1):68–82. https://doi.org/10.1016/j.molp.2014.12.007

Nunes IL, Mercadante AZ (2007) Encapsulation of lycopene using spray-drying and molecular inclusion processes. Braz Arch Biol Technol 50:893–900

Odriozola-Serrano I, Soliva-Fortuny R, Hernández-Jover T, Martín-Belloso O (2009) Carotenoid and phenolic profile of tomato juices processed by high intensity pulsed electric fields compared with conventional thermal treatments. Food Chem 112:258–266

Olmedilla-Alonso B (2017) Carotenoids: content in foods, in diet and bioavailability. COST Action EUROCAROTEN CA15136 Sci Newsl 2:1–9

Patel AR, Velikov KP (2011) Colloidal delivery systems in foods: a general comparison with oral drug delivery. LWT Food Sci Technol 44:1958–1964. https://doi.org/10.1016/j.lwt.2011.04.005

Perera CO, Yen GM (2007) Functional properties of carotenoids in human health. Int J Food Prop 10:201–230

Quan C, Johan C, Charlotta T (2009) Carotenoids particle formation by supercritical fluid technologies. Chin J Chem Eng 17:344–349

Rehman A, Tong Q, Jafari SM, Assadpour E, Shehzad Q, Aadil RM, Iqbal MW, Rashed MMA, Mushtaq BS, Ashraf W (2020) Carotenoid-loaded nanocarriers: a comprehensive review. Adv Colloid Interf Sci 275:Article 102048. https://doi.org/10.1016/j.cis.2019.102048

Ribeiro HS, Chu BS, Ichikawa S, Nakajima M (2008) Preparation of nanodispersions containing b-carotene by solvent displacement method. Food Hydrocoll 22(1):12–27. https://doi.org/10.1016/j.foodhyd.2007.04.009

Ribeiro D, Freitas M, Silva AMS, Carvalho F, Fernandes E (2018) Antioxidant and pro-oxidant activities of carotenoids and their oxidation products. Food Chem Toxicol 120(July):681–699. https://doi.org/10.1016/j.fct.2018.07.060

Roman MJ, Burri BJ, Singh RP (2012) Release and bioaccessibility of b-carotene from fortified almond butter during in vitro digestion. J Agric Food Chem 60(38):9659–9666. https://doi.org/10.1021/jf302843w

Saenz C, Tapia S, Chavez J, Robert P (2009) Microencapsulation by spray drying of bioactive compounds from cactus pear (Opuntiaficus-indica). Food Chem 114(2):616–622. https://doi.org/10.1016/j.foodchem.2008.09.095

Saini RK, Nile SH, Park SW (2015) Carotenoids from fruits and vegetables: chemistry, analysis, occurrence, bioavailability and biological activities. Food Res Int 76:735–750

Salter-Venzon D, Kazlova V, Izzy Ford S, Intra J, Klosner AE, Gellenbeck KW (2017) Evidence for decreased interaction and improved carotenoid bioavailability by sequential delivery of a supplement. Food Sci Nutr 5:424–433

Santos DT, Meireles MAA (2010) Carotenoid pigments encapsulation: fundamentals, techniques and recent trends. Open Chem Eng J 4(1):42–50

Santos DT, Martín Á, Meireles MAA, Cocero MJ (2012) Production of stabilized sub-micrometric particles of carotenoids using super critical fluid extraction of emulsions. J Supercrit Fluids 61:167–174

Schawartz SJ, Cooperstone JL, Cichon MJ, von Elbe JH, Giusti MM (2017) Colorants. In: Damodaran S, Parkin KL (eds) Fennema's food chemistry. CRC Press/Taylor& Francis, Boca Raton, p 681

Sen Gupta S, Ghosh M (2015) Synthesis, characterization, stability evaluation and release kinetics of fiber-encapsulated carotene nano-capsules. Grasas Aceites 66(4):e104. https://doi.org/10.3989/gya.0226151

Šeregelj VN, Cetkovic GS, Canadanovic-Brunet JM, Tumbas-Šaponjac VT, Vulic JJ, Stajˇcic SS et al (2017) Extraction and encapuslation of bioactive compounds from carrots. Acta PeriodicaTechnologica 2017:261–273. https://doi.org/10.2298/APT1748261S

Šeregelj V, Vulic J, Cetkovic G, Canadanovc-Brunet J, Šaponjac VT, Stajˇcic S et al (2020) Natural bioactive compounds in carrot waste for food applications and health benefits. Stud Nat Prod Chem 67:307–344. https://doi.org/10.1016/B978-0-12-819483-6.00009-6

Sessa M (2012) Nanoencapsulation of bioactive compounds for food applications; Universitá Degli Studi di Salerno: Fisciano, Italy, 2012.114

Silva EK, Meireles AA (2014) Encapsulation of food compounds using supercritical technologies: applications of supercritical carbon dioxide as an Antisolvent. Food Publ Health 4:247–258

Soh SH, Lee LY (2019) Microencapsulation and Nanoencapsulation using supercritical fluid (SCF) techniques. Pharmaceutics 11:21

Soukoulis C, Bohn T (2015) A comprehensive overview on the micro- and nano-technological encapsulation advances for enhancing the chemical stability and bioavailability of carotenoids. Crit Rev Food Sci Nutr 58(1):1–36. https://doi.org/10.1080/10408398.2014.971353

Soukoulis C, Bohn T (2018) A comprehensive overview on the micro- and nano-technological encapsulation advances for Enhancing the chemical stability and bioavailability of carotenoids. Crit Rev Food Sci Nutr 58:1–36

Sridhar K, Inbaraj BS, Chen B-H (2021) Recent advances on nanoparticle based strategies for improving carotenoid stability and biological activity. Antioxidants 10(5):713. https://doi.org/10.3390/antiox10050713

Steiner BM, McClements DJ, Davidov-Pardo G (2018) Encapsulation systems for lutein: a review. Trends Food Sci Technol 82(October):71–81. https://doi.org/10.1016/j.tifs.2018.10.003

Sy C, Dangles O, Borel P, Caris-Veyrat C (2015) Stability of bacterial carotenoids in the presence of iron in a model of the gastric compartment– comparison with dietary reference carotenoids. Arch Biochem Biophys 572:89–100. https://doi.org/10.1016/j.abb.2014.12.030

Tasch Holkem A, Franco Codevilla C, Bona D, da Silva C, Ragagnin de Menezes C (2015) Técnicas de preparação de sistemas nanotecnológico saplicadosaalimentos. Ciênc Nat 37:87

Thakur D, Jain A, Ghoshal G, Shivhare US, Katare OP (2017) Microencapsulation of _-carotene based on casein/guar GumBlend using zeta potential-yield stress phenomenon: an approach to enhance photo-stability and retention of functionality. AAPS PharmSciTech 18:1447–1459

Thurmann PA, Steffen J, Zwernemann C, Aebischer CP, Cohn W, Wendt G, Schalch W (2002) Plasma concentration response to drinks containing b-carotene as carrot juice or formulated as a water dispersible powder. Eur J Nutr 41:228–235. https://doi.org/10.1007/s00394-002-0381-3

Tiena L, Ispas-Szaboa MLP, Mateescu MA (2003) N-acylated chitosan: hydrophobic matrices for controlled drug release. J Control Release 93:1–13. https://doi.org/10.1016/S0168-3659(03)00327-4

Tirado DF, Palazzo I, Scognamiglio M, Calvo L, Della Porta G, Reverchon E (2019) Astaxanthin encapsulation in ethyl cellulose carriers by continuous supercritical emulsions extraction: A study on particle size, encapsulation efficiency, release profile and antioxidant activity. The Journal of Supercritical Fluids 150:128–36.

Walstra P, van Vliet T (2007) Dispersed systems: basic considerations. In: Damodaran S, Parkin KL, Fennema OR (eds) Fennema's food chemistry. Marcel Dekker, New York, pp 783–847. ISBN 978-0-8493-9272-6

Wang L, Bohn T (2012) Health-promoting food ingredients and functional food processing chapter 9. In: Bouayed J, Bohn T (eds) Nutrition, well-being and health. InTech Publisher, pp 201–212

Weiss J, Decker EA, McClements DJ, Kristbergsson K, Helgason T, Awad T (2008) Solid lipid nanoparticles as delivery systems for bioactive food components. Food Biophys 3:146–154

Williams AW, Boileau TW, Erdman JW (1998) Factors influencing the uptake and absorption of carotenoids. Exp Biol Med 218(2):106–108. https://doi.org/10.3181/00379727-218-44275

Xavier AA, Mercadante AZ (2019) The bio accessibility of carotenoids impacts the design of functional foods. Curr Opin Food Sci 26:1–8. https://doi.org/10.1016/j.cofs.2019.02.015

Xia Z, McClements DJ, Xiao H (2017) Influence of lipid content in a corn oil preparation on the bioaccessibility of _-carotene: a comparison of low-fat and high-fat samples: influence of lipid Conten. J Food Sci 82:373–379

Xianquan S, Shi J, Kakuda Y, Yueming J (2005) Stability of lycopene during food processing and storage. J Med Food 8(4):413–422. https://doi.org/10.1089/jmf.2005.8.413

Yoon G-A, Yeum K-J, Cho Y-S, Chen C-YO, Tang G, Blumberg JB, Russell RM, Yoon S, Lee-Kim YC (2012) Carotenoids and total phenolic contents in plant foods commonly consumed in Korea. Nutr Res Pract 6:481. [CrossRef] [PubMed]–490

Zhang L-H, Xu X-D, Shao B, Shen Q, Zhou H, Hong Y-M, Yu L-M (2015) Physicochemical properties and bioavailability of lutein microencapsulation (LM). Food Sci Technol Res 21(4):503–507. https://doi.org/10.3136/fstr.21.503

Zhao L, Temelli F, Curtis JM, Chen L (2017) Encapsulation of lutein in liposomes using supercritical carbon dioxide. Food Res Int 100:168–179. [CrossRef]

Zhong J, Dai LC (2011) Liposomal preparation by supercritical fluids technology. Afr J Biotechnol 10:16406–16413

Chapter 7
Recent Approaches for the Bioaccessibility and Bioavailability of Carotenoids

Caroline Ramos-Souza and Veridiana Vera De Rosso

1 Introduction

Widely spread in nature as a yellow, orange, or red lipophilic pigment, carotenoids are synthesized and stored in plastids, especially in chromoplasts and chloroplasts. In these organelles, carotenoid substructures are feasible as globular, tubular, crystalline, or membranous (Egea et al. 2010). In general, based on chemical structure, carotenoids are classified into nonpolar carotenes, which are composed exclusively of carbon and hydrogen, less nonpolar xanthophylls, possess at least one functional group containing oxygen, apo-carotenoids, C_{30} and C_{50} carotenoids (Britton and Khachik 2009). However, the human body cannot synthesize carotenoids, and almost 50 of the more than 1117 identified carotenoids are part of the human diet (Yabuzaki 2017). Generally, 80–90% of carotenoid intake comes from fruit and vegetables, of which 95% are found in human plasma represented by α-carotene, β-carotene, β-cryptoxanthin, lycopene, lutein, and zeaxanthin (Rodriguez-Amaya 2015a).

One of the most prominent carotenoids' health effects is their provitamin A activity. Vitamin A, an essential human micronutrient, can be enzymatically produced in the small intestine through the cleaved unsubstituted β-ionone terminal ring present mostly in carotenes (Rodriguez-Amaya 2015b; von Lintig et al. 2020). The biological effects of vitamin A are well-established, with an emphasis on cell differentiation, immunity, and light-dark visual adaptation (Eroglu et al. 2023). In addition, due to their accumulation in the retina, non-provitamin A carotenoids can be protective factors against the progression of age-related macular degenerative diseases by exerting effects against oxidative stress and the ability to scavenge free

C. Ramos-Souza · V. V. De Rosso (✉)
Nutrition and Food Service Research Center, Federal University of São Paulo (UNIFESP), São Paulo, Brazil
e-mail: veridiana.rosso@unifesp.br

R. Campos Chisté et al. (eds.), *Carotenoids*,
https://doi.org/10.1007/978-3-031-75322-0_7

radicals (Lombardo et al. 2022; Mrowicka et al. 2022). Furthermore, carotenoids have been associated with antioxidant and anti-inflammatory activity due to their ability to quench reactive oxygen or nitrogen species such as singlet oxygen or lipid peroxides and the carotenoids' ability to interact with transcription factors (Dingeo et al. 2020). Two transcription factors that stand out are nuclear factor erythroid 2-related factor 2 (Nrf2) related to physiological antioxidant homeostasis through gene expression of antioxidant enzymes such as catalase (CAT), superoxide dismutase (SOD) and glutathione peroxidase (GPx) and the pro-inflammatory nuclear transcription factor κB (NF-κB) which associated to the formation of pro-inflammatory cytokines (Kaulmann and Bohn 2014).

Epidemiological studies associated with a carotenoid-rich diet present a lower risk of developing diseases, highlighting research on the carotenoids' distribution and levels in different food matrices (Dias et al. 2018). Higher content of carotenoids in a food matrix does not necessarily show higher bioactivity, as carotenoids need to be released from the matrix and absorbed by the body to exert potential benefits, which means carotenoids need to be bioaccessible and bioavailable (Chacón-Ordóñez et al. 2019). Carotenoid bioaccessibility is defined as the proportion of carotenoids consumed and released from the food matrix during digestion in the luminal content and accessible for absorption in the small intestine through solubilization in mixed micelles (Rodrigues et al. 2022; Rodriguez-Amaya 2015a). The fraction of ingested that becomes available to the body for utilization in physiological functions or storage is carotenoid bioavailability (Rodriguez-Amaya 2015a).

2 Carotenoid Digestion: Where Do We Stand?

Carotenoid digestion is a complex, continuous, and dynamic process, and there are gaps in knowledge about it (Geng et al. 2023). After oral ingestion through the mouth, the carotenoid digestion process begins with the food matrix breakdown and the carotenoids released by mechanical and chemical steps (Kopec and Failla 2018). Chewing is the first mechanical step responsible for reducing the food particles' size and increasing the contact surface. As the oral phase lasts 1–2 min, the enzymatic effect on carotenoids is presumably small (Bohn 2018; Sun and Li 2020). However, saliva contains α-amylase, which is responsible for initiating starch digestion. Thus, when the carotenoid is in starchy matrices such as orange sweet potatoes, α-amylase participates in digestion and can be influenced by genetic factors (Bohn 2018). Although this influence has not been studied for carotenoid digestion, it has been shown that people with high consumption of starchy foods have higher levels of the enzyme in their saliva and express more copies of the salivary amylase gene (AMY1) (Perry et al. 2007). Food particles with saliva form the food bolus, which is conducted from the mouth to the esophagus to the stomach through peristaltic movements. The gastric phase can last from 1.5 to 2 h, and approximately three peristaltic contractions occur sequentially in a minute, promoting the gradual release of carotenoids (Kong and Singh 2010; Sun and Li 2020).

Food matrix carotenoid release is highly variable and depends on intrinsic and extrinsic factors, including the carotenoids' molecular structure and vegetable storage location, the food matrix constitution (as the presence of dietary fibers, proteins, fats, and minerals), food processing conditions (such as heating, cooking, bowling or steamed), and host related-factors such as single nucleotide polymorphisms (Fig. 7.1) (Bohn 2018; Castenmiller and West 1998; Molteni et al. 2022). Indeed, carotenoids located in chloroplasts in green-colored vegetables are associated with thylakoids due to their hydrophobic nature, resulting in rigid protein complexes and poor bioaccessibility. However, carotenoids located in chromoplasts, as in yellow, orange, or red fruits and vegetables, can be found in different substructures' physical forms (such as globular, tubular, membranous, crystalloid, or a combination between them) (Chacón-Ordóñez et al. 2019). These differences in carotenoid substructures can interfere with its bioaccessibility (Schweiggert and Carle 2015). Another fact is the presence of minerals in the food matrix, which can interfere with the solubilization and micellization processes, reducing the bioaccessibility of carotenoids (Shilpa et al. 2020).

After release from the food matrix, carotenoids are solubilized in lipid droplets in the stomach due to the high hydrophobicity (pKa ~ 10–15.6). This process can be influenced by the co-consumption of dietary fat, which is responsible for providing a solubilizing medium for carotenoids, facilitating their transfer to the aqueous micellar fraction, and stimulating gastric lipase release (Lemmens et al. 2014; Yao et al. 2022). Indeed, a meta-analysis of 27 *in vitro* studies and 12 randomized controlled trials showed a positive association with the bioaccessibility of carotenoids when co-consumed with dietary fat (Yao et al. 2022). However, the fat type strongly influences bioaccessibility, in which long-chain fatty acids presented better micellization when compared with medium-chain fatty and saturated fatty acids

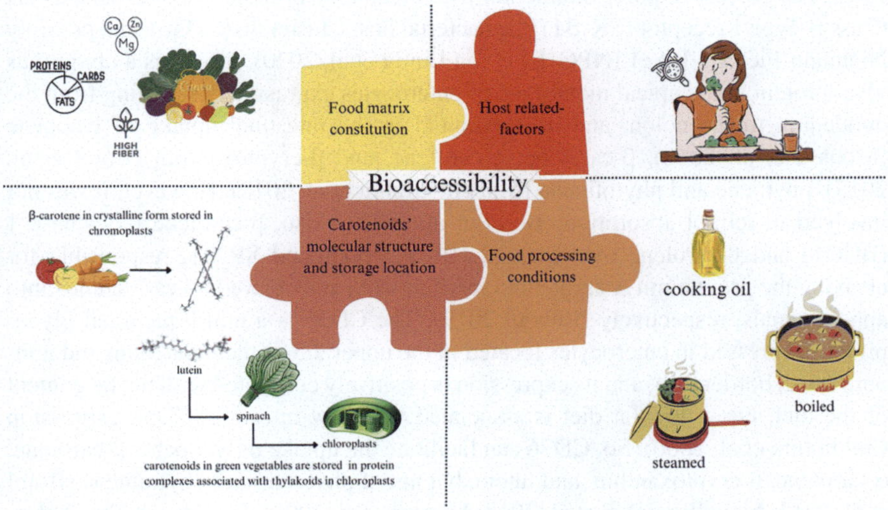

Fig. 7.1 Intrinsic and extrinsic factors that influenced the carotenoids' release in the food matrix

(Mashurabad et al. 2017). Furthermore, gastric lipase release has been shown to promote the cleavage of dietary triacylglycerols into diglycerides and free fatty acids, an essential step for facilitating emulsification (Brodkorb et al. 2019; Carriere et al. 1993).

In the intestinal phase, micellization is completed with the secretion of bile salts, pancreatic lipase, and additional enzymes such as amylase, nucleosidases, trypsinogen, chymotrypsinogen, carboxypeptidase, elastase, phospholipase, and carboxylic ester lipase. Bile salts aid in the formation and stability of micelles, and pancreatic lipase is the main enzyme that is required to cleave triglycerides into mono- and diglycerides, which are part of mixed micelles (Carriere et al. 1993; Eroglu et al. 2023). In addition to carotenoids, the 3–10 nm mixed micelles incorporated cholesterol, free fatty acids, monoacylglycerols, lysophospholipids, phospholipids, and bile salts (Zhou et al. 2023). The more lipid micronutrients solubilized by the micelles, it is assumed that the more efficient their absorption will be (von Lintig et al. 2020). However, host-related factors, such as the secretion of bile salts and pancreatic lipase, may influence carotenoid micellization. Indeed, *in vitro* studies demonstrate micellization is affected by the bile salts, pancreatic lipase, or both in the spinach carotenoid (β-carotene, lutein, and zeaxanthin) (Biehler et al. 2011). Also, the solubilization in mixed micelles differs according to the polarity of the carotenoids. Carotenes migrated to the core rich in triacylglycerol due to their polarity, and xanthophylls were found on the surface of the mixed micelles simultaneously with proteins, phospholipids, and partially ionized fatty acids (Tyssandier et al. 2001). Thus, free xanthophylls are more bioaccessible than carotenes (Petry and Mercadante 2017).

The mixed micelles dissociated by the pH effect when they come into contact with the absorptive epithelial cells of the mucous layer lining the brush border membrane of the duodenum, and part of carotenoids begin to be absorbed by enterocytes. The process requires interaction with membrane protein receptors such as the Class B Type 1 receptor (SR-B1), the bacterial host Cluster 36 (CD36), and possibly Niemann-Pick C1 Like1 (NPC1L1) (von Lintig et al. 2020). SR-B1 is a ubiquitous glycoprotein in the apical membrane of enterocytes expressed descending from the duodenum to the colon, and it is essential in the intestinal uptake of lycopene (Moussa et al. 2008), β-carotene, α-carotene, and β-cryptoxanthin (Borel et al. 2013), phytoene and phytofluene (Mapelli-Brahm et al. 2018). However, it was not involved in retinol absorption (Borel et al. 2013). Also, β-carotene oxygenase 1 (BCO1) and β-carotene oxygenase 2 (BCO2) regulated SR-B1, responsible for cleaving the provitamin A carotenoid in vitamin A, and cleaved all carotenoids into apocarotenals, respectively (Reboul 2019). The CD36 is a multifunctional glycoprotein expressed in enterocytes located in the upper 2/3 of the duodenum and jejunum brush border villi, and its expression is positively correlated with the fat content of the diet, i.e., a low-fat diet is associated with low intestinal CD36 expression (Sukhotnik et al. 2000). So, CD36 can facilitate the uptake of lycopene, β-carotene, α-carotene, β-cryptoxanthin, and lutein, but not of phytoene and phytofluene (Borel et al. 2013; Mapelli-Brahm et al. 2018; Moussa et al. 2008). In addition, the cholesterol membrane transporters NPC1L1 are involved in the intestinal absorption of highly lipophilic compounds, including α-carotene, β-carotene, β-cryptoxanthin, and lutein (During et al. 2005; Sato et al. 2012), but not for lycopene (Moussa et al. 2008), phytoene, and phytofluene (Mapelli-Brahm et al. 2018).

Competition with fat-soluble dietary compounds and products of lipid digestion for active site binding and cellular uptake of membrane proteins may occur, and depending on the concentration in the lumen, carotenoids' absorption by passive diffusion process is still possible through the brush border membrane of absorptive cells (Kopec and Failla 2018; Reboul 2019). As with bioaccessibility, the polarity and flexibility of carotenoids influence their absorption (Reboul 2019).

Only a part of the carotenoids is absorbed in the small intestine, and the remaining carotenoids are transmitted to the colon. It is unclear whether carotenoids can be absorbed in the colon or interact with the intestinal microbiota, but it appears that carotenoids can be partially degraded to unknown metabolites (Bohn 2018; Eroglu et al. 2023). These metabolites may include apocarotenoids, of which only traces have been detected by chromatography and mass spectrometry in human plasma. Apocarotenoids can increase aqueous solubility and electrophilicity due to their more hydrophilic nature and thus efficiently target transcription factors such as NF-κB, PPARγ, and RAR/RXRs (Varghese et al. 2023). However, the origin of apocarotenoids is not elucidated, if they are intake from the diet or are enzymatically synthesized from carotenoids in the digestion process (Eroglu et al. 2023). Recently, a systematic review conducted according to PRISMA (Preferred Reporting Items for Systematic Reviews and Meta-Analysis) investigated the effects of carotenoids on gut microbiota and barrier, and inflammation in healthy animals (Silva Meneguelli et al. 2023). Both *in vitro* and *in vivo* studies seem to agree and have suggested beneficial effects of carotenoids on gut microbiota, such as short-chain fatty acids production, stabilizing and expression of tight junction proteins and mucins, reduced proinflammatory activity, increased anti-inflammatory cytokines and changed the composition of the intestinal microbiota (Eroglu et al. 2023; Silva Meneguelli et al. 2023; Stevens et al. 2021). So, it seems that carotenoids act on gut microbiota, and the beneficial effects depend on the kind of carotenoid (Eroglu et al. 2023). However, despite studies suggesting the positive effects of carotenoids on the gut microbiota, there are still gaps that need to be filled for a holistic understanding of the process of digestion and absorption of carotenoids in the human body and thus, their potential health benefits.

After cellular absorption, carotenoids and retinyl esters were incorporated in chylomicrons for transport and distribution in the body. Chylomicrons were secreted into the lymph and converted into chylomicron remnants by lipoprotein lipase. Then, chylomicron remnants were absorbed by the liver where carotenoids were utilized, stored, and secreted into VLDL particles (von Lintig et al. 2020). Figure 7.2 briefly outlines the carotenoid digestion process.

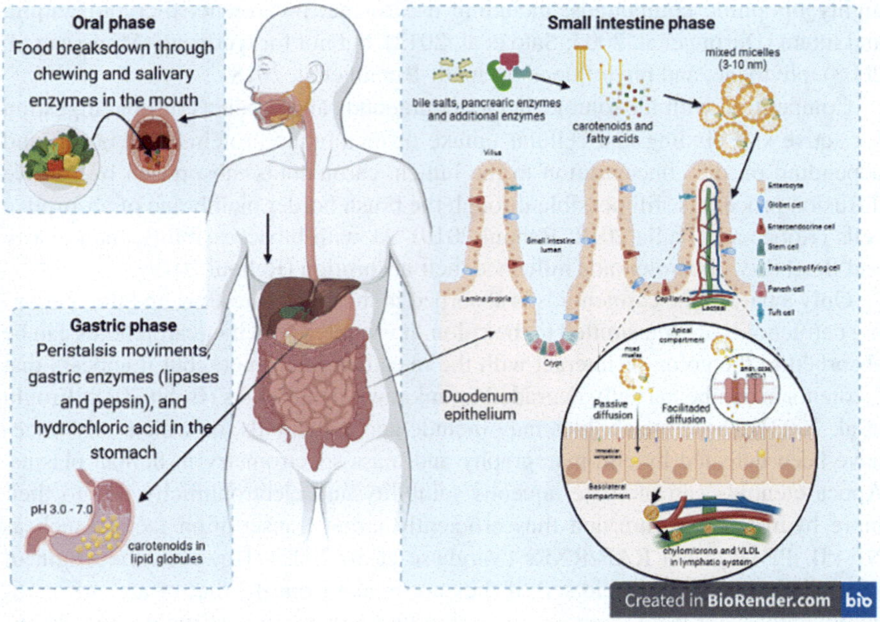

Fig. 7.2 Overview of the carotenoid digestion process. (Created in BioRender.com)

3 The INFOGEST Network Standardized Protocols

Food digestion has been investigated for more than 300 years (Bornhorst and Singh 2014). In search of detailed answers to the complex human digestive process, gastrointestinal digestion simulations are widespread in food and nutrition sciences. The use of *in vivo* simulations (human or animal) is laborious, time-consuming, resource-intensive, exhibits high inter-individual variability, and is limited by ethical considerations, as the use of animal models to understand human food digestion is questionable (Brodkorb et al. 2019). In addition, there is a tendency to limit studies involving complex living organisms as much as possible (Menard et al. 2023). Alternatively, *in vitro* simulations of human gastrointestinal food digestion have been widespread as less laborious, expensive, faster, without ethical considerations, and can be divided into static and dynamic models (Minekus et al. 2014).

Dynamic models are relatively complex, require substantial hardware and software, and are expensive to set up and maintain. Despite the dynamic models suitable for simulating and mimicking the digestion of foods with diverse end-points in different populations with host physiological interactions, these models have limited use and are not available for most researchers (Brodkorb et al. 2019). A wide variety of static digestion models with different parameters have been proposed due to the simplicity and prediction of *in vivo* digestion results (Bohn et al. 2017). A standardization strategy and the possibility of comparing results between research teams, an international consensus developed by the INFOGEST network proposed a harmonized and practical method of static *in vitro* simulation of gastrointestinal food digestion (Brodkorb et al. 2019; Minekus et al. 2014). The INFOGEST is a multidisciplinary, international, and dynamic research network in which scientists from different areas of knowledge from more than 45 countries intend to investigate food digestion (http://www.cost-infogest.eu).

The model developed and validated by the INFOGEST network evaluates *in vitro* static digestion in the oro-gastrointestinal systems and it was based on available relevant physiological data. The model can be applied to assess a wide range of food targets and constituents being suitable for mechanistic studies and hypothesis building due to its controlled conditions, reproducibility, and ease of sampling at the site of interest (Minekus et al. 2014). Thus, static models can simulate the oral, gastric, and intestinal phases through interconnected bioreactors that mimic the physical-chemical and enzymatic conditions, including temperatures, digestion transit times, enzyme activities, mineral compositions, pH, mucin and bile salt concentrations (Menard et al. 2023).

As the bioreactors are sequentially interconnected, gradual gastric emptying is not considered since the next digestive phase only starts when the previous one ends. The intestinal phase is usually composed of a single bioreactor and does not consider the physiological variations found in the three constituents of the small intestine (duodenum, jejunum, and ileum) (Brodkorb et al. 2019). Furthermore, pH, gastric fluid, enzyme activity, salt, and bile acids at each stage of digestion were kept constant which does not simulate the physiology of human digestion. So, the static models are not suitable for simulating the kinetics of digestion considering the different phases of human physiological digestion (Brodkorb et al. 2019). Despite the limitations of this simple *in vitro* static digestion model, it represents good intra and interlaboratory reproducibility, robustness with an easy evaluation of each digestion phase, and simplicity in execution. In addition, the method showed a good correlation with *in vivo* models (Bohn et al. 2018; Egger et al. 2016, 2017; Sanchón et al. 2018).

Recently, the standardized protocol developed in 2014 was updated for adults, which included the oral phase for all foods and gastric lipases (Brodkorb et al. 2019) and was adapted for the elderly population in general (Menard et al. 2023). Also, a static *in vitro* model was available to evaluate the digestion of commercial infant formulas based on infants born at term and aged 28 days (Ménard et al. 2018). As static models were restricted to describing kinetics and possible structural changes of the digestion process, and dynamic models were not accessible to all researchers besides the relevant physiological data, a low-cost semi-dynamic protocol was developed to fill this gap. This intermediate model includes a dynamic gastric phase considering crucial kinetic aspects, such as gradual acidification, gradual addition of enzymes and fluid secretion, and manual and gradual gastric emptying (Mulet-Cabero et al. 2020). The main modifications of the INFOGEST, INFOGEST 2.0, and INFOGEST semi-dynamic protocols are described in Fig. 7.3.

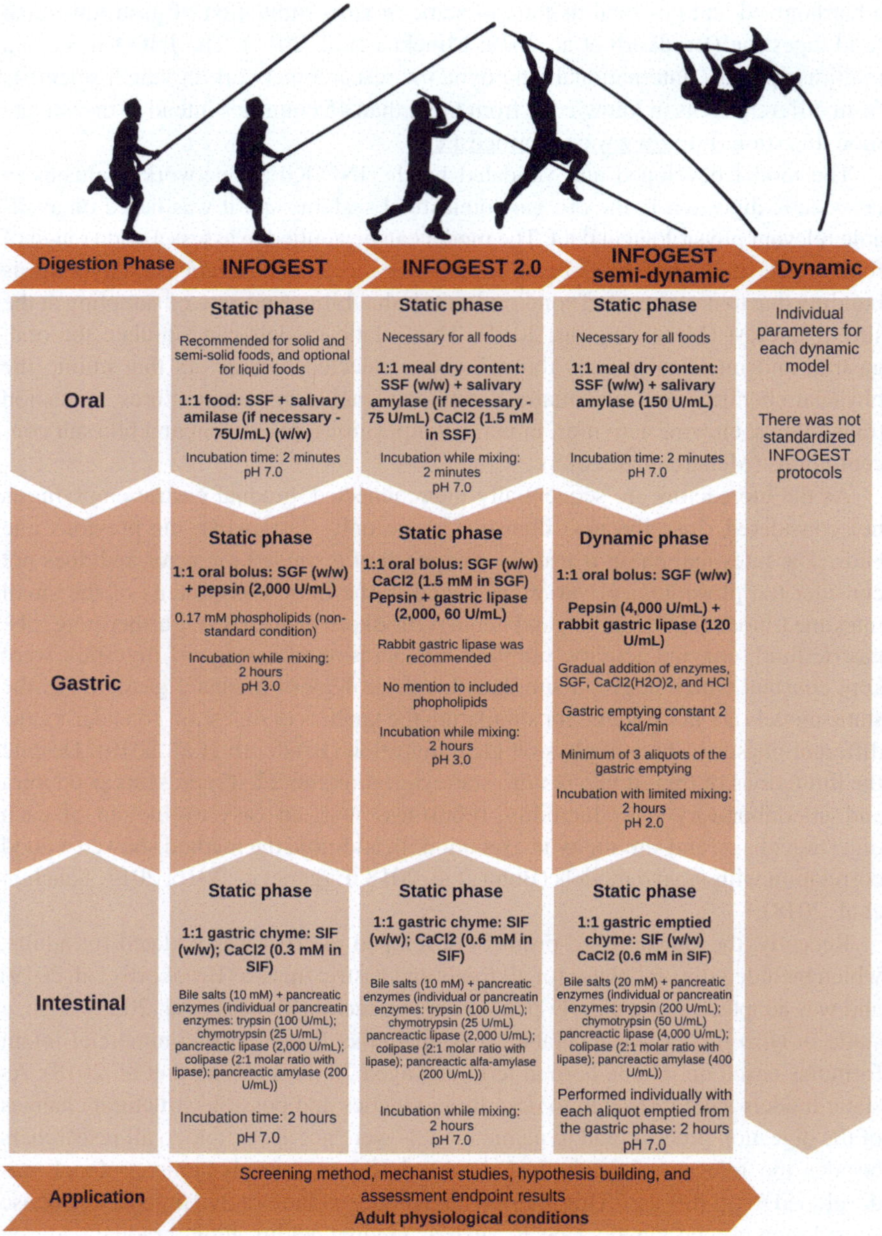

Fig. 7.3 Comparison among the currently available standardized protocols: INFOGEST, INFOGEST 2.0, and INFOGEST semi-dynamic. Salivary (SSF), gastric (SGF), and intestinal (SIF) simulated fluids

4 How to Simulate Carotenoid Digestion?

The INFOGEST method is not specific to any food component, and the steps of removing undigested material, extraction, and identification of the target compound after simulating oro-gastrointestinal digestion are not included in the protocol (Xavier and Mercadante 2022). Thus, although the standardized INFOGEST protocol allowed a release of carotenoids from the food matrix into the aqueous phase, additional steps were suggested (Brodkorb et al. 2019). Carotenoids present in food matrices such as fruits and vegetables usually have low bioaccessibility and bioavailability, due to their complex hydrophobic and conformational physical-chemical structures and low concentration of lipids (Estévez-Santiago et al. 2016; Failla et al. 2019). To understand the bioaccessibility of carotenoids, the INFOGEST protocol was adapted by different research groups (Eriksen et al. 2016; Estévez-Santiago et al. 2016; Hempel et al. 2017; Rodrigues et al. 2016).

The first model adapted from INFOGEST aimed to evaluate the bioaccessibility of the main contributors of provitamin A carotenoids considering fruit consumption in Spain. It has been proposed to include pancreatin cholesterol esterase in the intestinal phase to hydrolyze xanthophyll esters, centrifugation of the supernatant after the digestion process, and conventional extraction of carotenoids with acetone (Estévez-Santiago et al. 2016). The bioaccessibility found was surprisingly low for provitamin A carotenoids when compared to other sources of carotenoid provitamin A (Estévez-Santiago et al. 2016). The authors highlight the difficulty in comparing the results due to the non-standardization of the method and different ways to express the results (such as supernatant content, micellar phase content, intestinal phase with/without centrifugation, etc.), the influence of different matrices and processing, which was not always well described, and the addition of the saponification step. Indeed, discrepancies in the results of bioaccessibility were found due to the differences in the composition and amounts/activities of enzymes, salts, and bile employed, and the carotenoid bioaccessibility depends on the *in vitro* digestion protocol used (Rodrigues et al. 2016; Xavier and Mercadante 2022).

The bioaccessibility of carotenoids and carotenoid esters from goji berries (*Lycium barbarum* L.) was assessed using the standardized INFOGEST protocol. In addition, the authors investigated the location of carotenoid deposition and its influence on bioaccessibility (Hempel et al. 2017). It is suggested that the location and structural form of deposition impacts the efficiency of carotenoid release from the food matrix, and consequently affects its bioaccessibility and bioavailability (Schweiggert et al. 2012; Schweiggert and Carle 2015). After digestion, the samples were centrifuged and the supernatant was filtered through a membrane to obtain the fraction of micellarized carotenoids (Hempel et al. 2017). The release and bioaccessibility of zeaxanthin from goji berries were significantly higher than lutein from spinach, and this is due to the site of deposition of xanthophylls. Zeaxanthin in goji berries is deposited in liquid-crystalline form in tubular chromoplasts (Hempel et al. 2017), while lutein in spinach, used as a comparison standard, is typically incorporated into carotenoid-protein complexes within the thylakoid membrane of chloroplasts (Schweiggert and Carle 2015). Indeed, when comparing the bioaccessibility of carotenoids from carrots, mangoes, papayas, and tomatoes in an *in vitro* digestion model, it was observed that the physics form of carotenoids deposition in the chromoplast impacts the release efficiency of the food matrix (Schweiggert et al. 2012). In carrots and tomatoes, α-carotene, β-carotene, and lycopene exist in crystalline form, and in orange and yellowish fruits, such as mango and papaya, lycopene is dissolved in lipid droplets. Carotenoids dissolved in lipid droplets are more bioaccessible compared to crystalline forms and both are more bioaccessible compared to carotenoids in protein complexes in chloroplasts (Castenmiller and West 1998).

To evaluate the *in vitro* bioaccessibility of carotenoids and their carotenoid esters in a model of lipid-rich fruit (buriti, scientifically known as *Byrsonima crassifolia*) in the standardized INFOGEST protocol, two steps were added. The first one was the aqueous phase separation with mixed micelles after digestion by centrifugation and, the second step was the exhaustive extraction of carotenoids with diethyl ether, an organic solvent (Rodrigues et al. 2016). The extraction step of the mixed micelles was essential to obtain a clean and appropriate carotenoid extract for LC-PAD-MS analysis. The addition of these steps, in general, allowed a significant increase in bioaccessibility when compared to the simpler protocol previously used to study the bioaccessibility of carotenoids (Xavier et al. 2014). In this simpler protocol, the oral phase is absent, and the type and concentration of enzymes are different from those suggested in the modified INFOGEST protocol. In contrast, when evaluating the bioaccessibility of lutein in cupcakes fortified with a water-soluble lutein ester formulation with this protocol, the oral phase was included. Lutein fortification did not result in a net increase in bioaccessibility, moreover, only 30% of the lutein was hydrolyzed by pancreatic lipase. In dairy products (milk and yogurt), the same lutein formulation was applied to fortify the xanthophyll content and also showed incomplete hydrolysis of xanthophyll esters (Xavier et al. 2014). Thus, these results support the low affinity of lutein for pancreatic lipase and question the possibility of using other mucosal hydrolytic enzymes that can hydrolyze carotenoid esters (Xavier et al. 2018). In addition, the importance of studying the incorporation of carotenoids in functional food formulations is emphasized, and a need for future research on dynamic *in vitro* digestion models to more accurately estimate the bioaccessibility of carotenoids (Xavier et al. 2018).

Also, the carotenoids and the esters of carotenoids of mandarin pulps and cajá frozen pulp-based beverages were evaluated with the adapted INFOGEST protocols for carotenoids (Da Costa and Mercadante 2018; Petry and Mercadante 2017). The increases in the ratio of bile extract/food and the addition of rabbit gastric lipase contributed substantially to the improvement of the hydrolysis of β-cryptoxanthin esters, consequently, there was an increase in the concentration of bioaccessible carotenoids in fruit pulps during digestion *in vitro* (Petry and Mercadante 2019, 2020). Static *in vitro* models have used animal digestive enzyme extracts as substitutes for human gastric lipase and pancreatic secretions (Grundy et al. 2021). Among the available lipases, rabbit gastric extract was significantly the best contribution to the emulsification of carotenoids (Brodkorb et al. 2019; Iddir et al. 2021; Petry and

Mercadante 2020), and protease, amylase, lipase, and cholesterol esterase are commonly provided by porcine pancreatin (Grundy et al. 2021; Minekus et al. 2014). Carotenoids are usually found in their free and esterified form in fruits and vegetables, and saponification is typically used to remove interfering lipids and chlorophylls. However, on the INFOGEST protocol, carotenoid esters are poorly hydrolyzed and their identification in human plasma is infrequent. Thus, when a food matrix is intake, carotenoid esters are hydrolyzed in the gastrointestinal system (Petry and Mercadante 2017; Rodrigues et al. 2016). A low-cost modification to improve carotenoid ester hydrolysis is to increase the bile extract/food ratio (Xavier et al. 2014; Petry and Mercadante 2019) or include additional cholesterol esterase (CEL) (40 U/mL prepared in SIF) (Petry and Mercadante 2020).

Recently, a guide for evaluation of *in vitro* bioaccessibility of carotenoids was published which describes an adapted INFOGEST protocol to carotenoids and esters of carotenoids. In this guide, enzymatic activities were determined and calculated according to the INFOGEST protocol with modifications for carotenoids. After the digestion process, the mixed micelles with bioaccessible carotenoids are obtained through centrifugation and filtration to finally perform carotenoid extraction, identification, and quantification by liquid chromatography analysis (Xavier and Mercadante 2022). Understanding what limits the bioaccessibility of carotenoids is essential to suggest functionality claims for ingredients and foods, and is mandatory for their design (Xavier and Mercadante 2019). Gaps in knowledge have been filled simultaneously about carotenoids' fate through the digestive process, and their mechanisms of action with the standardization of the *in vitro* digestion process with modifications for the analysis of the bioaccessibility of carotenoids. In this way, bioaccessibility is a crucial factor for the success of a product, especially when the food matrix of the development of a product is different from the matrix that initially contained the carotenoid (Xavier et al. 2018). Although, the inside of what interferes with carotenoid bioaccessibility helps the food industry and people in their homes in the choice of processes of preparation and food components (as included dietary fat) in the preparation of the meal, to successfully lead to the bioaccessibility, bioavailability, and bioactivity of carotenoids in the human body.

5 Carotenoid's Bioaccessibility Using Dynamic Gastrointestinal Models

Although static *in vitro* digestion protocols are the most widely used approach to assess the bioaccessibility of carotenoids, some studies that have evaluated the bioaccessibility of carotenoids using dynamic gastrointestinal models are listed in Table 7.1.

The TNO Gastro-Intestinal Model (TIM-1) is a computer-controlled *in vitro* multicompartmental digestion model that precisely simulates the physiological processes in the upper gastrointestinal tract with a horizontal stomach and three

Table 7.1 Bioaccessibility and bioavailability of carotenoids studies employing dynamic digestion protocols

Source	Carotenoid	Food matrix	Dynamic model	Reference
Red tomatoes	β-Carotene, lutein, zeaxanthin, and lycopene	Western diet	TIM 1	Blanquet-Diot et al. (2009)
Yellow tomatoes	β-Carotene, lutein, zeaxanthin, and lycopene	Western diet	TIM 1	Blanquet-Diot et al. (2009)
Lycopene supplement (LycoVit 10%DC, BASF)	Lycopene	Western diet	TIM 1	Blanquet-Diot et al. (2009)
Tomatoes	Lycopene	Western diet	TIM 1 + Caco2 model	Déat et al. (2009)
Cooked fruits and vegetables	β-Carotene	Western diet	TIM 1	Loo-Bouwman et al. (2014)
Granules of a carotenoid mixture	β-Carotene, α-carotene, lutein, zeaxanthin, lycopene, and astaxanthin	Standard high-fat meal	TIM 1	Hu et al. (2021)
Eggs' yolk	Zeaxanthin and lutein	Cooking method (boiling, frying, and scrambling)	TIM 1	Nimalaratne et al. (2015)
Tomato sauce	Lycopene isomerization	Standard meal	TIM 1 + Caco2 model	Richelle et al. (2010)
Steamed or fried orange sweet potatoes	β-Carotene	–	Human Gastric Simulator (HGS)	Somaratne et al. (2020)
Commercial kale powder product	Lutein	–	DIDGI®	Lee et al. (2018)
Lutein supplement	Lutein	–	DIDGI®	Lee et al. (2018)
Raw pistachios, roasted salted pistachios, and muffins made with raw pistachios	Lutein	–	Mandalari protocol	Mandalari et al. (2013)
β-Carotene nanoemulsions	β-Carotene	–	Pinheiro protocol	Gomes et al. (2017)
Microemulsions of pitanga	β-Carotene and lycopene	–	Pinheiro protocol	Berni et al. (2020)
Microemulsions of buriti	β-Carotene and lycopene	–	Pinheiro protocol	Berni et al. (2020)

segments of the small intestine (duodenum, jejunum, and ileum) (thetimcompany.com). The main parameters of digestion are simulated such as pH, controlled with the addition of aqueous hydrochloric acid (HCl) solution or aqueous sodium bicarbonate ($NaHCO_3$) solution, body temperature, movement, and peristaltic transport. Also, the bile-secreted and pancreatic solutions happen at a true-to-life flow rate. In addition, it is possible to collect the sample at each level of the digestive process, as a bioaccessible fraction (Dupont et al. 2019).

The Human Gastric Simulator (HGS) reproduces gastric digestion and was designed at the University of California, Davis. The HGS operates at 37 °C with ~3 contractions per minute. It allows the study of food's physical and chemical breakdown by simulating continuous peristaltic contractions in a sample size range from 150 to 300 grams, closely mimicking the amplitude and frequency of human antral contractions (Brodkorb et al. 2019; Kong and Singh 2010). Also, the HGS enables control of gastric secretion, emptying rates, and pH based on specific physiological conditions, such as elderly or/and the uses of medication (Kong and Singh 2010). Furthermore, by assessing the breakdown of solid foods during gastric digestion, the HGS can aid in understanding the relationship between physical and chemical food breakdown and physiological outcomes such as gastric emptying, nutrient absorption, and satiety (Bornhorst 2017).

The development of the *in vitro* dynamic digestion system called DIGDI® occurs at the French National Institute of Agricultural Research (INRA, Rennes, France) and consists of two compartments simulating the stomach and the small intestine separated by a Teflon membrane with 2 mm holes. The objective was to monitor the kinetics of the deconstruction and hydrolysis of food in the digestion simulation (Ménard et al. 2014). The system was controlled by software called StoRM: stomach regulation and monitoring based on *in vivo* data, reproduced temperature, pH kinetics for the stomach and small intestine, the rate of secretions (such as HCl, NA2CO3, bile, and enzymes), the rate of gastric and intestinal emptying and peristaltic movements (Ménard et al. 2014). Thus, the DIDGI dynamic digestion model can be used to study and evaluate the stomach and small intestine under different conditions based on data obtained *in vivo*, such as preterm or term babies, adults, the elderly, and animals (Dupont et al. 2019).

Carotenoids' digestive stability in the red and yellow tomatoes and lycopene supplement was evaluated in a dynamic *in vitro* gastrointestinal tract model, TIM 1. The standard Western diet used consisted of 60 g of boiled egg white previously rinsed with water for complete egg yolk removal, 50 g of cooked pasta, 2.5 g of soy lecithin, and 20 g of sunflower oil with the addition of 100 g of puree red or yellow tomatoes double concentrate or lycopene beadlet with 78 mg of water-soluble powder (LycoVit 10%DC, BASF). The final homogenized sample adjusted with mineral water was 300 mL. During digestion and regardless of the meal, zeaxanthin and lutein were stable and resistant to gastrointestinal conditions. Β-carotene from the yellow tomato meal was stable under digestion conditions, while a 30% loss occurred for red tomato meals. In contrast, the all-*trans*-lycopene was degraded significantly during digestion for the evaluated matrices. Thus, the digestive stability of carotenoids depends on their nature, as β-carotene and lycopene (carotenes) seem more sensitive to digestive conditions in the tested samples than lutein or zeaxanthin (xanthophylls) (Blanquet-Diot et al. 2009).

TIM-1 model coupled with the Caco2 cellular model was used to study digestive stability and intestinal absorption of lycopene in tomatoes (Déat et al. 2009). The standard Western diet used was the same as described in Blanquet-Diot et al. (2009), used to evaluate the digestive stability of carotenoids in red and yellow tomatoes and lycopene supplements (Blanquet-Diot et al. 2009). Lycopene showed a recovery loss of 25% after 300 min in the digestion model, and a similar result was found by Blanquet-Diot et al. (2009). Furthermore, the lycopene tested by apical and basolateral application was absorbed in a time-dependent manner by the monolayer of Caco2 cells and detected after 30 min in the opposite compartment (Déat et al. 2009). Higher concentrations of lycopene were absorbed when applied to the apical chamber compared to the basolateral chamber. Thus, part of lycopene must be absorbed by the facilitated process mediated by a membrane transporter, the scavenger receptor class B type I (SR-BI) (Cai et al. 2004; Déat et al. 2009). Enterocyte-like human colonic adenocarcinoma (Caco2) together with HT-29 goblet cells are established as the most used intestinal cell model. They can be arranged in monolayers and adhered to the bottom of the culture vessels, allowing the study of carotenoid absorption. Cells can also be arranged in two or three compartments (apical, cellular, and basolateral) with the insertion of a semipermeable membrane. The bioaccessibility of β-carotene in two typical Western dietary samples was evaluated in TIM-1. The samples weighed 120.0 grams and contained 57% carbohydrate, 13% protein, and 30% fat. The mixed diet, representing a healthy diet high in cooked fruits and vegetables, contained 498 μg of β-carotene (90% as all-*trans*-β-carotene), and the oily diet contained 220 μg of β-carotene (98% as all-*trans*-β-carotene). Furthermore, the oily diet represents a diet low in vegetables and fruits but rich in β-carotene-enriched foods with salad dressing (30% vegetable oil suspension, Hoffmann-La Roche, Switzerland). A bioaccessibility of β-carotene was found in the mixed diet of 28% (28 μg/100 μg β-carotene) and, for the oily diet was found a bioaccessibility of 53% (53 μg/100 μg β-carotene). This 1.9:1 ratio was consistent with *in vivo* data, where apparent absorption was 1.9 times greater on the oil diet than on the mixed diet (Loo-Bouwman et al. 2014).

TIM-1 also was used to verify the bioaccessibility of the granules of a carotenoid mixture. The carotenoid mix was formulated with natural β-carotene, α-carotene, lutein, zeaxanthin, lycopene, and astaxanthin into granules with delayed release peak in the intestinal tract. Samples consisted of 480 mg granules along with 150 grams of a standard high-fat meal (composed of 50% fat, 20% protein, and 30% carbohydrates). As expected by the product design, lutein and zeaxanthin were released first, followed by lycopene and then α- and β-carotene. The bioaccessible fraction of α- and β-carotenes ranged from 7% to 8%, while lutein and zeaxanthin measured a cumulative bioaccessibility between 13% and 20%. However, the lycopene bioaccessibility was extremely low, as well as astaxanthin (Hu et al. 2021).

Zeaxanthin and lutein bioaccessibility were measured by the TIM-1 model for eggs' yolk. The qualitative chromatographic profile of carotenoids remained the same during digestion, with no significant trans-cis isomerization observed

(Nimalaratne et al. 2015). The isomerization was also investigated for all-*trans*-lycopene, and during its passage through the gastrointestinal tract, lycopene remained unchanged (Richelle et al. 2010). All-*trans*-lutein and all-*trans*-zeaxanthin were stable at 300 min with mean recoveries of 90% and 88%, respectively. The stability of lutein and zeaxanthin was also found in the study by Blanquet-Diot et al. (2009), described above. The bioaccessibility of lutein was 44%, 41%, and 39% in boiled, fried, and scrambled eggs, while for zeaxanthin was 45%, 42%, and 24%, respectively. Then, the differences in food preparations influenced the bioaccessibility, whose steamed eggs had the lowest (Nimalaratne et al. 2015).

Lycopene isomerization was investigated by *in vitro* static and dynamic (TIM-1) models, a cellular model with Caco2 cells to mimic the transport and metabolism of enterocytes, and an *in vivo* model in humans. Tomato sauce and the standard meal consisted of semolina (70 g) cooked in 200 ml of hot water, white bread (40 g), egg white (60 g), peanut oil (40 g), plain yogurt (125 g), sugar (5 g) and water (330 g) were used. The lycopene isomerization profile remained unchanged in static and dynamic gastrointestinal models and occurred within Caco2 cells. The absorption of lycopene located in the mixed micelles occurred in a time-dependent manner and was observed, and transferred of 13% of lycopene was from the mixed micelles to Caco2 or Caco2 SM cells while only 3% was transferred to Caco2 HTB37 after 6 h. Results in Caco2 cells correspond to *in vivo* results; thus, the lycopene that appears in chylomicrons and Caco2 cells corresponds to a mixture of *cis*-lycopene isomers (Richelle et al. 2010).

The influence of gastric juice diffusion on the release of β-carotene in steamed or fried orange sweet potatoes was analyzed in static and dynamic *in vitro* models in the Human Gastric Simulator (HGS). The analysis in the HGS lasted 10 min, as the study objective was not to reproduce the human digestion physicochemical reactions. The static model of *in vitro* bioaccessibility was performed according to the INFOGEST protocol (Minekus et al. 2014). After 4 h in the static model, the release of β-carotene was not affected by the structural differentiation suffered by the steaming and frying processes of orange sweet potato, releasing approximately 12% of the β-carotene. On the other hand, after *in vitro* digestion in the static model and 10 min in the HGS, where mechanical forces have been simulated, the cooking processes influence the sweet potato structure and, consequently, the release of β-carotene. There was a release of 35% of β-carotene for steamed sweet potato and 40% for fried sweet potato. Thus, mechanical forces influenced the modification of the sweet potato structure and, consequently, the release of β-carotene (Somaratne et al. 2020).

The lutein bioaccessibility in static and dynamic *in vitro* models was evaluated in a commercial kale powder product and a lutein supplement, composed of *Calendula officinalis* (calendula) flower extract with omega-3 fatty acids. Static digestion was performed according to the protocol by Garrett et al. (1999) with slight modifications, and dynamic digestion was used in the Digesteur Dynamique Gastro-Intestinale (DIDGI®) model. Kale powder obtained a maximum lutein bioaccessibility of approximately 60% after 120 min in a static digestion model nevertheless, in the dynamic model DIDGI it was necessary twice as long to reach values greater than

50%. For the lutein supplement, the maximum value of 67.88% was found in the dynamic digestion model, five times superior to static digestion, which was found at 12.34% in 120 min (Lee et al. 2018). Also, the quantification of lutein release during digestion was evaluated in a dynamic *in vitro* digestion model development for Mandalari et al. (2013) in samples of raw pistachios, roasted salted pistachios, and muffins made with raw pistachios. Lutein released in raw pistachios and salted and roasted pistachios occurred most in the gastric environment, and for pistachio muffins, total solubilization at the end of gastric digestion was observed (Mandalari et al. 2013).

B-carotene nanoemulsions were evaluated *in vitro* by static and dynamic models, following the protocol developed by (Pinheiro et al. 2013). For 100 g of nanoemulsion, 80 g of deionized water, 12 g of Cremophor RH 40, 8 g of Span 80, and 10 g of cupuaçu butter were used, in addition to β-carotene (0.6 g). In the static model, the bioaccessibility of β-carotene was 92%, calculated following Salvia-Trujillo et al. (2013), who propose using the β-carotene present in the raw digest as an initial amount for the calculation, and not the β-carotene encapsulated in the nanoparticle sample before the stomach phase. In contrast, the dynamic model had a maximum bioaccessibility of approximately 15% after 5 h, and the β-carotene release occurred in the duodenum (Gomes et al. 2017). The same protocol (Pinheiro et al. 2016) was used to analyze the bioaccessibility of β-carotene and lycopene in microemulsions of pitanga (*Eugenia uniflora*) and buriti (*Mauritia flexuosa*) fruits. The microemulsion was obtained by high-speed homogenization and ultrasound probe with the pulp, surfactant (Tween 80 or 2% Whey Protein Isolate), and 5% corn oil. Good recovery was obtained in the bioaccessibility of β-carotene and lycopene in microemulsions than in the whole pulp nonetheless the authors point out that there were losses of carotenoids to the stomach sac of the dynamic model developed and used (Berni et al. 2020).

6 Carotenoid's Bioaccessibility and Bioavailability: Gaps in Knowledge and Future Perspectives in Research

Carotenoid research groups propose their adaptation in INFOGEST protocol to assess bioaccessibility as the addition of enzymes enabling the hydrolysis of carotenoid and carotenoid esters, the undigested material isolation, and carotenoid extraction and identification after simulated oro-gastrointestinal digestion (Xavier and Mercadante 2022). INFOGEST 2.0 protocol recommended the use of gastric lipases (Brodkorb et al. 2019). However, human gastric lipase is not commercially available, and the available fungal lipases act at different pH and substrate sites compared to human lipase (Brodkorb et al. 2019; Iddir et al. 2021). Although rabbit gastric extracts showed satisfactory results in the hydrolysis of carotenoid esters, and their use is recommended for INFOGEST 2.0, the rabbit gastric extracts acquisition is not available for Lipolytech® (Petry and Mercadante 2019, 2020; Xavier and Mercadante 2022). Indeed, gastric lipase is essential to significantly increase lipolysis and the bioaccessibility of carotenoids from plant food matrices, especially the presence of carotenoid esters (Petry and Mercadante 2020). Thus, adaptations were made in the gastric phase in search of more physiological hydrolysis of carotenoids and their esters.

Another modification carried out by the carotenoid research groups in the standardized INFOGEST protocol is the inclusion of a porcine pancreas cholesterol esterase (CEL) solution to improve the hydrolysis of carotenoid esters in the intestinal phase (Petry and Mercadante 2017, 2020; Wen et al. 2018; Xavier and Mercadante 2022). Carotenoid esters are presumably hydrolyzed before their absorption in the human body, due to the lack of detection in the triacylglycerol-rich fraction of human plasma after consumption of foods rich in carotenoid esters (Failla et al. 2019; Hempel et al. 2017). It is not clearly defined how the process of hydrolysis of esters physiologically occurs, and it is assumed that CEL and pancreatic lipase may be enzymes responsible for their hydrolysis. Current *in vitro* digestion simulations do not efficiently reproduce the hydrolysis and even with the addition of CEL and pancreatic lipase, carotenoid esters are detected in a highly variable range in the bioaccessible fraction (Failla et al. 2019; Petry and Mercadante 2017; Rodrigues et al. 2016). Furthermore, a fraction of carotenoid esters consumed by humans was found in feces, suggesting that part of the ingested esters is not metabolized in the digestion process due to their presence in the lumen of the large intestine (Hernández-Alvarez et al. 2016). It is vital to understand the metabolic pathways of carotenoid esters, the action of digestive enzymes in their hydrolysis, and the possible impact of the presence of carotenoids and their esters on the intestinal microbiota (Failla et al. 2019). It is worth highlighting the high price of digestive enzymes such as rabbit gastric extracts and CEL, a relevant barrier to their addition in an *in vitro* model (Petry and Mercadante 2019). Also, the saponification of carotenoids and/or the bioaccessible fraction carried out by some studies neglects the understanding of carotenoid intake in human health. These protocol variations do not allow the comparison of the results among different studies (Xavier and Mariutti 2021). Thus, although there is a standardized INFOGEST protocol for in vitro static digestion, a consensus is needed on the conditions that mimic digestion, such as the type, amount, and release of digestive enzymes and the processes for isolating the bioaccessible fraction, identifying and extracting carotenoids to understand the health effects of consumed carotenoids.

Over the past 15 years, the research has included a more refined understanding of carotenoid bioavailability and the influences of intrinsic and extrinsic factors (Molteni et al. 2022; Shilpa et al. 2020), however, digestion remains a partially understood mechanical and chemical process and the impact of the food matrix deserves continued investigation (Hiolle et al. 2020). *In vitro* static digestion models study the impact of matrix constituents on the bioaccessibility of carotenoids, however, only a few studies have evaluated the role of biochemical and mechanical disintegration in carotenoid release. The decomposition of food is relevant for the release and absorption of nutrients during gastrointestinal digestion, requiring a thorough understanding of the impact of this process on nutrient digestibility

(Mennah-Govela and Bornhorst 2021). Indeed, the lack of mechanical forces generated by the stomach in the biochemical digestion of the food matrix of steamed or fried sweet potato did not affect the breakdown of cell walls, and the release of β-carotene was shown to be independent during gastric digestion between cooked samples (steamed and fried) (Somaratne et al. 2020). Furthermore, as shown in the **Sect. 5**, few studies have evaluated the in vitro digestion process using dynamic models. The widely used static models have the restriction of not simulating the dynamic process of digestion, and keeping pH levels, digestive enzymes, and bile concentrations constant. In contrast, dynamic systems when programmed with relevant physiological parameters can mimic the human digestive process (Dupont et al. 2019). Thus, it is essential to carry out studies on dynamic systems with carotenoids in their matrix and incorporate them into products for consumption.

A limitation of the digestion protocols is the absence of microbiota in the distal parts of the small intestine (Dupont et al. 2019) and the absorption of carotenoids by intestinal tissue is the essential step to understanding their bioavailability (Kalungwana et al. 2023). Studies performed in human volunteers continue to represent the "gold" standard for bioavailability, but challenges associated with host factors that influence aspects of digestion and carotenoid release from the food matrix and its bioaccessibility can lead to misinterpretation of results (Reboul 2019). In addition, studies with humans have high cost and ethical restrictions, and make it impossible to elucidate the whole processes and their regulation at the cellular and molecular level associated with the intestinal uptake, metabolism, and absorption of target compounds by the intestinal epidermis (Rodrigues and Failla 2021). Thus, the characteristics of the intestinal epithelium are mimicked through *in vitro* cell culture models that allow studying the uptake, metabolism, and absorption of target compounds. This model is a relatively simple, fast, and economical alternative, which supports the construction of reproducible hypotheses under well-controlled conditions and mechanistic investigations that cannot be conducted *in vivo* (Rodrigues and Failla 2021).

To better simulate the epithelial transport and absorption of carotenoids after the digestion process, cellular models (such as Caco-2, HT-29, IPEC-J2, or coculture of Caco-2 and HT-29MTX) can be coupled (Déat et al. 2009). The cell line widely accepted is human colon adenocarcinoma Caco-2 cells and evaluated the permeation after intestinal digestion (Rodrigues and Failla 2021). However, the model has limitations as the expression of enzyme and protein transporters responsible for carotenoid uptake and epithelial cell composition (Bohn 2018; Desmarchelier and Borel 2017; Failla et al. 2019; Kopec and Failla 2018). More realistic approaches are developed with the expectation of results of more excellent predictive reliability, such as the use of pluripotent stem cells for the generation of self-organized enteroids in 2D and 3D or cultivated in dynamic conditions and the use of technological devices, such as the chip intestinal (Rodrigues and Failla 2021). Gut-on-a-chip is a microfluidic technology for the development of the 3D topography of the organ and its physical-chemical characteristics. It is a technology that simulates the flow of fluids, mechanical forces, and oxygen gradients, correlating with *in vivo* data (Valiei et al. 2023). Another more physiological and recent approach is the use of the Ussing chamber technique to evaluate mouse intestinal tissue e*x vivo* (Kalungwana et al. 2023; Mulet-Cabero et al. 2020). This model has successfully been used to assess carotenoid absorption, and carotenoids were detected in the basolateral section, allowing a more realistic comparison of *ex vivo* intestinal permeability data with *in vivo* human data (Kalungwana et al. 2023). However, the use of fresh and viable mouse intestines and the technical difficulties related to the model represented relevant limitations (Kalungwana et al. 2023). The use of mathematical modeling is also noteworthy as a growing approach to the investigation of digestive processes. *In silico* models can interconnect molecular mechanisms with the system view approach and thus connect the available pieces to predict metabolic responses to consumed foods (Le Feunteun et al. 2020). While there is a long way to go toward a holistic understanding of carotenoid digestion and absorption, the progress has been astounding and the scientific efforts and proposed models stupendous.

7 Concluding Remarks

The consumption of carotenoids is associated with beneficial effects on human health, but there is a tortuous path to a holistic understanding of carotenoid digestion. Thus, the assessment of bioaccessibility and bioavailability is hampered by the limited knowledge of its absorption mechanisms. The use of animals is increasingly questioned for simulations of digestion processes due to physiological and dietary differences and *in vitro* studies stood out.

The standardization of the protocol for simulating *in vitro* digestion represented an important advance in the area of digestion and made it possible to compare results between different research groups. However, the INFOGEST protocol does not establish the steps after digestion, such as separation of the micellar fraction, extraction, isolation, and identification of carotenoids, being developed by each research group. Furthermore, the saponification process of carotenoids and the inclusion of carotenoid esters is another important point to be discussed. The researchers find it difficult to simulate the complete hydrolysis of carotenoid esters in the *in vitro* digestion process. However, carotenoid esters are poorly detected in human blood plasma, being hydrolyzed in the process of human digestion.

Thus, increasingly complex *in vitro* digestion models are being developed, such as semi-dynamic and dynamic models coupled to cells or fermenters to simulate the intestinal microbiota. These recent approaches are essential to better understanding the complexity of carotenoid digestion and absorption.

Acknowledgments The authors acknowledge "Fundação de Amparo à Pesquisa do Estado de São Paulo—FAPESP" for the funding support through fellowship (2021/00596-7) and grant (2021/02692-3). VV. De Rosso acknowledges the National Council for Scientific and Technological Development (CNPq) for their financial support (fellowship 315378/2021-2).

References

Berni P, Pinheiro AC, Bourbon AI, Guimarães M, Canniatti-Brazaca SG, Vicente AA (2020) Characterization of the behavior of carotenoids from pitanga (Eugenia uniflora) and buriti (Mauritia flexuosa) during microemulsion production and in a dynamic gastrointestinal system. J Food Sci Technol 57(2):650–662. https://doi.org/10.1007/s13197-019-04097-7

Biehler E, Kaulmann A, Hoffmann L, Krause E, Bohn T (2011) Dietary and host-related factors influencing carotenoid bioaccessibility from spinach (Spinacia oleracea). Food Chem 125(4):1328–1334. https://doi.org/10.1016/j.foodchem.2010.09.110

Blanquet-Diot S, Soufi M, Rambeau M, Rock E, Alric M (2009) Digestive stability of xanthophylls exceeds that of carotenes as studied in a dynamic in vitro gastrointestinal system. J Nutr 139(5):876–883. https://doi.org/10.3945/jn.108.103655

Bohn T (2018) Metabolic fate of bioaccessible and non-bioaccessible carotenoids. In: Non-extractable polyphenols and carotenoids: importance in human nutrition and health, vol 5. Royal Society of Chemistry, pp 165–200

Bohn T, Desmarchelier C, Dragsted LO, Nielsen CS, Stahl W, Rühl R, Keijer J, Borel P (2017) Host-related factors explaining interindividual variability of carotenoid bioavailability and tissue concentrations in humans. Mol Nutr Food Res 61(6):1600685. https://doi.org/10.1002/mnfr.201600685

Bohn T, Carriere F, Day L, Deglaire A, Egger L, Freitas D, Golding M, Le Feunteun S, Macierzanka A, Menard O, Miralles B, Moscovici A, Portmann R, Recio I, Rémond D, Santé-Lhoutelier V, Wooster TJ, Lesmes U, Mackie AR, Dupont D (2018) Correlation between in vitro and in vivo data on food digestion. What can we predict with static in vitro digestion models? Crit Rev Food Sci Nutr 58(13):2239–2261. https://doi.org/10.1080/10408398.2017.1315362

Borel P, Lietz G, Goncalves A, de Edelenyi FS, Lecompte S, Curtis P, Goumidi L, Caslake MJ, Miles EA, Packard C, Calder PC, Mathers JC, Minihane AM, Tourniaire F, Kesse-Guyot E, Galan P, Hercberg S, Breidenassel C, Gross MG et al (2013) CD36 and SR-BI are involved in cellular uptake of provitamin a carotenoids by Caco-2 and hek cells, and some of their genetic variants are associated with plasma concentrations of these micronutrients in humans. J Nutr 143(4):448–456. https://doi.org/10.3945/jn.112.172734

Bornhorst GM (2017) Gastric mixing during food digestion: mechanisms and applications. Annu Rev Food Sci Technol 8:523–542. https://doi.org/10.1146/annurev-food-030216-025802

Bornhorst GM, Singh RP (2014) Gastric digestion in vivo and in vitro: how the structural aspects of food influence the digestion process. Annu Rev Food Sci Technol 5(1):111–132. https://doi.org/10.1146/annurev-food-030713-092346

Britton G, Khachik F (2009) Carotenoids in foods. In: Britton G, Liaaen-Jensen S, Pfander H (eds) Carotenoids, vol 5. https://doi.org/10.15547/ictte.2018.06.006

Brodkorb A, Egger L, Alminger M, Alvito P, Assunção R, Ballance S, Bohn T, Bourlieu-Lacanal C, Boutrou R, Carrière F, Clemente A, Corredig M, Dupont D, Dufour C, Edwards C, Golding M, Karakaya S, Kirkhus B, Le Feunteun S et al (2019) INFOGEST static in vitro simulation of gastrointestinal food digestion. Nat Protoc 14(4):991–1014. https://doi.org/10.1038/s41596-018-0119-1

Cai L, Eckhardt ERM, Shi W, Zhao Z, Nasser M, De Villiers WJS, Van Der Westhuyzen DR (2004) Scavenger receptor class B type I reduces cholesterol absorption in cultured enterocyte CaCo-2 cells. J Lipid Res 45(2):253–262. https://doi.org/10.1194/jlr.M300303-JLR200

Carriere F, Barrowman JA, Verger R, Laugier R (1993) Secretion and contribution to lipolysis of gastric and pancreatic lipases during a test meal in humans. Gastroenterology 105:878–888

Castenmiller JJM, West CE (1998) Bioavailability and bioconversion of carotenoids. Annu Rev Nutr 18:19–38

Chacón-Ordóñez T, Carle R, Schweiggert R (2019) Bioaccessibility of carotenoids from plant and animal foods. J Sci Food Agric 99(7):3220–3239. https://doi.org/10.1002/jsfa.9525

Da Costa GA, Mercadante AZ (2018) In vitro bioaccessibility of free and esterified carotenoids in cajá frozen pulp-based beverages. J Food Compos Anal 68:53–59. https://doi.org/10.1016/j.jfca.2017.02.012

Déat E, Blanquet-Diot S, Jarrige JF, Denis S, Beyssac E, Alric M (2009) Combining the dynamic TNO-gastrointestinal tract system with a Caco-2 cell culture model: application to the assessment of lycopene and α-tocopherol bioavailability from a whole food. J Agric Food Chem 57(23):11314–11320. https://doi.org/10.1021/jf902392a

Desmarchelier C, Borel P (2017) Overview of carotenoid bioavailability determinants: from dietary factors to host genetic variations. Trends Food Sci Technol 69:270–280. https://doi.org/10.1016/j.tifs.2017.03.002

Dias MG, Olmedilla-Alonso B, Hornero-Méndez D, Mercadante AZ, Osorio C, Vargas-Murga L, Meléndez-Martínez AJ (2018) Comprehensive database of carotenoid contents in ibero-american foods. A valuable tool in the context of functional foods and the establishment of recommended intakes of bioactives. J Agric Food Chem 66(20):5055–5107. https://doi.org/10.1021/acs.jafc.7b06148

Dingeo G, Brito A, Samouda H, Iddir M, La Frano MR, Bohn T (2020) Phytochemicals as modifiers of gut microbial communities. Food Func 11(10):8444–8471. Royal Society of Chemistry. https://doi.org/10.1039/d0fo01483d

Dupont D, Alric M, Blanquet-Diot S, Bornhorst G, Cueva C, Deglaire A, Denis S, Ferrua M, Havenaar R, Lelieveld J, Mackie AR, Marzorati M, Menard O, Minekus M, Miralles B, Recio I, Van den Abbeele P (2019) Can dynamic in vitro digestion systems mimic the physiological reality? Crit Rev Food Sci Nutr 59(10):1546–1562. https://doi.org/10.1080/10408398.2017.1421900

During A, Dawson HD, Harrison EH (2005) Carotenoid transport is decreased and expression of the lipid transporters SR-BI, NPC1L1, and ABCA1 is downregulated in Caco-2 cells treated with ezetimibe. J Nutr 135:2305–2312. https://doi.org/10.1093/jn/135.10.2305

Egea I, Barsan C, Bian W, Purgatto E, Latché A, Chervin C, Bouzayen M, Pech JC (2010) Chromoplast differentiation: current status and perspectives. Plant Cell Physiol 51(10):1601–1611. https://doi.org/10.1093/pcp/pcq136

Egger L, Ménard O, Delgado-Andrade C, Alvito P, Assunção R, Balance S, Barberá R, Brodkorb A, Cattenoz T, Clemente A, Comi I, Dupont D, Garcia-Llatas G, Lagarda MJ, Le Feunteun S, JanssenDuijghuijsen L, Karakaya S, Lesmes U, Mackie AR et al (2016) The harmonized INFOGEST in vitro digestion method: from knowledge to action. Food Res Int 88:217–225. https://doi.org/10.1016/j.foodres.2015.12.006

Egger L, Schlegel P, Baumann C, Stoffers H, Guggisberg D, Brügger C, Dürr D, Stoll P, Vergères G, Portmann R (2017) Physiological comparability of the harmonized INFOGEST in vitro digestion method to in vivo pig digestion. Food Res Int 102:567–574. https://doi.org/10.1016/j.foodres.2017.09.047

Eriksen JN, Luu AY, Dragsted LO, Arrigoni E (2016) In vitro liberation of carotenoids from spinach and Asia salads after different domestic kitchen procedures. Food Chem 203:23–27. https://doi.org/10.1016/j.foodchem.2016.02.033

Eroglu A, Al'Abri IS, Kopec RE, Crook N, Bohn T (2023) Carotenoids and their health benefits as derived via their interactions with gut microbiota. Adv Nutr 14(2):238–255. https://doi.org/10.1016/j.advnut.2022.10.007

Estévez-Santiago R, Olmedilla-Alonso B, Fernández-Jalao I (2016) Bioaccessibility of provitamin A carotenoids from fruits: application of a standardised static in vitro digestion method. Food Func 7(3):1354–1366. https://doi.org/10.1039/c5fo01242b

Failla ML, Rodrigues DB, Chitchumroonchokchai C (2019) Bioavailability and metabolism of carotenoid esters: physical, chemical and biological properties. In: Food chemistry, function and analysis, vol 13. Royal Society of Chemistry, pp 390–420. https://doi.org/10.1039/9781788015851-00390

Garrett DA, Failla ML, Sarama RJ (1999) Development of an in vitro digestion method to assess carotenoid bioavailability from meals. J Agric Food Chem 47(10):4301–4309. https://doi.org/10.1021/jf9903298

Geng T, Bao S, Sun X, Ma D, Zhang H, Ge Q, Liu X, Ma T (2023) A clarification of concepts related to the digestion and absorption of carotenoids and a new standardized carotenoids bioavailability evaluation system. Food Chem 400(134060):1–13. https://doi.org/10.1016/j.foodchem.2022.134060

Gomes GVL, Sola MR, Marostegan LFP, Jange CG, Cazado CPS, Pinheiro AC, Vicente AA, Pinho SC (2017) Physico-chemical stability and in vitro digestibility of beta-carotene-loaded lipid nanoparticles of cupuacu butter (Theobroma grandiflorum) produced by the phase inversion temperature (PIT) method. J Food Eng 192:93–102. https://doi.org/10.1016/j.jfoodeng.2016.08.001

Grundy MML, Abrahamse E, Almgren A, Alminger M, Andres A, Ariëns RMC, Bastiaan-Net S, Bourlieu-Lacanal C, Brodkorb A, Bronze MR, Comi I, Couëdelo L, Dupont D, Durand A, El SN, Grauwet T, Heerup C, Heredia A, Infantes Garcia MR et al (2021) INFOGEST inter-laboratory recommendations for assaying gastric and pancreatic lipases activities prior to in vitro digestion studies. J Funct Foods 82:104497. https://doi.org/10.1016/j.jff.2021.104497

Hempel J, Schädle CN, Sprenger J, Heller A, Carle R, Schweiggert RM (2017) Ultrastructural deposition forms and bioaccessibility of carotenoids and carotenoid esters from goji berries (Lycium barbarum L.). Food Chem 218:525–533. https://doi.org/10.1016/j.foodchem.2016.09.065

Hernández-Alvarez E, Blanco-Navarro I, Pérez-Sacristán B, Sánchez-Siles LM, Granado-Lorencio F (2016) In vitro digestion-assisted development of a β-cryptoxanthin-rich functional beverage; in vivo validation using systemic response and faecal content. Food Chem 208:18–25. https://doi.org/10.1016/j.foodchem.2016.03.119

Hiolle M, Lechevalier V, Floury J, Boulier-Monthéan N, Prioul C, Dupont D, Nau F (2020) In vitro digestion of complex foods: how microstructure influences food disintegration and micronutrient bioaccessibility. Food Res Int 128:108817. https://doi.org/10.1016/j.foodres.2019.108817

Hu C, Salter Venzon D, Lange K, Maathuis A, Bellmann S, Gellenbeck K (2021) Evaluation of the bioaccessibility of a carotenoid beadlet blend using an in vitro system mimicking the upper gastrointestinal tract. Food Sci Nutr 9(6):3289–3296. https://doi.org/10.1002/fsn3.2295

Iddir M, Porras Yaruro JF, Larondelle Y, Bohn T (2021) Gastric lipase can significantly increase lipolysis and carotenoid bioaccessibility from plant food matrices in the harmonized INFOGEST static in vitro digestion model. Food Func 12(19):9043–9053. https://doi.org/10.1039/d1fo00786f

Kalungwana NA, Marshall L, Mackie A, Boesch C (2023) An ex vivo intestinal absorption model is more effective than an in vitro cell model to characterise absorption of dietary carotenoids following simulated gastrointestinal digestion. Food Res Int 166:112558. https://doi.org/10.1016/j.foodres.2023.112558

Kaulmann A, Bohn T (2014) Carotenoids, inflammation, and oxidative stress-implications of cellular signaling pathways and relation to chronic disease prevention. Nutr Res 34(11):907–929. Elsevier Inc. https://doi.org/10.1016/j.nutres.2014.07.010

Kong F, Singh RP (2010) A human gastric simulator (HGS) to study food digestion in human stomach. J Food Sci 75(9):1–9. https://doi.org/10.1111/j.1750-3841.2010.01856.x

Kopec RE, Failla ML (2018) Recent advances in the bioaccessibility and bioavailability of carotenoids and effects of other dietary lipophiles. J Food Compos Anal 68:16–30. https://doi.org/10.1016/j.jfca.2017.06.008

Le Feunteun S, Mackie AR, Dupont D (2020) In silico trials of food digestion and absorption: how far are we? Curr Opin Food Sci 31:121–125. https://doi.org/10.1016/j.cofs.2020.04.006

Lee EH, Cha KH, Vuong TT, Kim SM, Pan CH (2018) Comparison of static and dynamic in vitro digestion models to estimate the bioaccessibility of lutein in lutein-rich foods. Appl Biol Chem 61(4):441–447. https://doi.org/10.1007/s13765-018-0378-0

Lemmens L, Colle I, Van Buggenhout S, Palmero P, Van Loey A, Hendrickx M (2014) Carotenoid bioaccessibility in fruit- and vegetable-based food products as affected by product (micro)

structural characteristics and the presence of lipids: a review. Trends Food Sci Technol 38(2):125–135. https://doi.org/10.1016/j.tifs.2014.05.005

Lombardo M, Serrao S, Lombardo G (2022) Challenges in age-related macular degeneration: from risk factors to novel diagnostics and prevention strategies. Front Med 9(887104):1–17. https://doi.org/10.3389/fmed.2022.887104

Loo-Bouwman CA, Van Naber THJ, Minekus M, Van Breemen RB, Hulshof PJM, Schaafsma G (2014) Food matrix effects on bioaccessibility of β-carotene can be measured in an in vitro gastrointestinal model. J Agric Food Chem 62(4):950–955. https://doi.org/10.1021/jf403312v

Mandalari G, Bisignano C, Filocamo A, Chessa S, Sarò M, Torre G, Faulks RM, Dugo P (2013) Bioaccessibility of pistachio polyphenols, xanthophylls, and tocopherols during simulated human digestion. Nutrition 29(1):338–344. https://doi.org/10.1016/j.nut.2012.08.004

Mapelli-Brahm P, Desmarchelier C, Margier M, Reboul E, Meléndez Martínez AJ, Borel P (2018) Phytoene and phytofluene isolated from a tomato extract are readily incorporated in mixed micelles and absorbed by Caco-2 cells, as compared to lycopene, and SR-BI is involved in their cellular uptake. Mol Nutr Food Res 62(22):e1800703. https://doi.org/10.1002/mnfr.201800703

Mashurabad PC, Palika R, Jyrwa YW, Bhaskarachary K, Pullakhandam R (2017) Dietary fat composition, food matrix and relative polarity modulate the micellarization and intestinal uptake of carotenoids from vegetables and fruits. J Food Sci Technol 54(2):333–341. https://doi.org/10.1007/s13197-016-2466-7

Ménard O, Cattenoz T, Guillemin H, Souchon I, Deglaire A, Dupont D, Picque D (2014) Validation of a new in vitro dynamic system to simulate infant digestion. Food Chem 145:1039–1045. https://doi.org/10.1016/j.foodchem.2013.09.036

Ménard O, Famelart MH, Deglaire A, Le Gouar Y, Guérin S, Malbert CH, Dupont D (2018) Gastric emptying and dynamic in vitro digestion of drinkable yogurts: effect of viscosity and composition. Nutrients 10(9):1–13. https://doi.org/10.3390/nu10091308

Menard O, Lesmes U, Shani-Levi CS, Araiza Calahorra A, Lavoisier A, Morzel M, Rieder A, Feron G, Nebbia S, Mashiah L, Andres A, Bornhorst G, Carrière F, Egger L, Gwala S, Heredia A, Kirkhus B, Macierzanka A, Portman R et al (2023) Static in vitro digestion model adapted to the general older adult population: an INFOGEST international consensus. Food Func 14(10):4569–4582. https://doi.org/10.1039/d3fo00535f

Mennah-Govela YA, Bornhorst GM (2021) Breakdown mechanisms of whey protein gels during dynamic in vitro gastric digestion. Food Func 12(5):2112–2125. https://doi.org/10.1039/d0fo03325a

Minekus M, Alminger M, Alvito P, Ballance S, Bohn T, Bourlieu C, Carrière F, Boutrou R, Corredig M, Dupont D, Dufour C, Egger L, Golding M, Karakaya S, Kirkhus B, Le Feunteun S, Lesmes U, MacIerzanka A, MacKie A et al (2014) A standardised static in vitro digestion method suitable for food-an international consensus. Food Func 5(6):1113–1124. https://doi.org/10.1039/c3fo60702j

Molteni C, La Motta C, Valoppi F (2022) Improving the bioaccessibility and bioavailability of carotenoids by means of nanostructured delivery systems: a comprehensive review. Antioxidants 11(10). https://doi.org/10.3390/antiox11101931

Moussa M, Landrier J-FX, Reboul E, Ghiringhelli O, Coméra C, Collet X, Frö K, Bö V, Borel P (2008) Lycopene absorption in human intestinal cells and in mice involves scavenger receptor class B type I but not Niemann-Pick C1-Like 1. J Nutr 138:1432–1436. https://academic.oup.com/jn/article/138/8/1432/4750794

Mrowicka M, Mrowicki J, Kucharska E, Majsterek I (2022) Lutein and zeaxanthin and their roles in age-related macular degeneration— neurodegenerative disease. Nutrients 14(827):1–14. https://doi.org/10.3390/nu14040827

Mulet-Cabero AI, Egger L, Portmann R, Ménard O, Marze S, Minekus M, Le Feunteun S, Sarkar A, Grundy MML, Carrière F, Golding M, Dupont D, Recio I, Brodkorb A, Mackie A (2020) A standardised semi-dynamic: in vitro digestion method suitable for food-an international consensus. Food Func 11(2):1702–1720. https://doi.org/10.1039/c9fo01293a

Nimalaratne C, Savard P, Gauthier SF, Schieber A, Wu J (2015) Bioaccessibility and digestive stability of carotenoids in cooked eggs studied using a dynamic in vitro gastrointestinal model. J Agric Food Chem 63(11):2956–2962. https://doi.org/10.1021/jf505615w

Perry GH, Dominy NJ, Claw KG, Lee AS, Fiegler H, Redon R, Werner J, Villanea FA, Mountain JL, Misra R, Carter NP, Lee C, Stone AC (2007) Diet and the evolution of human amylase gene copy number variation. Nat Genet 39(10):1256–1260. https://doi.org/10.1038/ng2123

Petry FC, Mercadante AZ (2017) Impact of in vitro digestion phases on the stability and bioaccessibility of carotenoids and their esters in Mandarin pulps. Food Func 8(11):3951–3963. https://doi.org/10.1039/c7fo01075c

Petry FC, Mercadante AZ (2019) Bile amount affects both the degree of micellarization and the hydrolysis extent of carotenoid esters during in vitro digestion. Food Func 10(12):8250–8262. https://doi.org/10.1039/c9fo01453e

Petry FC, Mercadante AZ (2020) Addition of either gastric lipase or cholesterol esterase to improve both β-cryptoxanthin ester hydrolysis and micellarization during in vitro digestion of fruit pulps. Food Res Int 137:109691. https://doi.org/10.1016/j.foodres.2020.109691

Pinheiro AC, Coimbra MA, Vicente AA (2016) In vitro behaviour of curcumin nanoemulsions stabilized by biopolymer emulsifiers – Effect of interfacial composition. Food Hydrocoll, 52:460–467. https://doi.org/10.1016/j.foodhyd.2015.07.025

Pinheiro AC, Lad M, Silva HD, Coimbra MA, Boland M, Vicente AA (2013) Unravelling the behaviour of curcumin nanoemulsions during in vitro digestion: effect of the surface charge. Soft Matter 9(11):3147–3154. https://doi.org/10.1039/c3sm27527b

Reboul E (2019) Mechanisms of carotenoid intestinal absorption: where do we stand? Nutrients 11(4):1–12. https://doi.org/10.3390/nu11040838

Richelle M, Sanchez B, Tavazzi I, Lambelet P, Bortlik K, Williamson G (2010) Lycopene isomerisation takes place within enterocytes during absorption in human subjects. Br J Nutr 103(12):1800–1807. https://doi.org/10.1017/S0007114510000103

Rodrigues DB, Failla ML (2021) Intestinal cell models for investigating the uptake, metabolism and absorption of dietary nutrients and bioactive compounds. Curr Opin Food Sci 41:169–179. https://doi.org/10.1016/j.cofs.2021.04.002

Rodrigues DB, Mariutti LRB, Mercadante AZ (2016) An in vitro digestion method adapted for carotenoids and carotenoid esters: moving forward towards standardization. Food Func 7(12):4992–5001. https://doi.org/10.1039/c6fo01293k

Rodrigues DB, Marques MC, Hacke A, Loubet Filho PS, Cazarin CBB, Mariutti LRB (2022) Trust your gut: bioavailability and bioaccessibility of dietary compounds. Curr Res Food Sci 5:228–233. https://doi.org/10.1016/j.crfs.2022.01.002

Rodriguez-Amaya DB (2015a) Bioaccessibility and bioavailability. In: Food carotenoids: chemistry, biology and technology. Wiley Blackwell, pp 225–254

Rodriguez-Amaya DB (2015b) Provitamin A activity. In: Food carotenoids: chemistry, biology and technology. Wiley Blackwell, pp 255–281

Salvia-Trujillo L, Qian C, Martín-Belloso O, McClements DJ (2013) Modulating β-carotene bioaccessibility by controlling oil composition and concentration in edible nanoemulsions. Food Chem 139(1–4):878–884. https://doi.org/10.1016/j.foodchem.2013.02.024

Sanchón J, Fernández-Tomé S, Miralles B, Hernández-Ledesma B, Tomé D, Gaudichon C, Recio I (2018) Protein degradation and peptide release from milk proteins in human jejunum. Comparison with in vitro gastrointestinal simulation. Food Chem 239:486–494. https://doi.org/10.1016/j.foodchem.2017.06.134

Sato Y, Suzuki R, Kobayashi M, Itagaki S, Hirano T, Noda T, Mizuno S, Sugawara M, Iseki K (2012) Involvement of cholesterol membrane transporter Niemann-Pick C1-Like 1 in the intestinal absorption of lutein. J Pharm Pharmaceut Sci 15(2) www.cspsCanada.org

Schweiggert RM, Carle R (2015) Carotenoid deposition in plant and animal foods and its impact on bioavailability. Crit Rev Food Sci Nutr 57(9):1807–1830. https://doi.org/10.1080/10408398.2015.1012756

Schweiggert RM, Mezger D, Schimpf F, Steingass CB, Carle R (2012) Influence of chromoplast morphology on carotenoid bioaccessibility of carrot, mango, papaya, and tomato. Food Chem 135(4):2736–2742. https://doi.org/10.1016/j.foodchem.2012.07.035

Shilpa S, Shwetha HJ, Raju M, Lakshminarayana R (2020) Factors affecting bioaccessibility and bio-efficacy of carotenoids. In: Carotenoids: properties, processing and applications. Elsevier, pp 41–73. https://doi.org/10.1016/B978-0-12-817067-0.00002-6

Silva Meneguelli T, Duarte Villas Mishima M, Hermsdorff HHM, Martino HSD, Bressan J, Tako E (2023) Effect of carotenoids on gut health and inflammatory status: a systematic review of in vivo animal studies. Crit Rev Food Sci Nutr:1–16. https://doi.org/10.1080/10408398.2023.2234025

Somaratne G, Ye A, Nau F, Ferrua MJ, Dupont D, Paul Singh R, Singh J (2020) Role of biochemical and mechanical disintegration on β-carotene release from steamed and fried sweet potatoes during in vitro gastric digestion. Food Res Int 136(109481):1–9. https://doi.org/10.1016/j.foodres.2020.109481

Stevens Y, Pinheiro I, Salden B, Duysburgh C, Bolca S, Degroote J, Majdeddin M, Van Noten N, Gleize B, Caris-Veyrat C, Michiels J, Jonkers D, Troost F, Possemiers S, Masclee A (2021) Effect of a carotenoid-producing Bacillus strain on intestinal barrier integrity and systemic delivery of carotenoids: a randomised trial in animals and humans. J Funct Foods 80:104445. https://doi.org/10.1016/j.jff.2021.104445

Sukhotnik I, Gork S, Coran A, Chen M, Harmon C (2000) Effect of high fat diet on fatty acid translocase expression and fatty acid transport in isolated. J Am Coll Surg 191(4):S41–S42

Sun T, Li L (2020) Toward the 'golden' era: the status in uncovering the regulatory control of carotenoid accumulation in plants. Plant Sci 290(110331):1–11. https://doi.org/10.1016/j.plantsci.2019.110331

Tyssandier V, Lyan B, Borel P (2001) Main factors governing the transfer of carotenoids from emulsion lipid droplets to micelles. Biochim Biophys Acta 1533:285–292

Valiei A, Aminian-Dehkordi J, Mofrad MRK (2023) Gut-on-a-chip models for dissecting the gut microbiology and physiology. APL Bioeng 7(1):011502. American Institute of Physics Inc. https://doi.org/10.1063/5.0126541

Varghese R, Buragohain T, Banerjee I, Mukherjee R, Penshanwar SN, Agasti S, Ramamoorthy S (2023) The apocarotenoid production in microbial biofactories: an overview. J Biotechnol 374:5–16. https://doi.org/10.1016/j.jbiotec.2023.07.009

von Lintig J, Moon J, Lee J, Ramkumar S (2020) Carotenoid metabolism at the intestinal barrier. Biochim Biophys Acta Mol Cell Biol Lipids 1865(11):158580. https://doi.org/10.1016/j.bbalip.2019.158580

Wen X, Hempel J, Schweiggert RM, WangY NY, Carle R (2018) Screening of critical factors influencing the efficient hydrolysis of zeaxanthin dipalmitate in an adapted in vitro- digestion model. Food Chem 257:36–43. https://doi.org/10.1016/j.foodchem.2018.02.116

Xavier AAO, Mariutti LR (2021) Static and semi-dynamic in vitro digestion methods: state of the art and recent achievements towards standardization. Curr Opin Food Sci 41:260–273. https://doi.org/10.1016/j.cofs.2021.08.002

Xavier AAO, Mercadante AZ (2019) The bioaccessibility of carotenoids impacts the design of functional foods. Curr Opin Food Sci 26:1–8. https://doi.org/10.1016/j.cofs.2019.02.015

Xavier AAO, Mercadante AZ (2022) A guide for the evaluation of in vitro bioaccessibility of carotenoids. In: Methods in enzymology, vol 674. Academic Press Inc., pp 297–327. https://doi.org/10.1016/bs.mie.2022.06.002

Xavier AAO, Mercadante AZ, Garrido-Fernández J, Pérez-Gálvez A (2014) Fat content affects bioaccessibility and efficiency of enzymatic hydrolysis of lutein esters added to milk and yogurt. Food Res Int 65(PB):171–176. https://doi.org/10.1016/j.foodres.2014.06.016

Xavier AAO, Carvajal-Lérida I, Garrido-Fernández J, Pérez-Gálvez A (2018) In vitro bioaccessibility of lutein from cupcakes fortified with a water-soluble lutein esters formulation. J Food Compos Anal 68:60–64. https://doi.org/10.1016/j.jfca.2017.01.015

Yabuzaki J (2017) Carotenoids database: structures, chemical fingerprints and distribution among organisms. Database 2017(1). https://doi.org/10.1093/database/bax004

Yao Y, Tan P, Kim JE (2022) Effects of dietary fats on the bioaccessibility and bioavailability of carotenoids: a systematic review and meta-analysis of in vitro studies and randomized controlled trials. Nutr Rev 80(4):741–761. https://doi.org/10.1093/nutrit/nuab098

Zhou H, Tan Y, Mcclements DJ (2023) Applications of the INFOGEST in vitro digestion model to foods: a review. Annu Rev Food Sci Technol 14:135–156. https://doi.org/10.1146/annurev-food-060721

Chapter 8
Carotenoids from Marine Sources: General Aspects, Biological Properties and Industrial Applications

Pooja Bargali, Ravendra Kumar, Ayush Devrani, Himani Karakoti, Sonu Kumar Mahawer, Om Prakash, and Avdhesh Kumar

1 Introduction

In the different classes of naturally occurring compounds, terpene is one of the large classes of chemical compounds present in stems, leaves, flowers, roots, and other part of the plants. They are made up of isoprene units having a general formula $(C_5H_8)_n$, where (n) indicates the number of isoprene units in the molecule. Based on the value of *n* these compounds are categorized as monoterpenes, sesquiterpenes, diterpenes, triterpenes, tetraterpene, and polyterpenes (Mabou and Yossa 2021; Wagner and Elmadfa 2003). Tetraterpens consists of eight isoprene units with molecular formula $C_{40}H_{64}$. Among these natural fat-soluble pigments are called carotenoids.

Carotenoids are present either with oxygenated parts called xanthophylls, or unsaturated hydrocarbon chain named carotenes. Their eight isoprene units are connected in a head-to-tail fashion in lycopene to provide it with a conjugated system which leads to the chromophoric property of the molecule. Sometimes these structures get combined to form a bicyclic hydrocarbon (i.e., β-carotene) (Huang et al. 2012). Carotenoids, belonging to the tetraterpene family are synthesized by higher plants, algae, and photosynthetic bacteria and are not found in animals but can be

P. Bargali · R. Kumar (✉) · A. Devrani · H. Karakoti · O. Prakash
Department of Chemistry, College of Basic Sciences and Humanities, G.B. Pant University of Agriculture and Technology, Pantnagar, Uttarakhand, India

S. K. Mahawer
Department of Chemistry, College of Basic Sciences and Humanities, G.B. Pant University of Agriculture and Technology, Pantnagar, Uttarakhand, India

ICAR-Indian Grassland and Fodder Research Institute, Jhansi, Uttar Pradesh, India

A. Kumar
Department of Aquaculture, College of Fisheries, G.B. Pant University of Agriculture and Technology, Pantnagar, Uttarakhand, India

R. Campos Chisté et al. (eds.), *Carotenoids*,
https://doi.org/10.1007/978-3-031-75322-0_8

modified by them through metabolic reactions (Miyashita and Hosokawa 2014; Furr and Clark 2003).

Carotenoids show several significant beneficial effects on human health, such as anti-aging, anti-atherosclerosis, anti-tumor growth suppression, antioxidant, anti-inflammatory, and apoptotic induction (Mabou and Yossa 2021). Specifically, marine carotenoids, such as fucoxanthin, and astaxanthin can prevent cardiovascular complaints and cure illnesses that are very sensitive to light, they have also been shown to enhance immunological response and have powerful antioxidant effects on human health (Mathews 1964; Wollgast and Anklam 2000).

Except for autotrophic marine organisms, non-photosynthetic marine entities cannot synthesize carotenoids from de novo synthesis. Consequently, marine animals can acquire carotenoids directly from their diet or undergo partial modifications through metabolic processes. Common metabolic transformations observed in animals include translation, oxidative cleavage of double bonds, oxidation, reduction, and cleavage of epoxy bonds (Mata-Gómez et al. 2014; Corinaldesi et al. 2017). Notably, among heterotrophic organisms, bacteria, archaea, marine fungi (especially pigmented yeasts), and thraustochytrids (generally categorized as fungi-like protists) stand out as significant sources of carotenoids (Aasen et al. 2016). Marine carotenoids provide vibrant and stable colors to food products, replacing synthetic alternatives with natural and potentially health-promoting options. Astaxanthin lends a reddish hue to salmon and shrimp, while fucoxanthin gives brown seaweed its characteristic color. Fortifying food products with marine carotenoids like astaxanthin can enrich their nutritional profile and offer additional health benefits to consumers.

2 Source of Marine Carotenoids

Marine carotenoids, pigments found in various ocean dwellers like algae, fungi, and bacteria, offer a vibrant palette of colors and a treasure trove of potential applications. Generally, all the colored fruits and vegetables we consume are good sources of carotenoids (Galasso et al. 2017). Marine organisms producing carotenoids include algae, animals, and microorganisms such as fungi, microalgae, cyanobacteria, archaea, heterotrophic bacteria, etc. Among them, microalgae are the primary source of carotenoids used in industry applications while others are suitable for their widespread usage in pharmaceutical and cosmetic sectors (Corinaldesi et al. 2017).

2.1 *Marine Algae*

These photosynthetic microorganism uses carbon dioxide to produce bioactive compounds such as carbohydrates, proteins, vitamins, lipids, polyunsaturated fatty acids, and pigments (Xu et al. 2009; Walker et al. 2005). In marine ecosystems, microalgae are the major natural source of carotenoids which constitute about 8–14% of their dry weight. They used it as a necessary pigment for their survival, serving as structural and functional parts of the photosynthetic system within the cell as primary carotenoids or secondary carotenoids on exposure to particular environment stress.

Carotenoids produced by microalgae are xanthophylls such as violaxanthin, fucoxanthin, zeaxanthin, diadinoxanthin, lutein, dinoxanthin, ketocarotenoids, canthaxanthin, and asthaxanthin, and carotenoids such as α- and β-carotene (Raposo et al. 2015). Lutein, fucoxanthin, β-carotene, and siphonaxanthin are the significant carotenoids found in seaweeds. They are typically present in the chloroplast, gathered in vesicles, bound to membranes, and other macromolecules in the intracellular space. Fucoxanthin is the carotenoid present in several brown macroalgae, which are found worldwide in cold and temperate habitats (Galasso et al. 2017). *Haematococcus pluvialis* reigns supreme as a source of this potent antioxidant carotenoid astaxanthin. Other microalgae like *Chlorella minutissima* and *Scenedesmus obliquus* also contribute to the production of astaxanthins. *Dunaliella salina* shines as a rich source of β-carotene, and the precursor of vitamin A. Brown algae with significant quantities of fucoxanthin possesses anti-inflammatory and anti-diabetic properties (Peng et al. 2011). Some examples of marine carotenoids obtained from microalgae are shown in Fig. 8.1.

2.2 *Prokaryotes, Fungi and Cyanobacteria*

Bacteria from the genera *Agrobacterium* and *Paracoccus*, which are often isolated from coastal environments, are among the marine heterotrophic microorganism that mainly produces astaxanthin (Ambati et al. 2014; Vílchez et al. 2011). New strains of marine bacteria from the *Flavobacteriaceae* family have produced two uncommon carotenoids, saproxanthin, and myxol, which have significant antioxidant effects (Shindo et al. 2007). The predominant carotenoids produced by marine cyanobacteria include β-carotene, and several forms of xanthophylls, like synechoxanthin, canthaxanthin, caloxanthin, echinenone, myxoxanthophyll, nostoxanthin, and zeaxanthin (Zakar et al. 2016). There are several yeast species synthesizing astaxanthin belonging to particular genera such as *Xanthophyllomyces*, *Phaffia,* and *Rhodotorula* (Ambati et al. 2014).

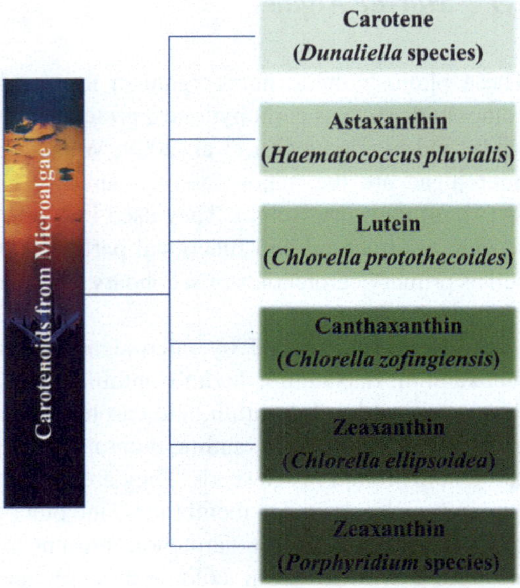

Fig. 8.1 Marine carotenoids found in microalgae

2.3 *Marine Animals*

A broad Carotenoids present in marine animals are metabolites of β-carotene, fucoxanthin, peridinin, alloxanthin, diatoxanthin, astaxanthin, etc. These animals contain carotenoids, which support both the metabolic processes and the food chain (Maoka et al. 2011). Interestingly, among the reported carotenoids allenic carotenoids, except for neoxanthin and its derivatives, all acetylenic carotenoids are gained from marine algae and animals (Britton et al. 2004). Through metabolic pathways such as oxidation, reduction, translation of double bonds, oxidative cleavage of double bonds, and cleavage of epoxy bonds animals primarily convert carotenoids (Maoka et al. 2011).

2.4 *Carotenoids in Porifera (Marine Sponges)*

Sponges possess bright colors due to the presence of carotenoids (Misawa 2021). Isorenieratene, renieratene, and renierapurpurin are examples of aryl carotenoids that are distinctive to sponges (Maoka et al. 2011).

2.5 *Carotenoids in Coelenterata (Sea Anemones)*

There were some jellyfish that had astaxanthin, which comes from dietary zooplankton. Certain corals were discovered to contain peridinin, pyrrhoxanthin, and diadinoxanthin originating from symbiotic dinoflagellates (Osinga et al. 2011; Matsuno 2001; Maoka et al. 2011). The sea anemones *Actinia equina* and *Tealia felina* have been shown to contain the rare carotenoids 2-nor-astaxanthin and actinioerythrin which are the metabolites of astaxanthin (Misawa 2021).

2.6 *Carotenoids in Mollusca (Mollusks)*

Chitons are a good source of carotenoids, specifically lutein, zeaxanthin, fucoxanthin, and their metabolites (Tsushima et al. 1989). Brown and red algae are the food source for abalone, *Haliotis discus*, and turban shell *Turbo cornutus,* and these shells contain the carotenoids lutein, fucoxanthin, zeaxanthin, and β-carotene. *Charonia sauliae*, a triton consumes starfish as food therfore the main carotenoids in triton were identified as astaxanthin, 7,8-didehydroastaxanthin, and 7,8,7′,8′-tetradehydroastaxanthin, which are distinctive carotenoids found in starfish. It has been discovered that astaxanthin, which comes from eating microcrustaceans, is a significant carotenoid in the whelk *Buccinum bayani*. On the other hand, *Drupella fragum* feeds on corals. Therefore, the two main carotenoids found in this sea snail are peridinin and diadinoxanthin. *Aplysia kurodai*, a sea hares reported with apocartenoids, a series of 8′-apocarotenal and 8′-apocarotenols derived from β-carotene, zeaxanthin, and lutein. Major carotenoids in octopus and cuttlefish species have been identified as astaxanthin and its esters (Galasso et al. 2017).

2.7 *Carotenoids in Arthopoda (Crustaceans)*

Astaxanthin is the main carotenoid found in crustaceans. Astaxanthin is found in crustaceans as carotenoproteins (i.e. crustacyanin) that have purple, blue, and yellow colour pigments (Matsuno 2001; Maoka et al. 2011). Adonirubin, echinenone, 3-hydroxyechinenone, canthaxanthin, and β-carotene which is consumed via dietary algae can be converted by a variety of crustaceans into astaxanthin. Salmon, shrimp, and lobster accumulate astaxanthin from their diet of microalgae, contributing to their coloration and potential health benefits.

2.8 Carotenoids in Echinodermata

Carotenoids such as asthaxanthin, fucoxanthin, β-carotene, echinenone, canthaxanthin are present in echinoderms like sea urchin, holoturians, and starfish (Matsuno 1989). The functions of carotenoids specifically in sea urchins are the production of eggs, biological, and developmental functions. Gonads of sea urchins contain a major carotenoid, β-echinenone (Tsushima 2007).

2.9 Carotenoids in Protochordata (Tunicates)

Filter feeders, and tunicates mostly include halocynthiaxanthin, fucoxanthin, mytiloxanthin, mytiloxanthinone, and alloxanthin, which may be metabolic metabolites of carotenoids presumed to be present in dietary phytoplankton. Tunicates contain peridinin and its metabolites as well. *Amaroucium pliciferum* is an ascidian, and from it two distinct marine carotenoids were isolated: amarouciaxanthin A and B (Matsuno et al. 1985).

2.10 Carotenoids in Pisces (Fish)

Several fish species store carotenoids in their tegument and gonads in an esterified state. All 40% of fish species, salmonid and perciformes cannot synthesize astaxanthin, they get it from consuming crustacean zooplankton. In the muscles of fish, astaxanthin is deposited. Astaxanthin (pink-red), canthaxanthin (orange-red), lutein (greenish-yellow), zeaxanthin (yellow-orange), β-carotene (orange), tunaxanthin, taraxanthin, and doradexanthins (yellow) are the carotenoids that have been isolated from fish shells (Maoka et al. 2011). Among these compounds astaxanthin and canthaxanthin are widely used to enhance the color of fish (Manikandan et al. 2020).

2.11 Carotenoids in Mammalia (Mammals)

Lutein, oxygenated carotenoid is found to be deposited in rodents, birds, frogs, and human retinas. It affects certain common ocular disorders brought on by long-time exposure to light and oxidative stress, such as cataracts and macular degeneration. Because of the antioxidant properties of lutein, it can decrease the amount of oxidative stress-induced damage by decreasing the amount of oxygen that can pass through the ocular membranes (Koutsos et al. 2013). Not all mammals store carotenoids and those who do, in them these pigments undergo metabolic changes linked to parturition (Goodwin 1950).

3 Chemical Composition and Biosynthesis of the Main Carotenoids in Marine Sources

Marine carotenoids belong to the broader class of isoprenoids, characterized by repeating five-carbon units, consisting of 40 carbon atoms with six-membered rings at each end of the polyprenoid chain in which lycopene is one of the exceptions. β-Carotene ($C_{40}H_{46}$, Molecular Weight = 536 g/mol), is one of the carotenoids with two β-ionone rings (end groups) joined by a conjugated polyene chain (Fig. 8.2). Due to the presence a number of conjugated carbon-carbon double bonds, which facilitates carotenoids to have its molar absorptivity:13200 M^{-1} cm^{-1}(i.e., high light absorbing range) and wavelength of maximum light absorption: 450–500 nm (i.e., intense red-yellow colors) (Furr and Clark 2003). The solvent's capacity to polarise substantially influences the energy of the electronic transition that results in a carotenoid's major absorption band. When the solvent's refractive index rises, the absorption moves to lower energy, or the spectrum shifts to longer wavelengths (Britton et al. 2004).

Carotenoids, vital pigments found throughout the marine world, play a key role in both the organisms that produce them and the ecosystems they inhabit. Marine carotenoid biosynthesis, similar to their terrestrial counterparts, begins with the formation of isoprenoid precursors, specifically isopentenyl pyrophosphate (IPP) and its isomer dimethylallyl pyrophosphate (DMAPP). These C5 building blocks arise through two independent pathways. The Mevalonate (MVA) pathway (common in eukaryotes and some bacteria) and the 2-C-methyl-D-erythritol 4-phosphate (MEP) pathway, also known as the non-MVA pathway, is dominant in archaea and some bacteria. In the formation of IPP/DMAPP, a series of enzymatic reactions are responsible for conversion into carotenoids. Condensation of three IPP units yields farnesyl pyrophosphate (C_{15}). Subsequently, one additional IPP combines with farnesyl pyrophosphate through the action of geranylgeranyl pyrophosphate synthase (CrtE) to form geranylgeranyl Pyrophosphate (GGPP) (C_{20}). Condensation of two GGPP molecules by Phytoene Synthase (CrtB) leads to the formation of the first carotene, phytoene (C_{40}). Phytoene undergoes a series of enzymatic modifications involving cyclization and desaturation reactions, mediated by various enzymes like lycopene cyclase (CrtY) and carotene desaturases (CrtI), resulting in the

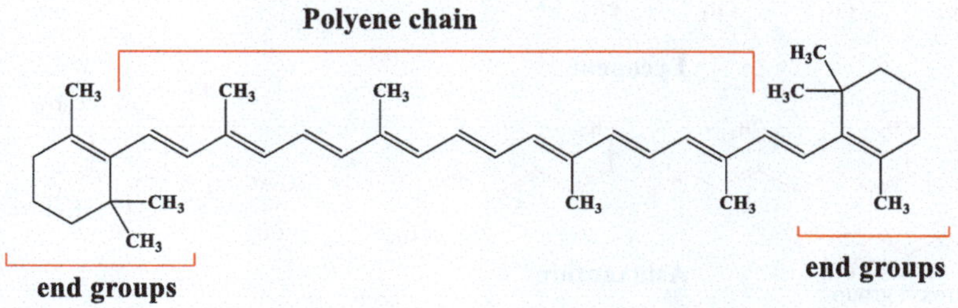

Fig. 8.2 Chemical structure of β-carotene

formation of various carotenoids like lycopene and β-carotene. Further modifications involving hydroxylation and other reactions catalyzed by enzymes like hydroxylases and epoxidases transform the basic carotene structure into xanthophylls (Britton et al. 2004; Takaichi 2011).

As previously stated, xanthophyll and carotenes are the two classes of carotenoids. Some examples of carotenes are β-carotene, α-carotenes, γ-carotene and lycopene. Lutein, astaxanthin, zeaxanthin, fucoxanthin, β-cryptoxanthin and peridinin are xanthophylls containing oxygen as a carotenoid containing functional groups such as hydroxyl, carbonyl, aldehyde, carboxylic, epoxide, and furanoxide (Furr and Clark 2003; Britton et al. 2004; Maoka 2009). Figure 8.3 illustrates the structural variation of different types of carotenoids which are obtained from different marine sources.

β-Carotene highly double bond conjugated red-colored carotenoid is richly present in green leafy plants such as parsley, spinach, and broccoli, as well as certain

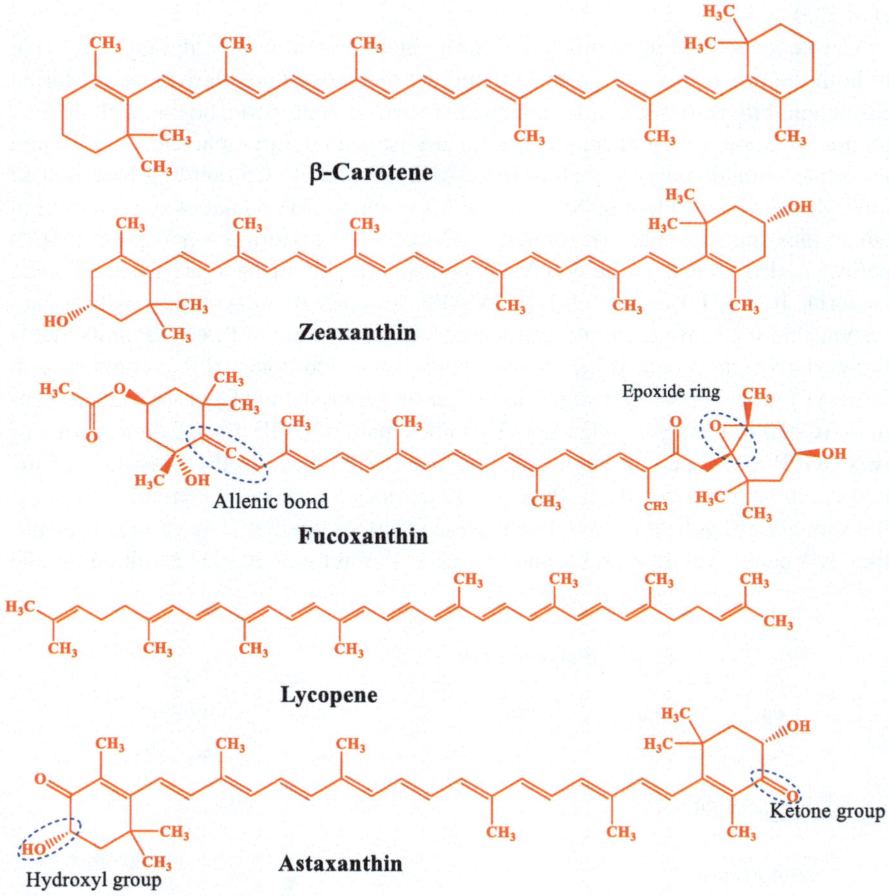

Fig. 8.3 Chemical structures of marine carotenoids

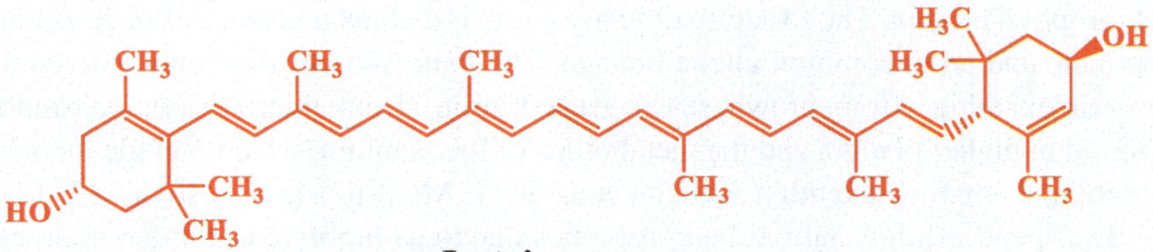

Lutein

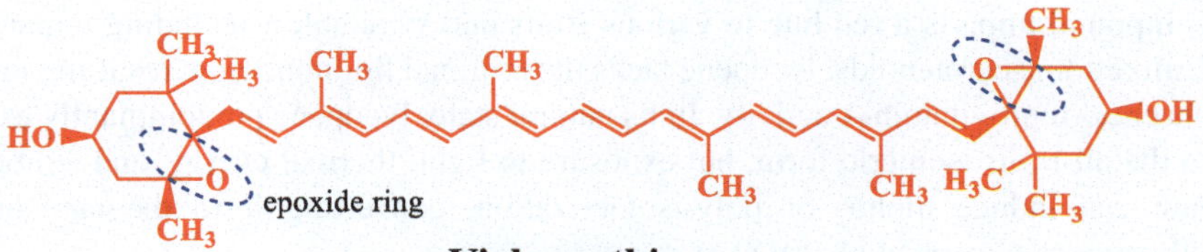

Violaxanthin

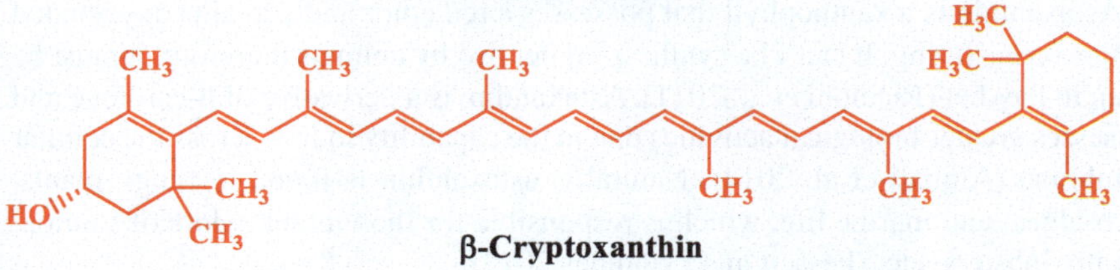

β-Cryptoxanthin

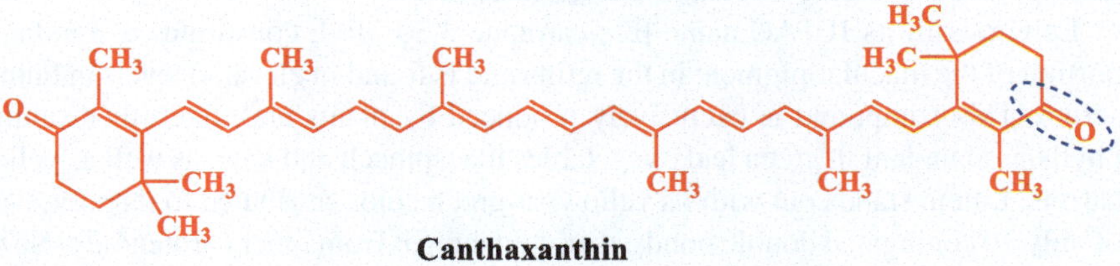

Canthaxanthin

Fig. 8.3 (continued)

fruits like mandarin and peach, and various vegetables including carrot and pumpkin (Del Campo et al. 2007). Its distinctive feature lies in the beta rings located at both ends of the molecule.

Zeaxanthin, a prevalent carotenoid alcohol in nature, is produced by plants and specific microorganisms. It constitutes one of the primary components of the macular pigment in the retina. Zeaxanthin shares isomerism with lutein, differing by the location of a single double bond. Unlike lutein, Zeaxanthin does not possess vitamin A activity. Both Zeaxanthin and lutein, play a crucial role in preventing age-related macular degeneration (AMD), a prominent cause of vision impairment and blindness (Tanaka et al. 2012).

Most naturally occurring marine species, such as seaweeds and microalgae, contain fucoxanthin, an orange pigment. It is a natural pigment that is dominant in brown microalgae (Phaeophyceae) and diatoms (Bacillariophyta), which makes up more than 10% of the estimated total natural synthesis of carotenoids, making it the most abundant of all carotenoids. It functions as a photoprotection for effective light utilization, photosynthesis upregulation, and light harvesting by forming a complex with chlorophyll protein. The structure of fucoxanthin is distinct and consists of 5,6-monoepoxide and an uncommon allene linkage. The main sources of allenic carotenoids are fucoxanthin (from brown seaweed), perinijin (from microalgae), neoxanthin (found in higher plants), and the metabolites of fucoxanthin, which include fucoxanthinol and amarouciaxanthin A (Peng et al. 2011; Miyashita and Hosokawa 2014).

Lycopene exhibits antioxidant properties due to its highly unsaturated hydrocarbon chain, comprising eleven conjugated and two unconjugated double bonds. This compound imparts a red hue to various fruits and vegetables, including tomatoes. Unlike some carotenoids, lycopene lacks the terminal β-ionone ring, resulting in the absence of provitamin A activity. In its natural state, lycopene predominantly exists in the all-*trans* isomeric form, but exposure to light, thermal energy, and temperature can induce mono- or poly-isomerization, converting it to the *cis*-isomer (Torregrosa-Crespo et al. 2018).

Astaxanthin is a xanthophyll that possesses a red color and contains oxygenated groups on each ring. It can't be synthesized denovo by animals therefore, it must be taken in the diet (Riccioni et al. 2011). Astaxanthin is a derivative of β-carotene and possesses greater biological activities due to the capability to link across the cellular membrane (Ambati et al. 2014). Naturally, astaxanthin is found in fungi, plants, microalgae, and marine life, which is responsible for the vibrant colors of salmon, shrimp, lobsters, etc. (Fassett and Coombes 2012).

Lutein, with its IUPAC name β, ε-carotene-3,30 -diol, constitutes a significant portion of the macular pigment in the retina and is found in elevated levels in human serum. This compound is exclusively produced by plants and, like other xanthophylls, is abundant in green leafy vegetables like spinach and kale, as well as yellow carrots. Lutein stands out with its yellowish-green color, attributed to its possession of only 10 conjugated double bonds, distinguishing it from other carotenoids (Naziri et al. 2014).

Violaxanthin, a natural xanthophyll pigment with an orange hue, is found in various brown algae and plants like pansies. It is biosynthesized from zeaxanthin through the process of epoxidation. The interconversions between violaxanthin and zeaxanthin are primarily facilitated by the violaxanthin de-epoxidase enzyme, as observed in studies by Yamamoto et al. (1972). The thermal stability of violaxanthin is attributed to the disulfide bridges inherent in its structure, as highlighted by Hallin et al. (2015).

β-Cryptoxanthin is recognized as a pro-vitamin A, serving as a valuable source of vitamin A. When derived from common food sources, it demonstrates notable bioavailability. Some foods rich in β-cryptoxanthin may be deemed comparable to those abundant in β-carotene as effective sources of retinol, as suggested studies by (Burri 2015).

Canthaxanthin, with the IUPAC name β, β-carotene-4,40 -dione, is a diketo carotenoid pigment characterized by its orange-red color. Initially discovered in edible mushrooms, it has since been identified in bacteria, crustaceans, and various fish species. In certain green and blue-green algae, canthaxanthin is produced as a secondary carotenoid towards the conclusion of the growth phase (Torregrosa-Crespo et al. 2018).

4 Biological Properties of Carotenoids from Marine Sources

Marine carotenoids were reported to possess relevant biological properties (Table 8.1), such as antioxidant and anti-inflammatory, anti-proliferative, antimicrobial and anticancer properties. Figure 8.4 illustrates the main investigated biological properties of marine carotenoids. They can serve as effective skin photo-protection against the harmful effects of solar UV radiation or be utilized in nutraceutical, and cosmeceutical formulations to prevent diseases related to oxidative stress (Nichols and Katiyar 2010; Berthon et al. 2017; Soliev et al. 2011; Aman Mohammadi et al. 2022).

Table 8.1 Some biological properties investigated for carotenoids from marine sources

Carotenoid	MW	Marine source	Biological property investigated	References
β-Carotene	$C_{40}H_{56}$	Microalgae: *Dunaliella salina,* fungi: *Thraustochytrium* strains, shellfish, cyanobacteria, sea urchin, starfish	Anticancer, antioxidant, anti-inflammatory, cure of erythema	Allard et al. (1994), Stahl and Sies (2012)
Astaxanthin	$C_{40}H_{52}O_4$	Chlorophyta: *Haematococcus pluvialis,* bacteria: *Agrobacterium,* yeast: *Rhodotorula,* sea snails: *Charonia sauliae,* sea urchin: *Arbacia lixula,* also found in: starfish, crabs, shrimp, lobsters, and shellfish	Antioxidant, antiproliferative, anti-obesity, cardiovascular protective, antihypertensive and neuroprotective effects	Al-Amin et al. (2015), Sharoni et al. (2022), Jyonouchi et al. (2000), Hussein et al. (2005)
Siphonaxanthin	$C_{40}H_{56}O_4$	Green algae	Antiproliferative effect	Ganesan et al. (2011)
Canthaxanthin	$C_{40}H_{52}O_2$	Found in sea urchin, starfish, crustaceans and algae; marine bacteria (*Micrococcus luteus*)	Immune system stimulation, antiproliferative, anticacer	Huang et al. (1992), Palozza et al. (1998), Shklar and Schwartz (1988)
Fucoxanthin	$C_{42}H_{58}O_6$	Seaweed (*Undaria pinnatifida*), *Laminalia japonica, Isochrysis aff. Galbana* and brown algae	Antiproliferative, antioxidant UV protection, anti-obesity	Hosokawa et al. (1999), Heo and Jeon (2009), Peng et al. (2011)
Zeaxanthin	$C_{40}H_{56}O_2$	Microalgae: *Dunaliella salina,* seagrass: *Posidonia oceanica, Cymodocea nodosa,* etc.	ADM (Acellular dermal matrix) prevention, antioxidant	Gao et al. (2011), Giblin (2000)

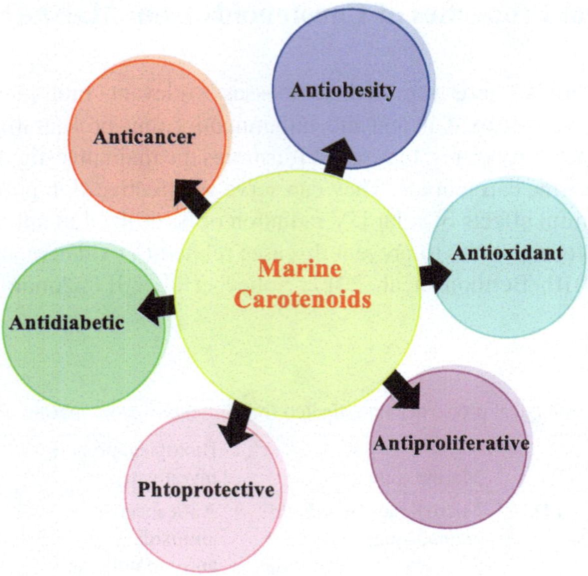

Fig. 8.4 Investigated biological properties of marine carotenoids

4.1 Antioxidant Properties

Natural carotenoids play various functions at different levels, including neutralizing singlet molecular oxygen, converting hydroperoxides to more stable compounds, preventing the generation of free radicals, and chelating pro-oxidant metals to less harmful forms (Pacher et al. 2007).

Astaxanthin, a red pigment from marine organisms, exhibits potent antioxidant activity, protecting the lipid bilayer from peroxidation due to its unique structure with non-polar carbon–carbon bonds and polar ionic rings. Astaxanthin surpasses other carotenoids in antioxidant potency by a factor of 10 (Gammone et al. 2015; Guerin et al. 2003). Beyond its neurological effects, astaxanthin plays a crucial role in safeguarding testicles from oxidative damage and potential testosterone production reduction by exerting strong antioxidant activity on Leydig cells (Wang et al. 2015). Astaxanthin's structure allows it to scavenge radicals both inside and outside cell membranes (Ambati et al. 2014).

In recent years, marine bacteria from Flavobacteriaceae family have yielded two rare marine carotenoids myxol and saproxanthin, which inhibit lipid peroxidation in the brain (Gammone et al. 2015; Shindo et al. 2004). Previous studies have suggested that fucoxanthin, obtained from *Sargassum siliquastrum*, possesses strong antioxidant properties and can protect against oxidative stress *in vitro* (Heo et al. 2008). To explore the *in vivo* characteristics, Sangeetha et al. (2009) conducted a study comparing fucoxanthin with β-carotene by administration of both compounds to rats with retinol deficiency.

Mytiloxanthin, fucoxanthin's metabolite found in shellfish, tunicates and exhibits high singlet oxygen quenching activity (62%). Maoka et al. (2011) reported that mytiloxanthin has a comparatively more potent inhibitory effect on lipid peroxidation at the same concentration of fucoxanthin, astaxanthin, and β-carotene. Zeaxanthin protects against oxidative damage and regulates antioxidant processes, including reducing oxidized glutathione and elevating reduced glutathione levels (Giblin 2000).

4.2 Role of Carotenoids in Provitamin A Activity and Eye Health

One crucial role of carotenoids in the human body is their capacity to convert provitamin A carotenoids, such as β-carotene, into retinol (vitamin A). This function is present in approximately 10% of carotenoids found in nature, as noted by García-González et al. (2005) Vitamin A holds significant importance for child health and survival, with its deficiency linked to vision disturbances and various respiratory and oral cavity pathologies, as highlighted by the World Health Organization in 1998. Unlike plants, humans cannot synthesize carotenoids de novo but can convert them into vitamin A. These precursor compounds primarily come from the diet, with fruits, vegetables, and microalgae being major sources. β-Carotene, in particular, contributes to 100% of provitamin A activity (Graham and Rosser 2000).

Research suggests that lutein and zeaxanthin are vital for maintaining normal visual function in the macula of the human eye, where they absorb blue light and protect against harmful photooxidative effects. Their antioxidant properties safeguard the eye from detrimental photochemical reactions as reported by Khachik et al. (2002), and Landrum and Bone (2001). Hammond et al. (1998) reported that in individuals aged above 64, in the retina the concentration of zeaxanthin and lutein directly influences visual sensitivity and low levels have been linked to a higher prevalence of cataracts (Olmedilla et al. 2001). Lutein and zeaxanthin absorb excess blue light before it comes in contact with photoreceptors, thus preventing damage to the eye macula (Greenstein et al. 2007).

4.3 Carotenoids in Cardiovascular Diseases

Specifically, the oxidative stress triggered by reactive oxygen species (ROS) can lead to the oxidation of low-density lipoproteins (LDL), and carotenoids exhibit a significant attribute in protecting LDL from oxidation (Dugas et al. 1999; Bub et al. 2000). Several epidemiological studies suggest that diets abundant in carotenoids could offer protection against specific cardiovascular ailments by mitigating the effects of oxidizing agents and oxidative stress in the advancement and clinical presentation of coronary heart disease, as indicated by Kohlmeier and Hastings

(1995). On the other hand, decreasing levels of lutein in plasma are associated with increased susceptibility to myocardial infarction (Street et al. 1994), while a lutein-rich diet has been associated with reduced risk of stroke (Ascherio et al. 1999).

4.4 Role of Carotenoids in Cancer Prevention

Lycopene has been extensively studied among the carotenoids, because of its strong anticancer properties (Giovannucci et al. 1995; Narisawa et al. 1996). Studies, such as those by Giovannucci et al. (1995), have shown an opposite relation between lycopene intake and prostate cancer. Astaxanthin has also been proposed as more effective against benign prostatic hyperplasia and prostate cancer by inhibiting the 5-α-reductase enzyme (Anderson 2001). Moreover, β-carotene, cantaxanthin, zeaxanthin, and astaxanthin have demonstrated encouraging outcomes in decreasing the size and quantity of liver neoplasms *in vivo*, as observed by Nishino et al. (2002).

While the anti-tumoral activities of carotenoids like zeaxanthin, criptoxanthin, and lutein have been minimally studied, the idea of using carotenoids as chemoprotective agents is not yet supported by clinical trials. Instead, it is suggested that the beneficial effects of carotenoids are derived from regular consumption in typical quantities, in line with dietary recommendations of five daily portions of fruits and vegetables to provide essential nutrients and phytochemical compounds, including carotenoids (Mayne 1996).

4.5 Antiobesity Effect of Fucoxanthin

For the first time, the antioxidant effect of fucoxanthin was studied using brown seaweed lipids from Wakame (*Undaria pinnatifida*) which contains 10% fucoxanthin. It is seen that the body weight of obese diabetic KK-Ay (where KK-A^y represents a cross between diabetic KK and lethal yellow (Ay) mice) mice was reduced compared with the control although no change was there in food intake. This reduction continues with a drop in perirenal, mesentery retroperitoneal, and uterine white adipose tissue corrected for body weight. Supplementing with fucoxanthin dramatically reduced body weight and reduced the build-up of extra fat in the abdomen WAT of obese mice fed a high-fat (HF) diet. In the mouse model, dietary delivery of the HF diet led to hyperglycemia, hyperinsulinemia, and hyperleptinemia. Supplementing with fucoxanthin normalized these aberrations (Miyashita and Hosokawa 2014; Woo et al. 2009; Maeda et al. 2007). Attempts to screen for anti-obesity activity in other carotenoids have been guided by the theory of antiobesity effect of fucoxanthin. So, screening of natural compounds for antiobesity effects on preadipocyte differentiation and adipogenesis in adipose cells is evaluated. Only fucoxanthin, neoxanthin, and two fucoxanthin metabolites (fucoxanthinol and amarouciaxanthin A) demonstrated an encouraging suppressive effect on the differentiation of adipose cells when various carotenoids were screened for potential suppression effects on adipocyte differentiation (Okada et al. 2008; Yim et al. 2011).

5 Biological Function of Carotenoids in Marine Life

Carotenoids from foods are converted by marine organisms and stored in their organs. Metabolic transformations enhance carotenoids' antioxidative and photoprotective properties. For example, in crustaceans, β-carotene is converted to astaxanthin and is stored in integuments, carapaces, ovaries, and eggs. The pigment undergoes a biochemical alternation that turns it from yellow (β-carotene) to red (astaxanthin). In marine invertebrates, astaxanthin can occasionally take the form of a red, blue, or purple carotenoid protein complex. It functions as a photoreceptor, protects from harmful rays, and also helps camouflage the animals under light conditions. This conversion is robust to the antioxidant effect of carotenoids such as the protection against photo-oxidation, lipid peroxidation, and quenching of singlet oxygen (Britton et al. 2008).

In another example, carotenoids produced by microalgae are made available to sea angels through *Limacina helicina* in the food chain. Without undergoing any metabolic changes, *L. helicina* directly absorbs carotenoids from food algae, including diatoxanthin, and accumulates them. Conversely, ingested diatoxanthin from *L. helicina* is oxidatively metabolized to pectenolone by the sea angel. A carbonyl group is added to diatoxanthin at C-4′, causing the carotenoid to become red and exhibit stronger antioxidant and photoprotective properties. As a result, the sea angel's gonads store pectenolone for use as a photo protector and antioxidant (Maoka et al. 2014).

6 Food Chain and Metabolism of Carotenoids in Aquatic Animals

As of now, we know that all aquatic animals containing carotenoids come mainly from algae or bacteria. Herbivorous animals intake carotenoids from the dietary algae and metabolize them. As seen in Fig. 8.5, carotenoids from diet herbivorous animals are ingested and metabolized by carnivorous species.

Example of Transfer of Carotenoid from Zooxanthellae (Dinoflagellate algae) to Starfish Through Corals Filter-feeding coral, *Acoropora japonica* is paired with zooxanthellae in a symbiotic (or interdependent) relationship. It's seen that absorption and accumulation of carotenoids take place by symbiotic zooxanthellae without going through metabolic changes in it. Thereafter, carnivorous, crown-of-thorns starfish, *Acanthaster planci* mostly feeds on coral. The two main carotenoids, astaxantin and 7,8- didehydroastaxanthin have been identified in this starfish and found to be oxidative metabolites of diatoxanthin and β-carotene, respectively. As repre-

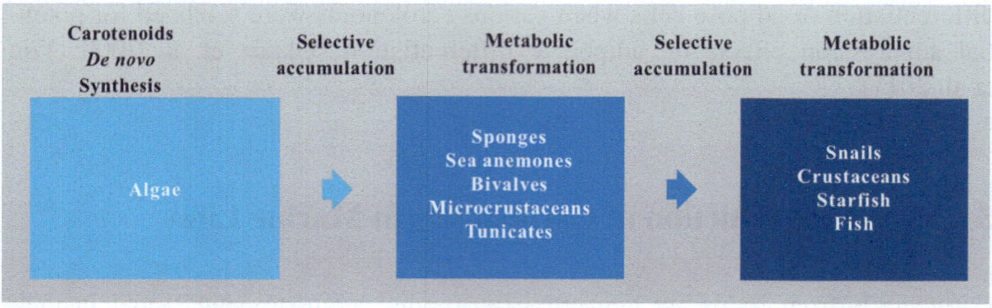

Fig. 8.5 Metabolism of carotenoids in marine organisms

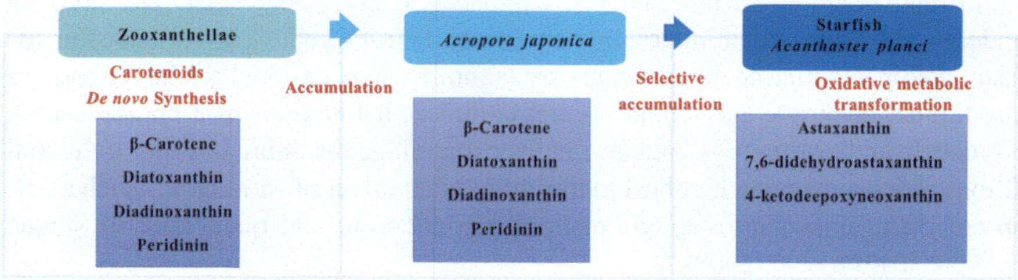

Fig. 8.6 Accumulation and metabolism of carotenoids

sented in Fig. 8.6, carotenoids derived from zooxanthellae are collected and oxidatively metabolized in the crown-of-thorns starfish through the coral.

Example of Transfer of Carotenoid from Diatoms to Sea Angel A Sea Angel, is a sea slug having a small, gelatinous, and transparent body. Because of the presence of carotenoids their gonads and viscera have a vivid orange-red colour. It is carnivorous and feeds on *Limacina helicina,* a small sea snail. *L. helicina* which is herbivorous further feeds on microalgae such as dinoflagellates and diatoms. And so, via *L. helicina* in the food chain, the sea angel get carotenoids produced by microalgae. Without going through any metabolic changes, *L. helicina* absorbs carotenoids from dietary algae. Figure 8.4 describes the oxidative metabolism of carotenoids from *L. helicina* (Misawa 2021).

7 Industrial Application of Marine Carotenoids

Nowadays, most carotenoids are chemically synthesized for industrial use and only a part is obtained directly by natural extraction from organisms. There is a huge demand for such natural compounds in the fields of cosmetics, nutraceuticals, and pharmaceuticals, and some examples of successful use of carotenoids in the global market can be mentioned.

Astaxanthin is produced by a green microalgae *Haematococcus pluvialis*, which is one of the main bioactive pigments (Misawa 2009). Patent applications related to astaxanthin are prevalent in the fields of food, feed, and nutraceuticals, serving as the primary driving force in the current market for this pigment (Torregrosa-Crespo et al. 2018). β-Carotene is mostly used by several industries as natural colorant in food and health care products, and as nutraceutical products in many countries (Guedes et al. 2011). Lutein is used as a food colorant, and neutraceutical (Galasso et al. 2017). Marine microalgae *Murielopsis* sp. (Del Campo et al. 2001), *Chlorella protothecoides* (Shi et al. 2000), *Scenedesmus almeriensis* is the source of lutein (Del Campo et al. 2007). Interestingly, the productivity of lutein in *C. protothecoides* is 100 times greater than the lutein from the marigold flower (Lin et al. 2015).

The health sector has become one of the fastest growing areas for the industrial application of carotenoids, but is still less in comparison to the industries interested in the coloring properties of these compounds. Many natural sources of carotenoids have colorant properties that are most used in the pharmaceutical and cosmetics sectors. Employment in these fields is quickly expanding because of their increased knowledge about the nutraceutical potential (Vílchez et al. 2011).

Fucoxanthin derived from seaweeds like *Laminar iajaponica* and other species, is primarily utilized as a supplement for diets that aim to reduce body weight. This may be because it can raise basal metabolism (Kanazawa et al. 2008). Astaxanthin products find widespread use in various commercial applications, taking the form of tablets, capsules, syrups, oils, soft gels, creams, biomass, and granulated powders.

Algal biotechnology has made its place in industry since the microorganism produces biomolecules that have several applications in the areas of food industry, pharmaceutical, health, agriculture, and as a source of production of biofuel (Borowitzka 2013). The presence of carotenoids in microalgae has significantly boosted their commercial value, leading to widespread applications in the food market. Various microalgae such as *Dunaliella*, *Chlorella*, *Arthrospira*, and *Spirulina* are now utilized as functional foods, available in the market in the form of pills, and tablets. These microalgae have been incorporated into the nutritional composition of products like drinks, sweets, etc. (Liang et al. 2004). Moreover, when carotenoids are present in the microalgal matrix they provide additional minerals inherent to the algal biomass in the formula which contributes positively to human health. Additionally, carotenoids find application as preservatives in cosmetics and solar protection products (Del Campo et al. 2001).

Regarding the environmental applications, microalgae have emerged as promising sources for biofuels, drawing attention due to their superior productivity in converting carbon dioxide into lipid-rich compounds compared to traditional oleaginous crops, and without the need for arable land competition. Recent progress in systems biology, genetic engineering, and utilization of biomass residues has introduced new possibilities, suggesting that biofuel production from microalgae could become economically viable within approximately 15 years. The incorporation of processes such as wastewater and flue gas treatments can enhance the cost-effectiveness and profitability of producing biodiesel and other bio-products from microalgae (Wijffels and Barbosa 2010; Mata et al. 2010).

References

Aasen IM, Ertesvåg H, Heggeset TMB, Liu B, Brautaset T, Vadstein O, Ellingsen TE (2016) Thraustochytrids as production organisms for docosahexaenoic acid (DHA), squalene, and carotenoids. Appl Microbiol Biotechnol 100:4309–4321. https://doi.org/10.1007/s00253-016-7498-4

Al-Amin MM, Akhter S, Hasan AT, Alam T, Nageeb Hasan SM, Saifullah ARM, Shohel M (2015) The antioxidant effect of astaxanthin is higher in young mice than aged: a regionspecific study on brain. Metab Brain Dis 30:1237–1246. https://doi.org/10.1007/s11011-015-9699-4

Allard JP, Royall D, Kurian R, Muggli R, Jeejeebhoy KN (1994) Effects of β-carotene supplementation on lipid peroxidation in humans. Am J Clin Nutr 59:884–890. https://doi.org/10.1093/ajcn/59.4.884

Aman Mohammadi M, Ahangari H, Mousazadeh S, Hosseini SM, Dufossé L (2022) Microbial pigments as an alternative to synthetic dyes and food additives: a brief review of recent studies. Bioprocess Biosyst Eng 45:1–12. https://doi.org/10.1007/s00449-021-02621-8

Ambati RR, Phang SM, Ravi S, Aswathanarayana RG (2014) Astaxanthin: sources, extraction, stability, biological activities and its commercial applications—a review. Mar Drugs 12:128–152. https://doi.org/10.3390/md12010128

Anderson M (2001) Inventor; Triarco Industries Inc, assignee. Method of inhibiting 5α-reductase with astaxanthin. United States patent US 6,277,417

Ascherio A, Rimm EB, Hernán MA, Giovannucci E, Kawachi I, Stampfer MJ, Willett WC (1999) Relation of consumption of vitamin E, vitamin C, and carotenoids to risk for stroke among men in the United States. Ann Intern Med 130:963–970. https://doi.org/10.7326/0003-4819-130-12-199906150-00003

Berthon JY, Nachat-Kappes R, Bey M, Cadoret JP, Renimel I, Filaire E (2017) Marine algae as attractive source to skin care. Free Radic Res 51:555–567. https://doi.org/10.1080/10715762.2017.1355550

Borowitzka MA (2013) High-value products from microalgae—their development and commercialisation. J Appl Phycol 25:743–746. https://doi.org/10.1007/s10811-013-9983-9

Britton G, Liaaen-Jensen S, Pfander H (2004) Introduction and guidelines on the use of the handbook. In: Carotenoids: handbook. Birkhäuser Basel, Basel, pp 1–33. https://doi.org/10.1007/978-3-0348-7836-4_1

Britton G, Liaaen-Jensen S, Pfander H (2008) Natural functions. In: Britton G, Liaaen-Jensen S, Pfander H (eds) Carotenoids – volume IV, 1st edn. Birkhäuser, Basel. https://doi.org/10.1007/978-3-7643-7499-0

Bub A, Watzl B, Abrahamse L, Delincee H, Adam S, Wever J, Muller H, Rechkemmer G (2000) Moderate intervention with carotenoid-rich vegetable products reduces lipid peroxidation in men. Nutr J 130:2200–2206. https://doi.org/10.1093/jn/130.9.2200

Burri BJ (2015) Beta-cryptoxanthin as a source of vitamin A. J Sci Food Agric 95:1786–1794. https://doi.org/10.1002/jsfa.6942

Corinaldesi C, Barone G, Marcellini F, Dell'Anno A, Danovaro R (2017) Marine microbial-derived molecules and their potential use in cosmeceutical and cosmetic products. Mar Drugs 15:118. https://doi.org/10.3390/md15040118

Del Campo JA, Rodríguez H, Moreno J, Vargas MÁ, Rivas J, Guerrero MG (2001) Lutein production by Muriellopsis sp. in an outdoor tubular photobioreactor. J Biotechnol 85:289–295. https://doi.org/10.1016/S0168-1656(00)00380-1

Del Campo JA, García-González M, Guerrero MG (2007) Outdoor cultivation of microalgae for carotenoid production: current state and perspectives. Appl Microbiol Biotechnol 74:1163–1174. https://doi.org/10.1007/s00253-007-0844-9

Dugas TR, Morel DW, Harrison EH (1999) Dietary supplementation with β-carotene, but not with lycopene, inhibits endothelial cell-mediated oxidation of low-density lipoprotein. Free Radic Biol Med 26:1238–1244. https://doi.org/10.1016/S0891-5849(98)00321-9

Fassett RG, Coombes JS (2012) Astaxanthin in cardiovascular health and disease. Molecules 17:2030–2048. https://doi.org/10.3390/molecules17022030

Furr HC, Clark RM (2003) Carotenoids physiology. In: Encyclopedia of food sciences and nutrition, pp 936–943. https://doi.org/10.1016/b0-12-227055-x/00177-2

Galasso C, Corinaldesi C, Sansone C (2017) Carotenoids from marine organisms: biological functions and industrial applications. Antioxidants 6:96. https://doi.org/10.3390/antiox6040096

Gammone MA, Riccioni G, D'Orazio N (2015) Marine carotenoids against oxidative stress: effects on human health. Mar Drugs 13:6226–6246. https://doi.org/10.3390/md13106226

Ganesan P, Noda K, Manabe Y, Ohkubo T, Tanaka Y, Maoka T, Sugawara T, Hirata T (2011) Siphonaxanthin, a marine algal carotenoids from green algae, effectively induces apoptosis in human leukemia (HL-60) cells. Biochim Biophys Acta 1810:497–503. https://doi.org/10.1016/j.bbagen.2011.02.008

Gao S, Qin T, Liu Z, Caceres MA, Ronchi CF, Chen CO, Yeum KJ, Taylor A, Blumberg JB, Liu Y, Shang F (2011) Lutein and zeaxanthin supplementation reduces H_2O_2-induced oxidative damage in human lens epithelial cells. Mol Vis 17:3180–3190

García-González M, Moreno J, Manzano JC, Florencio FJ, Guerrero MG (2005) Production of *Dunaliella salina* biomass rich in 9-*cis*-β-carotene and lutein in a closed tubular photobioreactor. J Biotechnol 115:81–90. https://doi.org/10.1016/j.jbiotec.2004.07.010

Giblin FJ (2000) Glutathione: A vital lens antioxidant. J Ocul Pharmacol Ther 16:121–135. https://doi.org/10.1089/jop.2000.16.121

Giovannucci E, Ascherio A, Rimm EB, Stampfer MJ, Colditz GA, Willett WC (1995) Intake of carotenoids and retino in relation to risk of prostate cancer. J Natl Cancer Inst 87:1767–1776. https://doi.org/10.1093/jnci/87.23.1767

Goodwin TW (1950) Carotenoids and reproduction. Biol Rev 25:391–413. https://doi.org/10.1111/j.1469-185X.1950.tb00765.x

Graham RD, Rosser JM (2000) Carotenoids in staple foods: their potential to improve human nutrition. Food Nutr Bull 21:404–409. https://doi.org/10.1177/156482650002100412

Greenstein VC, Chiosi F, Baker P, Seiple W, Holopigian K, Braunstein RE, Sparrow JR (2007) Scotopic sensitivity and color vision with a blue-light-absorbing intraocular lens. J Cataract Refract Surg 33:667–672. https://doi.org/10.1016/j.jcrs.2006.12.012

Guedes AC, Amaro HM, Malcata FX (2011) Microalgae as sources of high added-value compounds-a brief review of recent work. Biotechnol Prog 27:597–613. https://doi.org/10.1002/btpr.575

Guerin M, Huntley ME, Olaizola M (2003) Haematococcus astaxanthin: applications for human health and nutrition. Trends Biotechnol 21:210–216. https://doi.org/10.1016/S0167-7799(03)00078-7

Hallin EI, Guo K, Åkerlund HE (2015) Violaxanthin de-epoxidase disulphides and their role in activity and thermal stability. Photosynth Res 124:191–198. https://doi.org/10.1007/s11120-015-0118-9

Hammond BR, Wooten BR, Snodderly DM (1998) Preservation of visual sensitivity of older subjects: association with macular pigment density. Invest Ophthalmol Vis Sci 39:397–406

Heo SJ, Jeon YJ (2009) Protective effect of fucoxanthin isolated from *Sargassum siliquastrum* on UV-B induced cell damage. J Photochem Photobiol B Biol 95:101–107. https://doi.org/10.1016/j.jphotobiol.2008.11.011

Heo SJ, Ko SC, Kang SM, Kang HS, Kim JP, Kim SH, Lee KW, Cho MG, Jeon YJ (2008) Cytoprotective effect of fucoxanthin isolated from brown algae Sargassum siliquastrum against H_2O_2-induced cell damage. Eur Food Res Technol 228:145–151. https://doi.org/10.1007/s00217-008-0918-7

Hosokawa M, Wanezaki S, Miyauchi K, Kurihara H, Kohno H, Kawabata J, Odashima S, Takahashi K (1999) Apoptosis-inducing effect of fucoxanthin on human leukemia cell HL-60. Food Sci Technol Res 5:243–246. https://doi.org/10.3136/fstr.5.243

Huang DS, Odeleye OE, Watson RR (1992) Inhibitory effects of canthaxanthin on *in vitro* growth of murine tumor cells. Cancer Lett 65:209–213. https://doi.org/10.1016/0304-3835(92)90233-L

Huang M, Lu JJ, Huang MQ, Bao JL, Chen XP, Wang YT (2012) Terpenoids: natural products for cancer therapy. Expert Opin Investig Drugs 21:1801–1818. https://doi.org/10.1517/13543784.2012.727395

Hussein G, Nakamura M, Zhao Q, Iguchi T, Goto H, Sankawa U, Watanabe H (2005) Antihypertensive and neuroprotective effects of astaxanthin in experimental animals. Biol Pharm Bull 28:47–52. https://doi.org/10.1248/bpb.28.47

Jyonouchi H, Sun S, Iijima K, Gross MD (2000) Antitumor activity of astaxanthin and its mode of action. Nutr Cancer 36:59–65. https://doi.org/10.1207/S15327914NC3601_9

Kanazawa K, Ozaki Y, Hashimoto T, Das SK, Matsushita S, Hirano M, Okada T, Komoto A, Mori N, Nakatsuka M (2008) Commercial-scale preparation of biofunctional fucoxanthin from waste parts of brown sea algae *Laminalia japonica*. Food Sci Technol Res 14:573–582. https://doi.org/10.3136/fstr.14.573

Khachik F, de Moura FF, Zhao DY, Aebischer CP, Bernstein PS (2002) Transformations of selected carotenoids in plasma, liver, and ocular tissues of humans and in nonprimate animal models. Invest Ophthalmol Vis Sci 43:3383–3392

Kohlmeier L, Hastings SB (1995) Epidemiologic evidence of a role of carotenoids in cardiovascular disease prevention. Am J Clin Nutr 62:1370S–1376S. https://doi.org/10.1093/ajcn/62.6.1370S

Koutsos EA, Schmitt T, Colitz CM, Mazzaro L (2013) Absorption and ocular deposition of dietary lutein in marine mammals. Zoo Biol 32:316–323. https://doi.org/10.1002/zoo.21033

Landrum JT, Bone RA (2001) Lutein, zeaxanthin, and the macular pigment. Arch Biochem Biophys 385:28–40. https://doi.org/10.1006/abbi.2000.2171

Liang S, Liu X, Chen F, Chen Z (2004) Current microalgal health food R and D activities in China. In: Asian pacific phycology in the 21st century: prospects and challenges: proceeding of the second Asian Pacific phycological forum, held in Hong Kong, China. Springer, pp 45–48

Lin JH, Lee DJ, Chang JS (2015) Lutein production from biomass: marigold flowers versus microalgae. Bioresour Technol 184:421–428. https://doi.org/10.1016/j.biortech.2014.09.099

Mabou FD, Yossa IBN (2021) Terpenes: structural classification and biological activities. J Pharm Biol Sci 16:2319–7676. https://doi.org/10.9790/3008-1603012540

Maeda H, Hosokawa M, Sashima T, Miyashita K (2007) Ditary combination of fucoxanthin and fish oil attenuates the weight gain of white adipose tissue and decrease blood glucose in obese/diabetic KK-Ay mice. J Agric Biol Chem 55:7701–7706. https://doi.org/10.1021/jf071569n

Manikandan K, Felix N, Prabu E (2020) A review on the application and effect of carotenoids with respect to canthaxanthin in the culture of fishes and crustaceans. Int J Fish Aquat Stud 8:128–133. https://doi.org/10.22271/fsh.2020.v8.i5b.2314

Maoka T (2009) Recent progress in structural studies of carotenoids in animals and plants. Arch Biochem Biophys 483:191–195. https://doi.org/10.1016/j.abb.2008.10.019

Maoka T, Akimoto N, Tsushima M, Komemushi S, Mezaki T, Iwase F, Takahashi Y, Sameshima N, Mori M, Sakagami Y (2011) Carotenoids in marine invertebrates living along the Kuroshio Current Coast. Mar Drugs 9:1419–1427. https://doi.org/10.3390/md9081419

Maoka T, Kuwahara T, Narita M (2014) Carotenoids of sea angels *Clione limacina* and *Paedoclione doliiformis*, from the perspective of food chain. Mar Drugs 12:1460–1470. https://doi.org/10.3390/md12031460

Mata TM, Martins AA, Caetano NS (2010) Microalgae for biodiesel production and other applications: a review. Renew Sust Energ Rev 14:217–232. https://doi.org/10.1016/j.rser.2009.07.020

Mata-Gómez LC, Montañez JC, Méndez-Zavala A, Aguilar CN (2014) Biotechnological production of carotenoids by yeasts: an overview. Microb Cell Factories 13:1–11. https://doi.org/10.1186/1475-2859-13-12

Mathews MM (1964) Protective effect of β-carotene against lethal photosensitization by haematoporphyrin. Nature 203:1092. https://doi.org/10.1038/2031092a0

Matsuno T (1989) Animal carotenoids. In: Krinsky NI, Mathews-Roth MM, Taylor RF (eds) Carotenoids chemistry and biology. Plenum Press, New York, pp 59–74

Matsuno T (2001) Aquatic animal carotenoids. Fish Sci 67:771–783. https://doi.org/10.1046/j.1444-2906.2001.00323x

Matsuno T, Ookubo M, Komori T (1985) Carotenoids of tunicates, III. The structural elucidation of two new marine carotenoids, amarouciaxanthin A and B. J Nat Prod 48:606–613. https://doi.org/10.1021/np50040a015

Mayne ST (1996) Beta-carotene, carotenoids, and disease prevention in humans. FASEB J 10:690–701. https://doi.org/10.1096/fasebj.10.7.8635686

Misawa N (2009) Pathway engineering of plants toward astaxanthin production. Plant Biotechnol 26:93–99. https://doi.org/10.5511/plantbiotechnology.26.93

Misawa N (2021) Carotenoids: biosynthetic and biofunctional approaches, 1st edn. Springer, Singapore. https://doi.org/10.1007/978-981-15-7360-6

Miyashita K, Hosokawa M (2014) Carotenoids as a nutraceutical therapy for visceral obesity. In: Watson RR (ed) Nutrition in the prevention and treatment of abdominal obesity. Academic, Cambridge, pp 329–340. https://doi.org/10.1016/B978-0-12-407869-7.00030-1

Narisawa T, Fukaura Y, Hasebe M, Ito M, Aizawa R, Murakoshi M, Uemura S, Khachik F, Nishino H (1996) Inhibitory effects of natural carotenoids, α-carotene, β-carotene, lycopene and lutein, on colonic aberrant crypt foci formation in rats. Cancer Lett 107:137–142. https://doi.org/10.1093/jnci/87.23.1767

Naziri D, Hamidi M, Hassanzadeh S, Tarhriz V, Zanjani BM, Nazemyieh H, Hejazi MA, Hejazi MS (2014) Analysis of carotenoid production by Halorubrum sp. TBZ126; an extremely halophilic archeon from Urmia Lake. Adv Pharm Bull 4:61–67. https://doi.org/10.5681/apb.2014.010

Nichols JA, Katiyar SK (2010) Skin photoprotection by natural polyphenols: anti-inflammatory, antioxidant and DNA repair mechanisms. Arch Dermatol Res 302:71–83. https://doi.org/10.1007/s00403-009-1001-3

Nishino H, Murakoshi M, Ii T, Takemura M, Kuchide M, Kanazawa M, Yang Mou X, Wada S, Masuda M, Ohsaka Y, Yogosawa S (2002) Carotenoids in cancer chemoprevention. Cancer Metastasis Rev 21:257–264. https://doi.org/10.1023/A:1021206826750

Okada T, Nakai M, Maeda H, Hosokawa M, Sashima T, Miyashita K (2008) Suppressive effect of neoxanthin on the differentiation of 3T3-L1 adipose cells. J Oleo Sci 57:345–351. https://doi.org/10.5650/jos.57.345

Olmedilla B, Granado F, Blanco I, Vaquero M, Cajigal C (2001) Lutein in patients with cataracts and age-related macular degeneration: a long-term supplementation study. J Sci Food Agric 81:904–909. https://doi.org/10.1002/jsfa.905

Osinga R, Schutter M, Griffioen B, Wijffels RH, Verreth JA, Shafir S, Henard S, Taruffi M, Gili C, Lavorano S (2011) The biology and economics of coral growth. Mar Biotechnol 13:658–671. https://doi.org/10.1007/s10126-011-9382-7

Pacher P, Beckman JS, Liaudet L (2007) Nitric oxide and peroxynitrite in health and disease. Physiol Rev 87:315–424

Palozza P, Maggiano N, Calviello G, Lanza P, Piccioni E, Ranelletti FO, Bartoli GM (1998) Canthaxanthin induces apoptosis in human cancer cell lines. Carcinogenesis 19:373–376. https://doi.org/10.1093/carcin/19.2.373

Peng Y, Wu Z, Li Y (2011) Fucoxanthin, a marine carotenoid with promising anti-oxidative, anti-inflammatory and anti-diabetic properties. Mar Drugs 9:523–548. https://doi.org/10.3390/md9101806

Raposo MF, Morais AM, Morais RM (2015) Carotenoids from marine microalgae: a valuable natural source for the prevention of chronic diseases. Mar Drugs 13:5128–5155. https://doi.org/10.3390/md13085128

Riccioni G, D'Orazio N, Franceschelli S, Speranza L (2011) Marine carotenoids and cardiovascular risk markers. Mar Drugs 9:1166–1175. https://doi.org/10.3390/md9071166

Sangeetha RK, Bhaskar N, Baskaran V (2009) Comparative effects of β-carotene and fucoxanthin on retinol deficiency induced oxidative stress in rats. Mol Cell Biochem 331:59–67. https://doi.org/10.1007/s11010-009-0145-y

Sharoni Y, Agemy L, Giat Y, Kirilov E, Danilenko M, Levy J (2022) Lycopene and astaxanthin inhibit human prostate cancer cell proliferation induced by androgens. In: Proceedings of the 13th international symposium on carotenoids; Honolulu, HI, USA, 6–11 January 2002

Shi XM, Zhang XW, Chen F (2000) Heterotrophic production of biomass and lutein by *Chlorella protothecoides* on various nitrogen sources. Enzym Microb Technol 27:312–318. https://doi.org/10.1016/S0141-0229(00)00208-8

Shindo K, Kimura M, Iga M (2004) Potent antioxidative activity of cacalol, a sesquiterpene contained in Cacalia delphiniifolia Sleb et Zucc. Biosci Biotechnol Biochem 68:1393–1394. https://doi.org/10.1271/bbb.68.1393

Shindo K, Kikuta K, Suzuki A, Katsuta A, Kasai H, Yasumoto-Hirose M, Matsuo Y, Misawa N, Takaichi S (2007) Rare carotenoids, (3R)-saproxanthin and (3R,20S)-myxol, isolated from novel marine bacteria (Flavobacteriaceae) and their antioxidative activities. Appl Microbiol Biotechnol 74:1350–1357. https://doi.org/10.1007/s00253-006-0774-y

Shklar G, Schwartz J (1988) Tumor necrosis factor in experimental cancer regression with alpha tocopherol, beta-carotene, canthaxanthin and algae extract. Eur J Cancer Clin Oncol 24:839–850. https://doi.org/10.1016/0277-5379(88)90192-7

Soliev AB, Hosokawa K, Enomoto K (2011) Bioactive pigments from marine bacteria: applications and physiological roles. Evid Based Complement Alternat Med 2011:560–566. https://doi.org/10.1016/B978-012374473-9.00772-4

Stahl W, Sies H (2012) β-Carotene and other carotenoids in protection from sunlight. Am J Clin Nutr 96:1179S–1184S. https://doi.org/10.3945/ajcn.112.034819

Street DA, Comstock GW, Salkeld RM, Schüep W, Klag MJ (1994) Serum antioxidants and myocardial infarction. Are low levels of carotenoids and alpha-tocopherol risk factors for myocardial infarction? Circulation 90:1154–1161. https://doi.org/10.1161/01.CIR.90.3.1154

Takaichi S (2011) Carotenoids in algae: distributions, biosyntheses and functions. Mar Drugs 9:1101–1118. https://doi.org/10.3390/md9061101

Tanaka T, Shnimizu M, Moriwaki H (2012) Cancer chemoprevention by caroteno. Molecules 17:3202–3242. https://doi.org/10.3390/molecules17033202

Torregrosa-Crespo J, Montero Z, Fuentes JL, Reig García-Galbis M, Garbayo I, Vílchez C, Martínez-Espinosa RM (2018) Exploring the valuable carotenoids for the large-scale production by marine microorganisms. Mar Drugs 16:203. https://doi.org/10.3390/md16060203

Tsushima M (2007) Carotenoids in sea urchins. Dev Aquac Fish Sci 37:159–166. https://doi.org/10.1016/S0167-9309(07)80072-X

Tsushima M, Maoka T, Matsuno T (1989) Comparative biochemical studies of carotenoids in marine invertebrates. The first positive identification of ε,ε-carotene derivatives and isolation of two new carotenoids from Chitons. Comp Biochem Physiol 93B:665–671

Vílchez C, Forján E, Cuaresma M, Bédmar F, Garbayo I, Vega JM (2011) Marine carotenoids: biological functions and commercial applications. Mar Drugs 9:319–333. https://doi.org/10.3390/md9030319

Wagner KH, Elmadfa I (2003) Biological relevance of terpenoids. Ann Nutr Metab 47:95–106. https://doi.org/10.1159/000070030

Walker TL, Purton S, Becker DK, Collet C (2005) Microalgae as bioreactors. Plant Cell Rep 24:629–641. https://doi.org/10.1007/s00299-005-0004-6

Wang JY, Lee YJ, Chou MC, Chang R, Chiu CH, Liang YJ, Wu LS (2015) Astaxanthin protects steroidogenesis from hydrogen peroxide-induced oxidative stress in mouse Leydig cells. Mar Drugs 13:1375–1388. https://doi.org/10.1016/S0167-7799(03)00078-7

Wijffels RH, Barbosa MJ (2010) An outlook on microalgal biofuels. Science 329:796–799. https://doi.org/10.1126/science.11890

Wollgast J, Anklam E (2000) Polyphenols in chocolate: is there a contribution to human health? Food Res Int 33:449–459. https://doi.org/10.1016/S0963-9969(00)00069-7

Woo MN, Jeon SM, Shin YC, Lee MK, Kang MA, Choi MS (2009) Anti-obese property of fucoxanthin is partly mediated by altering lipid-regulating enzymes and uncoupling proteins of visceral adipose tissue in mice. Mol Nutr Food Res 53:1603–1611. https://doi.org/10.1002/mnfr.200900079

Xu L, Weathers PJ, Xiong XR, Liu CZ (2009) Microalgal bioreactors: challenges and opportunities. Eng Life Sci 9:178–189. https://doi.org/10.1002/elsc.200800111

Yamamoto HY, Kamite L, Wang YY (1972) An ascorbate-induced absorbance change in chloroplasts from violaxanthin de-epoxidation. Plant Physiol 49:224–228. https://doi.org/10.1104/pp.49.2.224

Yim MJ, Hosokawa M, Mizushina Y, Yoshida H, Saito Y, Miyashita K (2011) Suppressive effects of amarouciaxanthin A on 3T3- L1 adipocyte differentiation through down-regulation of PPARγ and C/EBP mRNAα expression. J Agric Food Chem 59:1646–1652. https://doi.org/10.1021/jf103290f

Zakar T, Laczko-Dobos H, Toth TN, Gombos Z (2016) Carotenoids assist in cyanobacterial Photosystem II assembly and function. Front Plant Sci 7:295. https://doi.org/10.3389/fpls.2016.00295

Chapter 9
Uses of Microorganisms for Carotenoid Production: Contribution of Modern Technologies in Food Sustainability

T. C. Nascimento, P. P. Nass, L. C. Schetinger, M. L. Nörnberg, P. A. Caetano, R. R. Dias, E. Jacob-Lopes, and L. Q. Zepka

1 Introduction

Carotenoids correspond to a class of compounds with diverse natural nonpolar structures responsible for the yellow, orange, or red color of many foods (Rodriguez-Amaya 2015). These are structures with multiple technological, biotechnological, and bioactive functions. Ubiquitous naturally occurring secondary metabolites, carotenoids, can protect organs, tissues, and cells from the adverse effects of free radicals and lipid peroxides (Britton 2020). As a result, they have become a focal point of extensive research, exploring their potential applications as dyes, supplements, nutraceuticals, and prevention of various diseases (Rodriguez-Amaya 2019; Britton 2020).

Photosynthetic microorganisms, such as microalgae and cyanobacteria, and non-photosynthetic microorganisms, such as bacteria and fungi, are capable of synthesizing (naturally or through genetic intervention) various carotenoids under different cultivation conditions and media (Foong et al. 2021). The microbial pathways involved in carotenoid biosynthesis are (I) the mevalonate pathway (MVA) and (II) the 2-C-methyl-D-erythritol-4-phosphate (MEP) pathway. While the first is normally used by *Archaea* bacteria, fungi, and microalgae, the second is the preferred route for prokaryotic bacteria (Lichtenthaler 2010; Disch et al. 1998; Schwender et al. 1997).

In recent years, these microscopic beings have become protagonists in the natural and sustainable obtaining of these pigments, as they can be cultivated in controlled and optimized environments, allowing the development of efficient processes

T. C. Nascimento · P. P. Nass · L. C. Schetinger · M. L. Nörnberg · P. A. Caetano
R. R. Dias · E. Jacob-Lopes · L. Q. Zepka (✉)
Department of Food Science and Technology, Federal University of Santa Maria (UFSM), Santa Maria, RS, Brazil
e-mail: jacoblopes@pq.cnpq.br

R. Campos Chisté et al. (eds.), *Carotenoids*,
https://doi.org/10.1007/978-3-031-75322-0_9

that are independent of non-renewable resources. Approximately a decade ago, sustainability began to stand out globally, under the principle of reducing dependence on fossil resources for the fundamental needs of planet Earth, including the food industry (Philp and Atlas 2017; Sundararajan and Ramasamy 2024). As a result, the enormous metabolic and genetic diversity of microorganisms can be taken advantage of, since there are clear routes for the production of natural compounds of commercial and technological interest such as carotenoids.

Although most microorganisms are not genuinely intended for industrial carotenoid production, metabolomics tools provide optimistic prospects for meeting global demand for more natural and sustainable nutritional products (Li et al. 2020). Strategies based on metabolic and genetic engineering have been explored to consolidate the overproduction of this bioproduct. Modern technologies related to enzyme overexpression, gene manipulation, insertion of carotenoid genes into non-carotenogenic microorganisms, and improvement of cellular storage have actively contributed to making microorganisms exceptional carotenoid biofactories (Sundararajan and Ramasamy 2024).

In this context, this Chapter brings together a comprehensive description of microbial carotenoid production, including aspects related to chemical structure, biosynthesis, and bioactivity, main microbial sources, modern strategies to increase sustainable production, as well as downstream processing and sustainability metrics.

2 Chemical, Bioactivity, and Biosynthesis of Carotenoids

Carotenoids, lipophilic natural pigments, contribute to yellow, orange, or red coloring. Typically, these compounds are C40 tetraterpenes or tetraterpenoids, formed by connecting eight C5 isoprenoid units head-to-tail, with a central tail-to-tail linkage reversing the order (Rodriguez-Amaya et al. 2008).

The predominant feature is a centrally positioned and elongated double-bond system. This polyene chain forms the light-absorbing chromophore, imparting color to carotenoids and playing a crucial role in their unique properties and functions (Rodriguez-Amaya 2015).

The bioactivity of these compounds is basically due to their antioxidant properties. Such properties of these isoprenoid compounds are intricately linked to their chemical structure, encompassing factors such as the count of conjugated double bonds (CDBs), the nature of structural end groups, and the presence of oxygen-containing substituents (Rodrigues et al. 2015; Rodriguez-Amaya et al. 2008).

The correlation between the expansion of the conjugated double bond (CBD) system and the bioactive potential can be elucidated through the preferential antioxidant mechanism of carotenoids, involving the donation of electrons from the polyenic chain (Rodrigues et al. 2012; El-Agamey et al. 2004).

The basic structure of the carotenoid can be synthesized using two pathways: the MVA and MEP pathway. Archaea and eukaryotic bacteria typically employ the MVA pathway, prokaryotic bacteria utilize the MEP pathway, and plants utilize both pathways for isoprenoid production (Lichtenthaler 2010; Disch et al. 1998; Schwender et al. 1997).

The central phosphate and carbon pathways play pivotal roles in providing essential compounds such as glyceraldehyde-3-phosphate (G3P), pyruvate, and acetyl-CoA, serving as substrates for both the MVA and MEP pathways (Egeland 2016; Lohr et al. 2012).

The MEP pathway initiates through the condensation of equal amounts of glyceraldehyde-3-phosphate (G3P) and pyruvate, ultimately giving rise to dimethylallyl diphosphate (DMAPP) and isopentenyl diphosphate (IPP) in a ratio of approximately 5:1, achieved through a sequence of seven steps (Fernandes et al. 2017; Egeland 2016).

The MVA pathway encompasses six steps in the synthesis of isopentenyl diphosphate (IPP) from acetyl-CoA. Subsequently, IPP undergoes conversion to its isomer, dimethylallyl diphosphate (DMAPP), facilitated by isopentenyl-pyrophosphate delta isomerase (IDI) (Fernandes et al. 2017; Egeland 2016).

Geranylgeranyl diphosphate (GGPP) synthase catalyzes the addition of three molecules of isopentenyl diphosphate (IPP) to one molecule of dimethylallyl diphosphate (DMAPP), resulting in the formation of C20 geranylgeranyl diphosphate (GGPP). Subsequently, two GGPP molecules undergo a tail-to-tail combination, leading to the synthesis of phytoene (C40), a colorless molecule (Rodriguez-Amaya 2015).

The generation of phytoene marks a pivotal step in carotenoid biosynthesis. Phytoene desaturase catalyzes the initial desaturation, resulting in the formation of ζ–carotene. ζ–carotene then undergoes two additional desaturation steps, culminating in the production of lycopene, the first colored product in this pathway. Lycopene serves as a crucial precursor in numerous reactions, contributing to the synthesis of diverse types of carotenoids (Fig. 9.1) (Cunningham and Gantt 2001).

Lycopene can follow two distinct metabolic pathways, giving rise to the synthesis of both α-carotene and β-carotene (Fig. 9.1). These compounds then undergo further transformations, contributing to the diverse array of carotenoids in subsequent metabolic processes.

Lycopene ε-cyclase (LCYE) and lycopene β-cyclase (LCYB) are enzymes that facilitate the cyclization of lycopene, resulting in the formation of α-carotene. Following a sequential process, α-carotene undergoes conversion into lutein through two successive hydroxylation reactions, catalyzed by cytochrome P450 enzymes: cytochrome P450 beta-hydroxylase (CYP97A) and cytochrome P450 epsilon hydroxylase (CYP97C) (Takaichi 2020).

In contrast, lycopene undergoes conversion to β-carotene through the activity of lycopene β-cyclase (LCYB). The ensuing β-carotene can follow diverse pathways. Specifically, β-carotene may undergo initial conversion into zeaxanthin catalyzed by β-carotene hydroxylase. Subsequently, zeaxanthin can undergo further transformation into astaxanthin, facilitated by β-carotene ketolase. Alternatively, β-carotene can be converted into canthaxanthin through the action of β-carotene ketolase. Following this, canthaxanthin can then undergo additional modification into

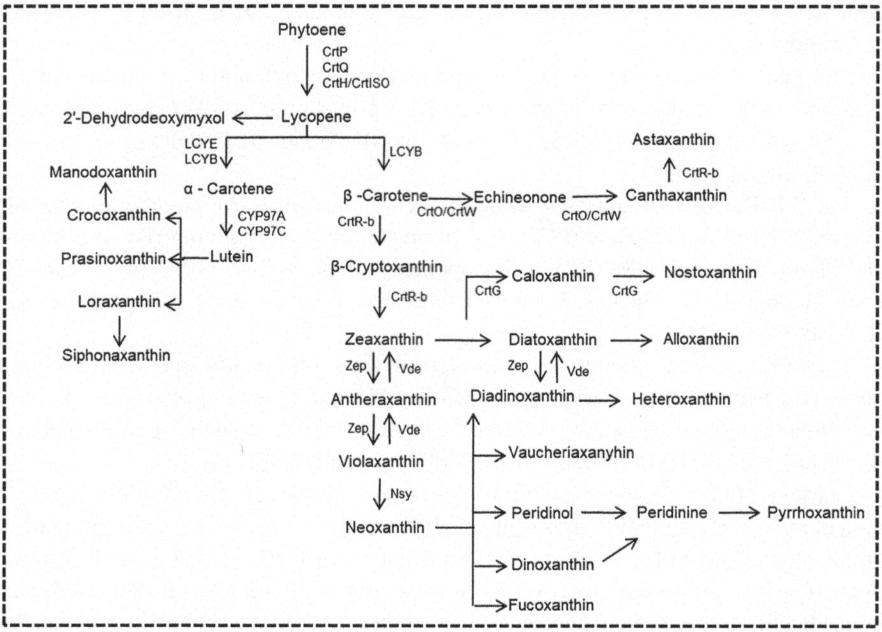

Fig. 9.1 Diagram of the carotenoid biosynthesis pathway

astaxanthin through the activity of β-carotene hydroxylase (Patias et al. 2017; Cunningham and Gantt 2001).

Finally, the biosynthesis of carotenoids by microorganisms is influenced by genetic factors, environmental conditions, and the cultural stress experienced during their growth.

3 Fungi Sources of Carotenoids

The fungal microorganisms have the potential to produce relevant amounts of carotenoids intracellularly, under high tolerance to stress, withstanding extreme pH and temperatures (Joshi et al. 2023). Some of the factors that can affect the production of carotenoids in fungi are the intensity of light and the culture medium (Foong et al. 2021).

Cultivation in media with a high sugar content makes these microorganisms industrially interesting. The fungi *Neurospora crassa*, and *Blacklea trispora,* as well as the yeasts *Rhodotorula* sp., *Rhodotorula mucilaginosa*, and *Rhodozyma phaffia* (*Xanthophyllomyces dendrorhous*), are some examples that are being studied to maximize and/or optimize the bioproduction of carotenoids, aiming to commercial use (Aksu and Eren 2005; Foong et al. 2021; Wang et al. 2021; Dyaa et al.

2022; Joshi et al. 2023; Sundararajan and Ramasamy 2024), as presented in Table 9.1.

The fungi *Neurospora crassa* has been exploited for the production of lycopene. The carbon flow directed to the mevalonate metabolic pathway enabled greater lycopene synthesis. Furthermore, exogenous oleic acid boosted intracellular triacylglycerol production, contributing to increased lycopene production through lipid storage capacity (Wang et al. 2021).

Another carotenogenic fungus, *Blakeslea trispora*, stands out for its production of β-carotene, in addition to the ability to be transformed by cyclization of lycopene. Lycopene biosynthesis in *B. trispora* can be stimulated (Liu et al. 2021). Fermentation of *B. trispora* at neutral to alkaline pH favors the production of lycopene, while β-carotene is formed as the predominant pigment at more acidic pH (Foong et al. 2021). In relation to this strain (*B. trispora*), both aeration and agitation influenced the concentration and productivity of β-carotene. The highest concentration and

Table 9.1 Production of carotenoids from some fungi sources

Fungi	Carotenoid	Culture method
Neurospora cassa[a]	Lycopene[a]	Optimized production with carbon flow through the mevalonate metabolic pathway; and with the addition of oleic acid through the production of triacylglycerols and lipid storage[a]
Blakeslea trispora[b,c,d]	β-carotene[b,c,d], Lycopene[b,c]	Production of β-carotene and transformation by cyclization of lycopeno[b] Optimization of lycopene production at neutral to alkaline pH and β-carotene predominantly at more acidic pH[c] Higher concentration and productivity of β-carotene at low rotation speed (150 rpm) and high aeration rate (1.5 vvm)[d]
Rhodotorula sp.[b,e,f]	β-carotene[b,e], Torulene[b,e], Torularhodin[b,e]	Optimization of carotenoids with malt extract, fructose, urea, yeast extract, 3.5% NaCl, 27.5 °C, pH 6.7, 180 rpm for 98 h[f]
Rhodotorula mucilaginosa[g,h,i]	β-carotene[g,h,i]	Optimization with sucrose as a carbon source and greater yield using lactose[g] Optimization in complex media and stress conditions, with UV light, magnesium sulfate, light/dark photoperiod[h] Optimization of β-carotene production using agro-industrial waste, pH 6.1, 25.8 °C, 119.6 rpm for 84 h[i]
Rhodozyma phaffia[e,j,k,l]	Astaxanthin[e,j,k,l]	Optimization with saccharified sugarcane bagasse, 55 °C, pH 4.8 for 12 h[k] Optimization with glucose and corn grain distillation residue as a nitrogen source, initial pH 6.7. Increased production by combining stimulation with TiO_2 and cultivation in a fermenter[l]

[a]Wang et al. (2021), [b]Liu et al. (2021), [c]Foong et al. (2021), [d]Mantzouridou et al. (2002), [e]Joshi et al. (2023), [f]Dyaa et al. (2022), [g]Aksu and Eren (2005), [h]Cortes et al. (2020), [i]Sharma and Ghoshal (2020), [j]Sundararajan and Ramasamy (2024), [k]Rangel-Basto et al. (2022), [l]Guan et al. (2023)

productivity of β-carotene were obtained at low rotation speed (150 rpm) and high aeration rate (1.5 vvm) (Mantzouridou et al. 2002).

The yeast species *Rhodotorula* sp. produces high concentrations of carotenoids, with β-carotene being the majority, in addition to being known for producing torulene and torularhodine (Liu et al. 2021; Joshi et al. 2023). The highest production of carotenoids from *Rhodotorula* sp. was achieved when the yeast was cultivated in a production medium composed of 3.7 g L^{-1} of malt extract, 7.7 g L^{-1} of fructose, 9 g L^{-1} of urea, 35 g L^{-1} of NaCl and 1 g L^{-1} of yeast extract at 27.5 °C, pH 6.7 and 180 rpm. The concentration of carotenoids increased from 21.5 mg L^{-1} to 229.9 mg L^{-1} after 98 h of incubation (Dyaa et al. 2022).

Regarding *Rhodotorula mucilaginosa*, the increase in sugar concentration accelerated yeast growth and total carotenoid production. The highest concentration of carotenoids (89 mg L^{-1} of fermentation solution) was obtained when 20 g L^{-1} of sucrose was used as a carbon source, while the highest yield (35 mg g^{-1} of dehydrated biomass) was achieved when 13.2 g L^{-1} of lactose was used as a carbon source in the fermentation solution (Aksu and Eren 2005).

Furthermore, using a complex medium, *R. mucilaginosa* was subjected to varied stress conditions, including osmotic stress using different salts, physical stress by ultraviolet (UV) light, and light stress using different photoperiods. The ideal growth conditions for carotenoid production were determined to be 1 min of UV light, 0.5 mg L^{-1} magnesium sulfate an 18:6 h light/dark period, which resulted in a yield of carotenoids of 118.3 μg g^{-1} from yeast (Cortes et al. 2020). Using agro-industrial residues, the best conditions for obtaining maximum carotenoids were at pH 6.1 incubated at 25.8 °C with agitation at 119.6 rpm for 84 h, which resulted in 717.35 μg g^{-1} of β-carotene, 7.33 g L^{-1} of cell mass, with aeration having a significant role in the carotenogenesis of *R. mucilaginosa*. The maximum amount of carotenoids obtained under aerobic fermentation conditions was 819.23 μg g^{-1} compared to 717.35 μg g^{-1} in a shake flask under similar conditions (Sharma and Ghoshal 2020).

Rhodozyma phaffia stands out for producing astaxanthin, one of the few natural alternatives for producing this pigment (Joshi et al. 2023; Sundararajan and Ramasamy 2024). The enzymatic saccharification process of sugar cane bagasse (enzymatic units, temperature, pH, and reaction time) resulted in holocellulose conversion of 68.1% ± 2.6 with 2.000 UEG.g^{-1}, at 55 °C, pH 4.8 for 12 h. The results confirm that saccharified sugarcane bagasse is a potential substrate for astaxanthin production (Rangel-Basto et al. 2022).

Furthermore, the effects of the carbon source, distillate residue of corn grains, and initial pH on the growth and astaxanthin production of *R. phaffia* were studied. The ideal composition of the medium was 32 g L^{-1} of glucose, 12 g L^{-1} of distillate residue of corn grains (distillery grains) as a nitrogen source, and an initial pH of 6.7. *R. phaffia* was cultivated in a shaking flask under optimized conditions, the biomass was 6.47 g L^{-1} and the astaxanthin content was 1.41 mg g^{-1}. The astaxanthin content was increased to 4.70 mg g^{-1} by combining TiO_2 stimulation and cultivation in a 5 L fermentor, which was 2.81 times higher than the control group (Guan et al. 2023).

The diversity of carotenoids produced is variable, which is related to the species, composition of the cultivation medium (composition of substrates—source of carbon and nitrogen, presence of metals, salts, and chemical agents), and environmental conditions (temperature, light, agitation, aeration and pH) (Liu et al. 2021). Studies suggest that the ideal conditions for carotenoid production vary according to environmental stress conditions (Igreja et al. 2021).

Temperature can affect fungal carotenoid production. Fungi can adapt to low temperatures with distinct molecular mechanisms, commonly known as the heat shock response. Furthermore, based on the difference in pH value, the productivity of carotenoids, among others, differs, compared to a fixed pH, while, the production of carotenoids differs significantly at controlled pH compared to uncontrolled pH values. Salt stress induces carotenoid production in carotenogenic species. Supplementation of metal ions positively interferes with the production of carotenoids, increasing markedly with a reduction in lipid concentration with the addition of aluminum sulfate in an optimized medium (Joshi et al. 2023).

The nature of the carbon and nitrogen sources also affects carotenoid formation in yeast and fungi in unpredictable ways. A high carbon-to-nitrogen ratio promotes carotenoid biosynthesis in yeast. There are several fungi that respond to more acidic pH conditions with pigment accumulation in submerged cultivation (Foong et al. 2021).

Microorganisms such as yeast have the advantage of a controlled fermentation process. The application of different conditions for personalized fermentation, as well as selected substrates, can make these yeasts attractive for commercial use. This process can work with the original strains or with those developed to use a specific substrate. This includes the production of carotenoids through the cultivation of agro-industrial waste (Sandmann 2022).

The ability of microorganisms to digest low-cost agro-industrial waste as substrates reduces the overall cost of carotenoid production. Furthermore, the microbial carotenoid fermentation process is relatively simple to manage and monitor, minimizing processing time. Other production challenges include low carotenoid content and safety concerns for certain opportunistic pathogens. Therefore, optimization of the carotenoid production process on an industrial scale is necessary for the commercial production of microbial carotenoids (Foong et al. 2021).

4 Bacteria Sources of Carotenoids

Various bacteria exhibit the ability to produce carotenoids and other pigments, especially when faced with challenging environmental conditions (Table 9.2). Noteworthy is the observation that specific symbiotic bacteria generate carotenoids, acting as a defensive shield against sunlight, oxidative stress, and potentially harmful bacteria. This phenomenon contributes significantly to the overall well-being of the symbiotic organism (Kusmita et al. 2023; Young and Lowe 2018).

Table 9.2 Production of carotenoids from some bacteria sources

Bacteria	Carotenoid	Culture method
Paracoccus marcusii[a]	Astaxanthin, Zeaxanthin, ζ-carotene and β-zeacarotene[a]	Optimized with inoculum size, nitrogen source, pH, temperature, and agitation speed[a]
Brevibacterium linens[b]	β-Cryptoxanthin[b]	Medium composed with glucose and sodium. PH adjusted to 6.9, 72 h culture, at 25 °C, 145 rpm agitation[b]
Bradyrhizobium sp.[c]	Canthaxanthin[c]	Photosynthesis with light and dark cycles[c]
Agrobacterium aurantiacum[d]	Astaxanthin[d]	Medium composed with glucose and sodium[d]
Paracoccus carotinifaciens[e]	Astaxanthin[e]	Culture medium with (in g L^{-1}): glucose (5), Na2HPO4 (3.6), NaCl (8), NaH2PO4 (1.2), ammonium chloride (2.5) and yeast extract (2.5) in 1.20 L. In a bioreactor at 22 °C, 600 rpm of agitation speed and air flow rate of 1.8 L min^{-1}[e]
Sphingomonas sp.[f,g,j]	Nostoxanthin[j], Astaxanthin[f,g], β- carotene[f,g]	Sea Water[k]. Plating in agar (5 g L^{-1} of HiVeg hydrolysate, 2.5 g L^{-1} of yeast extract, 1 g L^{-1} of dextrose, 9 g L^{-1} of agar) followed by incubation at 7 °C for 10 days[f,g]
Rhodospirillum rubrum[h,i]	Lycopene[h,i]	Dark, semi aerobic conditions[k]. Fed- batch cultivation with controlled acetate input, syngas devoid of carbon monoxide and phosphorous
Flavobacterium sp.	Zeaxanthin	Tryptic Soy Broth (Merck) supplemented with 1% Sea Salt (Sigma) for 24 h at 20 °C in an orbital shaker at 200 rpm[j]

[a]Naik and Gupte (2023), [b]Guyomarc'h et al. (2000), [c]Lorquin et al. (1997), [d]Yokoyama et al. (1994), [e]Katsumata et al. (2014), [f]Ram et al. (2020), [g]Mageswari et al. (2015), [h]George et al. (2020), [i]Karmann et al. (2019), [j]Kikukawa et al. (2021), [k]Wang et al. (2012)

Paracoccus marcusii can improve its carotenoid production with enhancement techniques. For instance, Naik and Gupte (2023) achieved optimization of bacterial pigment production by adjusting key parameters, including inoculum size, nitrogen source, pH, temperature, and agitation speed. With the improved approach, they were able to reach a yield of 7240 ± 41 μg L^{-1}. *Brevibacterium linens* is a bacterial strain employed as an adjunct culture in the cheese ripening process, as the French cheese variety *Fourme de Montbrison*. These bacteria can produce a mix of carotenoids, especially β-Cryptoxanthin under optimal conditions (Giuffrida et al. 2020; Guyomarc'h et al. 2000). β-Cryptoxanthin is a pro-vitamin A carotenoid pigment characterized by its yellow color, found in food sources like carrots, oranges, persimmons, papayas, pumpkins, and red sweet peppers. It is particularly abundant in citrus fruits such as satsuma and oranges, contributing to the prevention of diseases such as cancer and osteoporosis (Brahma and Dutta 2022).

Bradyrhizobium sp. has the potential to produce Canthaxanthin under altered light and dark cycle conditions. Canthaxanthin finds extensive application as a feed additive for enhancing skin and yolk color in poultry. Due to its insolubility in water and susceptibility to oxidation, commercial canthaxanthin is frequently microencapsulated with wall materials. This microencapsulation serves to enhance solubility and stability, ensuring its effectiveness in poultry feed (Lorquin et al. 1997; Wen et al. 2022). Inoculating seeds with *Bradyrhizobium* sp. resulted in organisms showing increased levels of total soluble sugars and ureides compared to control plants. In the current investigation, peanut plants originating from seeds inoculated with *Bradyrhizobium* sp. exhibited elevated leaf concentrations of bioactive compounds, pigments, carotenoids, nitrate, ammonia, and amino acids (Gericó et al. 2020).

Agrobacterium aurantiacum is a species of Gram-negative bacteria that has a known potential of producing astaxanthin (Goswami et al. 2010; Yokoyama et al. 1994). Recently, *Agrobacterium aurantiacum* has been utilized as a genic base for increasing astaxanthin yield in microorganisms, such as yeast. In an experiment, (Jin et al. 2018) extracted a gene from *Agrobacterium aurantiacum* and applied it to *Saccharomyces cerevisiae*. In comparison to the stress induced by β-carotene hydroxylase (CrtZ) from Alcaligenes sp. strain PC-1, the utilization of the CrtZ from *Agrobacterium aurantiacum* has demonstrated a notable enhancement in astaxanthin production, resulting in a 1.78-fold increase in yield and an elevated astaxanthin ratio of 88.7% compared to the initial 66.6%. Further augmentation of astaxanthin yield, reaching 10.1 mg g^{-1} DCW, was achieved through ARTP mutagenesis, meaning a high improvement in astaxanthin production. Also, the gene crtW from Agrobacterium aurantiacum was imported to drive astaxanthin production in *Escherichia coli* (Chou et al. 2019).

Sphingomonas sp. is an outstanding producer of yellow carotenoids, and it is naturally present in deep-sea sediment. Kikukawa et al. (2021) isolated these bacteria from Suruga Bay, in Shizuoka, Japan, thus, *Sphingomonas* sp. SG73 generated Nostoxanthin with a high level of purity, comprising nearly 97% (area%) of the overall carotenoid production. Notably, this strain exhibited halophilic traits and demonstrated tolerance to a salt concentration 1.5 times higher than that of seawater. Cultivating *Sphingomonas* sp. SG73 in the presence of 1.8% synthetic ocean salt. This resulted in a notable increase, with Nostoxanthin production rising by 2.5 times compared to cultivation without artificial sea salt. These results confirm the effectiveness of *Sphingomonas* sp. SG73 in producing Nostoxanthin and highlighting its appropriateness for synthesizing carotenoids using marine water, owing to its compatibility with sea salt.

Nostoxanthin is a poly-hydroxy yellow carotenoid obtained from certain photosynthetic bacteria, including specific Cyanobacteria species. The yellow carotenoid nostoxanthin ((2,3,2′,3′)-β,β-carotene-2,3,2′,3′-tetrol) is synthesized from zeaxanthin through the addition of two hydroxyl groups. Based on the chemical structure of nostoxanthin, it is anticipated to exhibit a similarly high antioxidative activity as zeaxanthin. However, more studies on its antioxidant activity are required to unravel its potential. This carotenoid was identified in certain prokaryotes, including specific Cyanobacteria such as the bacteriochlorophyll-a containing bacterium strain *Sandarakinorhabdus limnophila*, photosynthetic bacteria, the non-marine *Brevundimonas* sp., and most *Sphingomonas* species. Nevertheless, the concentration of nostoxanthin in known nostoxanthin-producing bacteria is reported to be low

(Gich and Overmann 2006; Jenkins et al. 1979; Kikukawa et al. 2021; Zhu et al. 2012).

Lycopene serves a crucial function as both an antioxidative and anticancer agent, gaining increasing significance as a valuable commodity in the global market. The production of this carotenoid utilized industrially as a natural colorant and food additive, is currently a subject of considerable interest (Caseiro et al. 2020; Su et al. 2018). Historically, microbial production metabolic strategies (see Sect. 6) have employed *Escherichia coli* as a host strain. Thus, Wang et al. (2012) conducted a study with *Rhodospirillum rubrum*, to better understand its industrial applications. Quantitative growth experiments were carried out in dark, semi-aerobic conditions, utilizing a growth medium designed for high cell density and elevated intracellular membrane levels, conducive to conventional industrial production in the absence of light. It resulted in lycopene yields of up to 2 mg g^{-1} (dry weight) of cells or up to 15 mg g^{-1} of culture. These findings align with the lycopene production levels observed in multiple *E. coli* strains previously engineered for similar purposes.

Regarding lycopene production, Qu et al. (2021) achieved a high lycopene yield with gene modifications. They utilized *Rhodobacter sphaeroides*, with targeted modifications. *Rhodobacter sphaeroides* is a non-sulfur purple bacterium characterized by significant metabolic versatility, demonstrating the ability to produce various valuable compounds, including carotenoids. Thus, genetic modifications are useful alternatives to enable the industrial production of carotenoids in bacterial strains (see Sect. 6).

5 Microalgae Sources of Carotenoids

Microalgae incorporate a diverse group, with about 72,500 species cataloged. Among the species of commercial relevance are Spirulina (*Arthrospira*) sp., *Chlorella* sp., *Crypthecodinium cohnii*, *Porphyridium cruentum*, *Haematococcus pluvialis*, *Phaeodactylum tricornutum*, *Nannochloropsis* sp., and *Dunaliella* sp. (Jacob-Lopes et al. 2019; Jacob-Lopes and Zepka 2023). However, *Haematococcus pluvialis*, and *Dunaliella*, stand out for their industrial output of carotenoids (Borowitzka 2013).

A significant number of microalgae carotenoids are used in commercial food (Fernandes et al. 2020; Nass et al. 2022a, b), cosmetics (Martínez Andrade et al. 2018), and pharmaceuticals (Khan et al. 2018). Due to their robustness and simple nutritional requirements, rapid growth, and substantial content of compounds to be exploited (Jacob-Lopes et al. 2019), which opens a field of exploitation with high technical-economic potential biotechnological processes (Ashokkumar et al. 2023).

As microalgae culture is ecofriendly and renewable, there is increasing affinity to use them as a source of bioactive pigments, popular carotenoids such as astaxanthin, β-carotene, zeaxanthin, lutein, violaxanthin, echinenone, and canthaxanthin, can be isolated, some of which are produced exclusively by microalgae, and finds application as antioxidant, anti-inflammatory, immunoprophylactic, antitumor activities among others (Nascimento et al. 2021).

The most commercialized carotenoid is β-carotene. The *Dunaliella salina* species, known for its β-carotene production, is colored in the reddish-orange spectrum. The carotenoid content in this species ranges from 3% to 13%, showcasing its significance in producing this particular carotenoid (Wu et al. 2016; Hotos and Avramidou 2021). The microalgae were found on Lake Hillier, which explains its colour (Pink color Lake). The pink color is believed to be due to the presence of organisms such as *Dunaliella salina* and *Salinibacter ruber* (red bacterium) and environmental conditions (Teller 1987).

D. salina exhibits an exceptionally broad tolerance to salinity, yet its optimal growth occurs at approximately 22% w/v NaCl, roughly seven times the concentration found in seawater. Notably, this microalgal species can thrive even in salinities reaching NaCl saturation, which is around 35% NaCl. Additionally, it demonstrates a remarkable resistance to higher temperatures, with an optimum temperature exceeding 28 °C. Furthermore, the presence of high light is crucial for facilitating the rapid accumulation of β-carotene within the cells, as highlighted by Borowitzka (2018).

The other well-known compound is astaxanthin from *Haematococcus sp*., which accumulates up to 1–7% of its dry cell weight. Production of astaxanthin derived from *Haematococcus sp*. and *Chlorella zofingiensis* are ideal strains for commercial astaxanthin production at an industrial scale (Xie et al. 2021; Joshi et al. 2023; Wood et al. 2023). According to Kang et al. (2005), their study comparing heterotrophic and photoautotrophic production of astaxanthin by Haematococcus pluvialis indicated that photoautotrophic means were more efficient in terms of greater productivity. The research found that under photoautotrophic induction, specifically through N-deprivation followed by supplementation with bicarbonate or CO_2, *Haematococcus* cells exhibited higher astaxanthin content compared to cells induced through heterotrophic.

Another bioactive carotenoid from microalgae is lutein. A study has estimated the annual production of microalgae at 350–750 kg ha^{-1} of microalgae (Kadri et al. 2023). In addition, the lutein production rate is 3–6 times higher for microalgae compared with that for marigold flowers, for the reason that microalgae (70–150 tons ha^{-1}) grow much faster than marigold flowers (6 tons/ha) (Fu et al. 2023). The strategies of cultivation adopted for lutein production along with the metabolic and genetic design help for lutein industrial production from microalgae primarily from the *Chlorella vulgaris* and *Chlorella sorokiniana* strain (McClure et al. 2019; Li et al. 2022).

The synthesis of echinenone and canthaxanthin, exclusive to these microorganisms, is characterized by the formation of ketocarotenoids, distinguishing them as supercarotenoids owing to their unique chemical structure. Echinenone, serving as a precursor to the compound, manifests a structural configuration that includes a carbonyl group intricately linked to the conjugated double bond system. This intricacy extends the chromophoric moiety, comprising 12 conjugated double bonds

Table 9.3 Production of carotenoids from some microalgae sources

Microalgae	Carotenoids	Culture method
Dunaliela Salina[a]	all-*trans*-β-carotene[a]	Mixotrophic: various salinities and light intensities
Chlorella zofingiensis[b]	all-*trans*-astaxanthin[b]	Photoautotrophic for mixotrophic in an up-scaled
Haematococcus sp[c]	all-*trans*-astaxanthin[c]	Photoautotrophic and heterotrophic
Chlorella sorokiniana[d]	all-*trans*-lutein, all-*trans*-β-carotene[d]	Cultivation modes: photoautotrophic, heterotrophic, and mixototrophic
Chlorella vulgaris[e]	all-*trans*-lutein, all-*trans*-β-carotene[e]	Mixotrophic: cultivations of light and nitrogen coupling strategies
Scenedesmus obliquus[f]	all-*trans*-β-carotene, all-*trans*-lutein, all-*trans*-zeaxanthin[f]	Photoautotrophic
Aphanotece microscopica[g]	all-*trans*-β-carotene, *cis* and *trans*-echinenone, all-*trans*-zeaxanthin[g]	Photoautotrophic
Desertifilum sp.[g]	all-*trans*-β-carotene, *cis* and *trans*-echinenone[g]	Photoautotrophic
Desertifilum sp.[h]	β-carotene, lutein, zeaxanthin, echinenone and canthaxanthin[h]	Photoautotrophic
Desertifilum sp.[i]	β-carotene, zeaxanthin, lutein, echinenone[i]	Heterotrophic using slaughterhouse wastewater

[a]Borowitzka (2018), [b]Wood et al. (2023), [c]Kang et al. (2005), [d]Patel et al. (2022), [e]Li et al. (2022), [f]Nascimento et al. (2021), [g]Rodrigues et al. (2015), [h]Maroneze et al. (2019), and [i]Rodrigues et al. (2014)

(Rodrigues et al. 2015; Maroneze et al. 2019; Nascimento et al. 2021; Nass et al. 2022b).

Following this, the quality of cultive medium shows a relevant role in microalgae growth and producing bioactive pigments (Dai et al. 2016). This information is depicted in the Table 9.3 illustrating the production of predominant carotenoids from various microalgae species and the cyanobacteria *Aphanotece microscopica* and *Desertifilum sp.*Notably, contingent upon the cultivation medium employed, the synthesis of principal carotenoids transpires. For comprehensive details, one is encouraged to refer to Table 9.3 for a thorough examination of the documented outcomes.

The metabolic versatility of these microorganisms is incredible. When living in hostile environments in extreme environments accumulate carotenoids during stress conditions to ensure survival (Jacob-Lopes et al. 2019; Jacob-Lopes and Zepka 2023).

6 Engineer Metabolic and Genetic Applied to Increase Microbial Carotenoid Production

Biotechnological production of carotenoids has grown year after year due to the global trend toward natural and sustainable food products. However, as previously demonstrated, only the production of astaxanthin and β-carotene from microalgae is a commercial reality to date. For this reason, several strategies have been explored to consolidate microbial production of carotenoids (Sundararajan and Ramasamy 2024). Among them, genetic and metabolic engineering stand out, allowing for the increasingly feasible modification, design, and optimization of microorganisms to serve as exceptional carotenoid biofactories (Li et al. 2020).

In essence, metabolic engineering aims to maximize the flow of carbon from substrates to the target compound while minimizing the flow to dispensable byproducts (Li et al. 2020). On the other hand, genetic engineering refers to techniques that employ the principles of metabolic engineering to artificially modify an organism, enabling it to produce a desired compound that is not naturally produced (Joshi et al. 2023).

The biosynthetic efficiency of most microorganisms is often affected by the limited number of precursors and the high competitiveness of branches during compound formation (Xu et al. 2023). However, advances in genetic and metabolic engineering allow for increased carotenoid synthesis through optimization or construction of artificial biosynthetic pathways.

Advances include (I) overexpression of key enzymes, (II) manipulation of specific genes, (III) mutation of carotenoid genes in non-carotenogenic microorganisms, and (IV) insertion of a storage membrane (Agarwal et al. 2023). Based on these advances, it is possible to optimize the levels of isoprenoid precursors and ensure efficient conversion into products of interest, consequently achieving satisfactory yields (Sundararajan and Ramasamy 2024).

Overexpression of key enzymes is based on increasing the expression level of enzymes at rate-limiting steps in the pathway. For example, the action of hydroxymethylglutaryl-CoA reductase is one of the rate-limiting steps in MVA. However, the application of optimized expression aids in the overproduction of the target compound (Li et al. 2020).

Furthermore, specific mutations also contribute to overproduction. β-carotene and ε-carotene derivatives increased by 300% and 200% through artificial fusions into CCD1 (carotenoid cleavage dioxygenase 1) and OfCCD1 (Osmanthus fragans carotenoid cleavage dioxygenase), respectively (Werner et al. 2019; Chen et al. 2019). To enhance lycopene production, the photosynthetic gene cluster repressor ppsR was removed from lycopene-producing Rb. *sphaeroides* strain (RL1) constructed in previous studies, aiming to disrupt the control of carotenoid synthesis by oxygen levels. Additionally, lycopene production was elevated through the overexpression of the activator prrA. The resulting superior lycopene producer, DppsR/OprrA, exhibited a high growth rate and achieved a lycopene production of 150.15 mg L^{-1} with a yield of 21.45 mg g^{-1} dry cell weight (DCW) under high oxygen conditions. Thus, specific genic modifications stand as useful techniques to improve carotenoid production in targeted bacterial strains, under industrial conditions (Chou et al. 2019; Jin et al. 2018; Qu et al. 2021; Zhu et al. 2012).

Referring to non-carotenogenic microorganisms, hosts such as *Saccharomyces cerevisiae*, and *E. coli* are frequently explored. However, *E. coli* stands out as the prominent host for constructing recombinant biosynthetic pathways (Ashokkumar et al. 2023). These hosts require heterologous genes to establish an isoprenoid biosynthesis pathway, as they lack them naturally. For example, the modification of the crtE/B/I genes from *Pantoea ananatis* to *Pantoea agglomerans* practically doubled lycopene production in *E. coli* (Yoon et al. 2007).

Considering that carotenoid productivity can be limited by the intracellular storage inability of these host microorganisms, as well as the toxicity that high intracellular concentrations of carotenoids can cause, the insertion of a storage membrane appears as an alternative solution (Li et al. 2020). A study conducted by Wu et al. (2017) showed that overexpression of enzymes that enhance membrane synthesis improved β-carotene production by 43% (glycerol-3-phosphate O-acyltransferase and a 1-acyl-sn-glycerol-3-phosphate acyltransferase) and lycopene by 37.5% (monoglucosyldiacylglycerol synthase).

In addition to the aforementioned advances, the use of CRISPRi/a (a bacterial adaptive immune system) for dynamic regulation of carotenoid production has also contributed to increasing production efficiency (Kampmann 2019). For example, recent enhancements in the synthetic biology toolbox for microalgae have been highlighted, potentially leading to significant improvements in their productivity (Naduthodi et al. 2021).

Additionally, genetic improvement of strains, a strategy that has existed for decades, continues to contribute to satisfactory results in increasing the production of microbial carotenoids (Ashokkumar et al. 2023). For instance, in a study conducted by Heider et al. (2012), genetic improvement of the bacterium *Corynebacterium glutamicum* resulted in a concentration 80 times higher compared to the wild-type strain.

Finally, as a result of increased productivity, these metabolic strategies yield immediate results contributing to the sustainable industrial production of pigments with commercial interest, as can be seen in the following Sect. 7.

7 Processing Downstream and Sustainability Metrics

The microbial production of carotenoids is intracellular, and after the upstream processes, a series of downstream unit operations are employed to recover these natural pigments. In the first stage, after cultivation, centrifugation, filtration, sedimentation, and/or flocculation operations are used to harvest the cellular biomass that contains the intracellular carotenoids. After this operation, drying using methods such as spraying and freeze-drying and cell disruption using mechanical and/or non-mechanical methods correspond to the pre-treatment process. Subsequently, extraction, which follows the pretreatment process, is commonly carried out using the supercritical fluid technique and, finally, purification applied to compounds used in high-value products is achieved through various chromatographic methods (Mussagy et al. 2019; Dias et al. 2020).

The extraction of microbial carotenoids is, in particular, a step that deserves attention. The good result of this process is intrinsically linked to the careful selection of the solvent(s), which can affect both the solubility and stability of the biological activities of carotenoids. Furthermore, regarding the choice of solvent(s), it is essential to consider attributes such as low toxicity, environmental friendliness, and high yield in terms of recovery and recyclability. Traditionally, carotenoid extraction from microbial biomasses has been carried out using conventional methods that involve the use of hazardous solvents such as acetone, chloroform, and hexane. These commonly used solvents pose a health and environmental risk due to toxicity, flammability, and pollution potential. In that regard, to overcome these bottlenecks, unconventional methods such as supercritical fluid extraction and enzyme-assisted extraction, which do not require dangerous solvents, have been promising alternatives to mitigate these concerns (Sharayei et al. 2021; Soldan et al. 2021; Sarkarat et al. 2023).

Still, about the step of purifying microbial carotenoids, depending on the commercial purpose, before applying chromatographic techniques, some attention must be paid to the presence of ester compounds. Removal of esters through saponification can improve the stability, solubility, or carotenoid absorption. In addition, after the saponification and purification step by chromatographic methods, normally, a subsequent step is still necessary where the product is formulated and conditioned to preserve the stability of the compounds. At this stage, drying (spray drying or freeze-drying) and crystallization are the two most commonly used methods (Mussagy et al. 2023).

In sum, it is noteworthy that there is no universal approach to carotenoid production, as this always depends on a combination of factors such as the origin of the microbial biomass and the nature of carotenoids. Each technique has advantages and disadvantages and the choice is determined especially based on the cost-benefit ratio and the desired final product. However, in any case, it is important to prioritize ecologically sustainable methods for the industrial recovery of carotenoids. Environmental sustainability is becoming imperative across all sectors (Efroymson et al. 2021).

In contemporary times, environmental sustainability is no longer seen as an option—it is a necessity—transcending the condition of choice to configure yourself as essential in building a sustainable tomorrow for future generations (Gelderman et al. 2021). In this sense, sustainable actions are gradually becoming an integral part of business practices, covering several sectors, such as the production of natural pigments. This transition to more sustainable practices meets the growing consumer demand for eco-friendly products and reflects a general awareness of the importance of preserving the environment. This benefits companies regarding reputation and market acceptance and contributes to a more positive impact on the planet (Ogrean and Herciu 2020).

Of note, to measure and report environmental performance, companies use various indicators, such as global warming potential and other specific indicators. Reports on the metrics of these indicators are often part of corporate social responsibility reports or sustainability reports. However, these reports are not always widely disseminated to the public. However, regardless, there is a growing literature on the environmental sustainability metrics of biological processes and products in favor of a circular bioeconomy (Kyriakopoulou et al. 2015; Onorato and Rösch 2020).

Comparatively, according to the results presented by Aldaghi et al. (2023), the chemical synthesis of astaxanthin—a carotenoid from the xanthophyll group—is the best option in environmental terms for producing the pigment, followed by production by the bacteria *Corynebacterium glutamicum* and microalgae *Haematococcus pluvialis*. When comparing chemical synthesis with biological processes, the first presents lower environmental impacts in the categories of abiotic depletion, acidification, eutrophication, freshwater aquatic ecotoxicity, global warming, human toxicity, marine aquatic ecotoxicity, photochemical oxidation, and terrestrial ecotoxicity. The ozone depletion potential category is the only one among those evaluated by the authors for which chemical synthesis presents—with little disparity—a greater environmental impact. The findings of this research align with the conclusions presented by Deprá et al. (2020), who, when evaluating the environmental sustainability of microalgae-based pigments, also conclude that synthetic pigments represent, to date, the most environmentally friendly route for carotenoid production.

However, while this is true, it is important to point out that microbial pigments may be superior to the synthetic form. Microbial pigments have been the subject of study due to their bioactive properties and the potential to offer functional benefits to human health. Furthermore, when migrating to a nature-centered production model, such as the bioeconomy, it is essential to highlight the role of biological products, such as microbial carotenoids, that are aligned with the principles of sustainability and renewability. Therefore, although they may not yet represent the best option, they are a crucial part of this transition. Besides that, it is noteworthy that interest in the production of microbial pigments, especially for industrial applications, is relatively more recent. Consequently, it is doable to foresee a future in which continuous developments and innovations improve the production of microbial carotenoids, establishing it as a more robust practice, with more sophisticated methods and techniques for a sustainable, reliable, and environmentally benign operation (Zepka et al. 2021; Nass et al. 2022a, b).

In the state-of-the-art, according to the study by Aldaghi et al. (2023), both the carotenoid production by bacteria and microalgae is energy intensive. The eco-smart replacement of fossil energy consumption with clean and renewable energy in unit operations can improve the environmental footprint of biological processes and products. Aldaghi et al. (2023) reported that the environmental impact of carotenoid production by bacteria and microalgae when replacing the conventional energy matrix with renewable energy from hydroelectric plants decreases by an average of 96.9% and 77.1%, respectively. Dias et al. (2022a), in proposing a roadmap for net-zero carbon emissions in commercial biologics, including carotenoids, concluded that the energy transition is the backbone for carbon-neutral operations in energy-intensive facilities. At the same time, considering the high energy consumption, the findings make it clear that the adoption of a renewable energy system, such as photovoltaic, could not only meet a portion of the facilities' energy demand but also would reduce the environmental footprint and electricity costs, thus improving the global sustainability of bio-based products (Dias et al. 2022b).

In sum, there are several opportunities for improvements in bio-based technology by mapping hot spots in production processes. By mapping hot spots, i.e., inefficient ones, it is possible to identify areas for enhancement. Some of these areas include: (i) enhancement of microorganism strains by genetic engineering, biochemical engineering, and transcription factor engineering to improve carotenoid production; (ii) advanced monitoring to control key variables of the biological processes; (iii) modeling computational to simulate and optimize biological processes; (iv) automation and robotics to ensure a consistent production environment, and (v) development of more efficient and flexible advanced bioreactors. These improvements can significantly contribute to the overall sustainability of biological products industrial. Finally, it is worth emphasizing that what emerges on the horizon is a new paradigm of goods and services production, oriented towards valuing nature and repositioning it back to the center of the economy. Therefore, bio-based products, such as carotenoids derived from microorganisms, are part of a flourishing sector of the current economy. Although, in some cases, they may not compete with their counterparts, they will continue to be actively pursued. The interest reflects growing awareness of the need to reduce dependence on non-renewable resources in line with a more eco-conscious vision.

8 Concluding Remarks

Finally, microorganisms such as fungi, bacteria, and microalgae are capable of synthesizing a variety of bioactive carotenoids, some of which are structurally unique (e.g., canthaxanthin and echinenone) using different metabolic pathways (MEP or MVA). These microorganisms have become essential in carotenoid production, aligning with the principles of global sustainability. This contribution is a result of metabolic and genetic engineering strategies that enable the overproduction of these compounds at an industrial level, in parallel with the exploration of green technologies in downstream processing. Although there is still progress to be made in improving sustainability metrics, bio-based products will continue to be actively demanded, reflecting the growing awareness of the need to reduce dependence on non-renewable resources, aligning with a more eco-friendly-conscious global vision.

References

Agarwal H, Bajpai S, Mishra A, Kohli I, Varma A, Fouillaud M, Dufossé L, Joshi NC (2023) Bacterial pigments and their multifaceted roles in contemporary biotechnology and pharmacological applications. Microorganisms 11(3):614. https://doi.org/10.3390/microorganisms11030614

Aksu Z, Eren AT (2005) Carotenoids production by the yeast *Rhodotorula mucilaginosa*: use of agricultural wastes as a carbon source. Process Biochem 40(9):2985–2991. https://doi.org/10.1016/j.procbio.2005.01.011

Aldaghi SA, Ubais R, Schmitt I, Wendisch VF, Costamagna M, Perucca M (2023) Life cycle assessment of bacterial, algal, and synthetic approaches for Astaxanthin production at a laboratory scale: comparative environmental analysis and sensitivity of energy sources. PRO 11(10):2911

Ashokkumar V, Flora G, Sevanan M, Sripriya R, Chen WH, Park JH, Kumar G (2023) Technological advances in the production of carotenoids and their applications–a critical review. Bioresour Technol 367:128215

Borowitzka M (2013) High-value products from microalgae-their development and commercialization. J Appl Phycol 25:743–756. https://doi.org/10.1007/s10811-013-9983-9

Borowitzka M (2018) Commercial-scale production of microalgae for bioproducts. In: Blue biotechnology: production and use of marine molecules, vol 1, pp 33–65. https://doi.org/10.1002/9783527801718.ch2

Brahma D, Dutta D (2022) Antioxidant property of beta-cryptoxanthin produced by Kocuria marina DAGII. Mater Today Proc 57:1833–1837. https://doi.org/10.1016/J.MATPR.2022.01.027

Britton G (2020) Carotenoid research: history and new perspectives for chemistry in biological systems. Biochim Biophys Acta Mol Cell Biol Lipids 1865:158699. https://doi.org/10.1016/j.bbalip.2020.158699

Caseiro M, Ascenso A, Costa A, Creagh-Flynn J, Johnson M, Simões S (2020) Lycopene in human health. LWT 127:109323. https://doi.org/10.1016/j.lwt.2020.109323

Chen X, Shukal S, Zhang C (2019) Integrating enzyme and metabolic engineering tools for enhanced α-ionone production. J Agric Food Chem 67(49):13451–13459. https://doi.org/10.1021/acs.jafc.9b00860

Chou YL, Ko CY, Yen CC, Chen LFO, Shaw JF (2019) Multiple promoters driving the expression of astaxanthin biosynthesis genes can enhance free-form astaxanthin production. J Microbiol Methods 160:20–28. https://doi.org/10.1016/J.MIMET.2019.03.012

Cortes AG, Vásquez JAG, Díaz YCA, Castrillón MR (2020) Effects of cellular stress on pigment production in *Rhodotorula mucilaginosa/alborubescens* AJB01 strain from the Caribbean region of Colombia. bioRxiv, 2020-05. https://doi.org/10.1101/2020.05.20.107201

Cunningham FX Jr, Gantt E (2001) One ring or two? Determination of ring number in carotenoids by lycopene ε-cyclases. Proc Natl Acad Sci 98:2905–2910

Dai R, Wang P, Jia P, Zhang Y, Chu X, Wang Y (2016) A review on factors affecting microcystins production by algae in aquatic environments. World J Microbiol Biotechnol 32:1–7

Deprá MC, dos Santos AM, Jacob-Lopes E (2020) Sustainability metrics in the microalgae-based pigments production: a life cycle assessment approach. In: Pigments from microalgae handbook, pp 363–390

Dias RR, Severo IA, Deprá MC, Maroneze MM, Zepka LQ, Jacob-Lopes E (2020) The next-generation of microalgae-based products. In: Microbial enzymes and biotechniques: interdisciplinary perspectives, pp 15–42

Dias RR, Deprá MC, Zepka LQ, Jacob-Lopes E (2022a) Roadmap to net-zero carbon emissions in commercial microalgae-based products: environmental sustainability and carbon offset costs. J Appl Phycol 34(3):1255–1268

Dias RR, Deprá MC, Severo IA, Zepka LQ, Jacob-Lopes E (2022b) Smart override of the energy matrix in commercial microalgae facilities: a transition path to a low-carbon bioeconomy. Sustain Energy Technol Assess 52:102073

Disch A, Schwender J, Müller C, Lichtenthaler HK, Rohmer M (1998) Distribution of the mevalonate and glyceraldehyde phosphate/pyruvate pathways for isoprenoid biosynthesis in unicellular algae and the cyanobacterium *Synechocystis* PCC 6714. Biochem J 333:381–388

Dyaa A, Soliman H, Abdelrazak A, Samra BN, Khojah E, Ahmed AF, El-Esawi MA, Elsayed A (2022) Optimization of carotenoids production from *Rhodotorula* sp. strain ATL72 for enhancing its biotechnological applications. J Fungi 8(2):160. https://doi.org/10.3390/jof8020160

Efroymson RA, Jager HI, Mandal S, Parish ES, Mathews TJ (2021) Better management practices for environmentally sustainable production of microalgae and algal biofuels. J Clean Prod 289:125150

Egeland ES (2016) Carotenoids. In: Borowitzka MA, Beardall J, Raven JA (eds) The physiology of microalgae. Springer, New York/Dordrecht/London, pp 507–563

El-Agamey A, Lowe GM, McGarvey DJ, Mortensen A, Phillip DM, Truscott TG et al (2004) Carotenoid radical chemistry and antioxidant/pro-oxidant properties. Arch Biochem Biophys 430(1):37–48

Fernandes AS, Pinheiro PN, Deprá MC, Jacob-Lopes E, Zepka LQ (2017) Carotenoids in microalgae. In: Carotenoids biosynthesis, properties and physiological effects. Nova Science Publishers

Fernandes AS, Petry FC, Mercadante AZ, Jacob-Lopes E, Zepka LQ (2020) HPLC-PDA-MS/MS as a strategy to characterize and quantify natural pigments from microalgae. Curr Res Food Sci 3:100–112

Foong LC, Loh CWL, Ng HS, Lan JC-W (2021) Recent development in the production strategies of microbial carotenoids. World J Microbiol Biotechnol 37:12. https://doi.org/10.1007/s11274-020-02967-3

Fu Y, Wang Y, Yi L, Liu J, Yang S, Liu B et al (2023) Lutein production from microalgae: a review. Bioresour Technol 376:128875

Gelderman CJ, Schijns J, Lambrechts W, Vijgen S (2021) Green marketing as an environmental practice: the impact on green satisfaction and green loyalty in a business-to-business context. Bus Strateg Environ 30(4):2061–2076

George DM, Vincent AS, Mackey HR (2020) An overview of anoxygenic phototrophic bacteria and their applications in environmental biotechnology for sustainable resource recovery. Biotechnol Rep 28:e00563. https://doi.org/10.1016/J.BTRE.2020.E00563

Gericó TG, Tavanti RFR, de Oliveira SC, Lourenzani AEBS, de Lima JP, Ribeiro RP, dos Santos LCC, dos Reis AR (2020) Bradyrhizobium sp. enhance ureide metabolism increasing peanuts yield. Arch Microbiol 202(3):645–656. https://doi.org/10.1007/s00203-019-01778-x

Gich F, Overmann J (2006) Sandarakinorhabdus limnophila gen. nov., sp. nov., a novel bacteriochlorophyll a-containing, obligately aerobic bacterium isolated from freshwater lakes. Int J Syst Evol Microbiol 56(4):847–854. https://doi.org/10.1099/ijs.0.63970-0

Giuffrida D, Monnet C, Laurent F, Cacciola F, Oteri M, Le Piver M, Caro Y, Donato P, Mondello L, Roueyre D, Dufossé L (2020) Carotenoids from the ripening bacterium Brevibacterium linens impart color to the rind of the French cheese, Fourme de Montbrison (PDO). Nat Prod Res 34(1):10–15. https://doi.org/10.1080/14786419.2019.1622107

Goswami G, Chaudhuri S, Dutta D (2010) The present perspective of astaxanthin with reference to biosynthesis and pharmacological importance. World J Microbiol Biotechnol 26(11):1925–1939. https://doi.org/10.1007/s11274-010-0373-z

Guan X, Zhang J, Xu N, Cai C, Lu Y, Liu Y, Dai W, Wang X, Nan B, Li X, Wang Y (2023) Optimization of culture medium and scale-up production of astaxanthin using corn steep liquor as substrate by response surface methodology. Prep Biochem Biotechnol 53(4):443–453. https://doi.org/10.1080/10826068.2022.2098324

Guyomarc'h F, Binet A, Dufossé L (2000) Production of carotenoids by Brevibacterium linens: variation among strains, kinetic aspects and HPLC profiles. J Ind Microbiol Biotechnol 24(1):64–70. https://doi.org/10.1038/sj.jim.2900761

Heider SA, Peters-Wendisch P, Wendisch VF (2012) Carotenoid biosynthesis and overproduction in *Corynebacterium glutamicum*. BMC Microbiol 10(12):198. https://doi.org/10.1186/1471-2180-12-198

Hotos GN, Avramidou D (2021) The effect of various salinities and light intensities on the growth performance of five locally isolated microalgae [*Amphidinium carterae*, *Nephroselmis sp.*,

Tetraselmis sp.(var. red pappas), *Asteromonas gracilis* and *Dunaliella sp.*] in laboratory batch cultures. J Marine Sci Eng 9(11):1275
Igreja WS, Maia FDA, Lopes AS, Chisté RC (2021) Biotechnological production of carotenoids using low cost-substrates is influenced by cultivation parameters: a review. Int J Mol Sci 22(16):8819. https://doi.org/10.3390/ijms22168819
Jacob-Lopes E, Zepka LQ (2023) Food and feed from microalgae: A historical perspective to future directions. In: Handbook of food and feed from microalgae. Academic, pp 3–7
Jacob-Lopes E, Maroneze MM, Deprá MC, Sartori RB, Dias RR, Zepka LQ (2019) Bioactive food compounds from microalgae: An innovative framework on industrial biorefineries. Curr Opin Food Sci 25:1–7
Jenkins CL, Andrewes AG, McQuade TJ, Starr MP (1979) The pigment of Pseudomonas paucimobilis is a carotenoid (Nostoxanthin), rather than a brominated aryl-polyene (xanthomonadin). Curr Microbiol 3(1):1–4. https://doi.org/10.1007/BF02603124
Jin J, Wang Y, Yao M, Gu X, Li B, Liu H, Ding M, Xiao W, Yuan Y (2018) Astaxanthin overproduction in yeast by strain engineering and new gene target uncovering. Biotechnol Biofuels 11(1):230. https://doi.org/10.1186/s13068-018-1227-4
Joshi K, Kumar P, Kataria R (2023) Microbial carotenoid production and their potential applications as antioxidants: a current update. Process Biochem 128:190–205. https://doi.org/10.1016/j.procbio.2023.02.020
Kadri MS, Singhania RR, Anisha GS, Gohil N, Singh V, Patel AK, Patel AK (2023) Microalgal lutein: Advancements in production, extraction, market potential, and applications. Bioresour Technol 129808 https://doi.org/10.1016/j.biortech.2023.129808
Kampmann M (2019) CRISPRi and CRISPRa screens in mammalian cells for precision biology and medicine. ACS Chem Biol 13(2):406–416. https://doi.org/10.1021/acschembio.7b00657
Kang CD, Lee JS, Park TH, Sim SJ (2005) Comparison of heterotrophic and photoautotrophic induction on astaxanthin production by *Haematococcus pluvialis*. Appl Microbiol Biotechnol 68:237–241
Karmann S, Panke S, Zinn M (2019) Fed-batch cultivations of rhodospirillum rubrum under multiple nutrient-limited growth conditions on syngas as a novel option to produce poly(3-hydroxybutyrate) (PHB). Front Bioeng Biotechnol 7(APR). https://doi.org/10.3389/fbioe.2019.00059
Katsumata T, Ishibashi T, Kyle D (2014) A sub-chronic toxicity evaluation of a natural astaxanthin-rich carotenoid extract of Paracoccus carotinifaciens in rats. Toxicol Rep 1. https://doi.org/10.1016/j.toxrep.2014.08.008
Khan MI, Shin JH, Kim JD (2018) The promising future of microalgae: current status, challenges, and optimization of a sustainable and renewable industry for biofuels, feed, and other products. Microb Cell Factories 17(1):1–21
Kikukawa H, Okaya T, Maoka T, Miyazaki M, Murofushi K, Kato T, Hirono-Hara Y, Katsumata M, Miyahara S, Hara KY (2021) Carotenoid nostoxanthin production by sphingomonas sp. Sg73 isolated from deep sea sediment. Mar Drugs 19(5). https://doi.org/10.3390/md19050274
Kusmita L, Nur Prasetyo Edi A, Dwi Franyoto Y, Mutmainah, Haryanti S, Dwi Retno Nurcahyanti A (2023) Sun protection and antibacterial activities of carotenoids from the soft coral Sinularia sp. symbiotic bacteria from Panjang Island, North Java Sea. Saudi Pharm J 31(8):101680. https://doi.org/10.1016/J.JSPS.2023.06.013
Kyriakopoulou K, Papadaki S, Krokida M (2015) Life cycle analysis of β-carotene extraction techniques. J Food Eng 167:51–58
Li C, Swofford SA, Sinskey AJ (2020) Modular engineering for microbial production of carotenoids. Metab Eng Commun 10:e00118. https://doi.org/10.1016/j.mec.2019.e00118
Li Y, Qi Z, Fan Y, Tang Y, Zhou R (2022) The concurrent production of lipid and lutein in Chlorella vulgaris triggered by light coupling nitrogen tactics. Biochem Eng J 182:108435
Lichtenthaler HK (2010) The non-mevalonate DOXP/MEP (deoxyxylulose 5- phosphate/methylerythritol 4-phosphate) pathway of chloroplast isoprenoid and pigment biosynthesis. In: Rebeiz CA, Benning C, Bohnert HJ, Daniell H, Hoober JK, Lichtenthaler HK, Portis AR, Tripathy BC (eds) The chloroplast. Basics and applications, advances in photosynthesis and respiration, vol 31. Springer, Dordrecht, pp 95–118

Liu C, Hu B, Cheng Y, Guo Y, Yao W, Qian H (2021) Carotenoids from fungi and microalgae: a review on their recent production, extraction, and developments. Bioresour Technol 337:125398. https://doi.org/10.1016/j.biortech.2021.125398

Lohr M, Schwender J, Polle JEW (2012) Isoprenoid biosynthesis in eukaryotic phototrophs: a spotlight on algae. Plant Sci 185–186:9–22

Lorquin J, Molouba F, Dreyfus BL (1997) Identification of the carotenoid pigment canthaxanthin from photosynthetic Bradyrhizobium strains. Appl Environ Microbiol 63(3):1151–1154. https://doi.org/10.1128/aem.63.3.1151-1154.1997

Mageswari A, Subramanian P, Srinivasan R, Karthikeyan S, Gothandam KM (2015) Astaxanthin from psychrotrophic Sphingomonas faeni exhibits antagonism against food-spoilage bacteria at low temperatures. Microbiol Res 179:38–44. https://doi.org/10.1016/J.MICRES.2015.06.010

Mantzouridou F, Roukas T, Kotzekidou P (2002) Effect of the aeration rate and agitation speed on β-carotene production and morphology of *Blakeslea trispora* in a stirred tank reactor: mathematical modeling. Biochem Eng J 10(2):123–135. https://doi.org/10.1016/s1369-703x(01)00166-8

Maroneze MM, Jacob-Lopes E, Zepka LQ, Roca M, Pérez-Gálvez A (2019) Esterified carotenoids as new food components in cyanobacteria. Food Chem 287:295–302

Martínez Andrade KA, Lauritano C, Romano G, Ianora A (2018) Marine microalgae with anti-cancer properties. Mar Drugs 16(5):165

McClure DD, Nightingale JK, Luiz A, Black S, Zhu J, Kavanagh JM (2019) Pilot-scale production of lutein using Chlorella vulgaris. Algal Res 44:101707

Mussagy CU, Winterburn J, Santos-Ebinuma VC, Pereira JFB (2019) Production and extraction of carotenoids produced by microorganisms. Appl Microbiol Biotechnol 103:1095–1114

Mussagy CU, Kot A, Dufossé L, Gonçalves CN, Pereira JF, Santos-Ebinuma VC et al (2023) Microbial astaxanthin: from bioprocessing to the market recognition. Appl Microbiol Biotechnol 107:1–17

Naduthodi MIS, Claassens NJ, D'Adamo S, van Oost J, Barbosa MJ (2021) Synthetic biology approaches to enhance microalgal productivity. Trends Biotechnol 39(10):1019–1036. https://doi.org/10.1016/j.tibtech.2020.12.010

Naik R, Gupte S (2023) Characterization of pigment produced by high carotenoid yielding bacteria Paracoccus marcusii RSPO1 and evaluation of its anti-diabetic, anti-microbial and antioxidant properties. Nat Prod Res 38:968–977. https://doi.org/10.1080/14786419.2023.2208358

Nascimento TC, Pinheiro PN, Fernandes AS, Murador DC, Neves BV, de Menezes CR et al (2021) Bioaccessibility and intestinal uptake of carotenoids from microalgae *Scenedesmus obliquus*. LWT 140:110780

Nass PP, do Nascimento TC, Fernandes AS, Caetano PA, de Rosso VV, Jacob-Lopes E, Zepka LQ (2022a) Guidance for formulating ingredients/products from *Chlorella vulgaris* and *Arthrospira platensis* considering carotenoid and chlorophyll bioaccessibility and cellular uptake. Food Res Int 157:111469

Nass PP, Do Nascimento TC, Fernandes AS, Zepka LQ, Jacob-Lopes E (2022b) Food bioactive compounds from microalgae. In: Algal genetic resources: cosmeceuticals, nutraceuticals, and pharmaceuticals from Algae, p 45

Ogrean C, Herciu M (2020) Business models addressing sustainability challenges—towards a new research agenda. Sustain For 12(9):3534

Onorato C, Rösch C (2020) Comparative life cycle assessment of astaxanthin production with Haematococcus pluvialis in different photobioreactor technologies. Algal Res 50:102005

Patel AK, Albarico FPJB, Perumal PK, Vadrale AP, Nian CT, Chau HTB et al (2022) Algae as an emerging source of bioactive pigments. Bioresour Technol 351:126910

Patias LD, Fernandes A, Petry FC, Mercadante AZ, Jacob-Lopes E, Zepka LQ (2017) Carotenoid profile of three microalgae/cyanobacteria species with peroxyl radical scavenger capacity. Food Res Int 100:260–266

Philp J, Atlas R (2017) Microbial resources for global sustainability. In: Microbial resources from functional existence in nature to applications, pp 77–101. https://doi.org/10.1016/B978-0-12-804765-1.00004-7

Qu Y, Su A, Li Y, Meng Y, Chen Z (2021) Manipulation of the regulatory genes ppsR and prrA in Rhodobacter sphaeroides enhances lycopene production. J Agric Food Chem 69(14):4134–4143. https://doi.org/10.1021/acs.jafc.0c08158

Ram S, Mitra M, Shah F, Tirkey SR, Mishra S (2020 Bacteria as an alternate biofactory for carotenoid production: A review of its applications, opportunities and challenges. J Funct Foods 67:103867. https://doi.org/10.1016/J.JFF.2020.103867

Rangel-Basto YA, Ramos-Valdivia AC, Rojas CMCG, Ponce-Noyola T (2022) Saccharified sugarcane bagasse as a substrate for astaxanthin production by *Xanthophyllomyces dendrorhous*. Biomass Convers Biorefinery 14:8071–8079. https://doi.org/10.1007/s13399-022-02934-y

Rodrigues E, Mariutti LRB, Chisté RC, Mercadante AZ (2012) Development of a novel microassay for evaluation of peroxyl radical scavenger capacity: application to carotenoids and structure-activity relationship. Food Chem 135(3):2103–2111

Rodrigues DB, Flores ÉM, Barin JS, Mercadante AZ, Jacob-Lopes E, Zepka LQ (2014) Production of carotenoids from microalgae cultivated using agroindustrial wastes. Food Res Int 65:144–148

Rodrigues DB, Menezes CR, Mercadante AZ, Jacob-Lopes E, Zepka LQ (2015) Bioactive pigments from microalgae *Phormidium autumnale*. Food Res Int 77:273–279

Rodriguez-Amaya DB (2015) Food carotenoids: chemistry, biology and technology. IFT Press/Wiley, Oxford

Rodriguez-Amaya DB (2019) Bioactive carotenes and Xanthophylls in plant foods. Encycl Food Chem:260–266. https://doi.org/10.1016/B978-0-08-100596-5.22359-7

Rodriguez-Amaya DB, Kimura M, Godoy HT, Amaya-Farfan J (2008) Updated Brazilian database on food carotenoids: factors affecting carotenoid composition. J Food Compos Anal 21:445–463

Sandmann G (2022) Carotenoids and their biosynthesis in fungi. Molecules 27(4):1431. https://doi.org/10.3390/molecules27041431

Sarkarat R, Mohamadnia S, Tavakoli O (2023) Recent advances in non-conventional techniques for extraction of phycobiliproteins and carotenoids from microalgae. Braz J Chem Eng 40(2):321–342

Schwender J, Zeidler J, Gröner R, Müller C, Focke M, Braun S, Lichtenthaler FW, Lichtenthaler HK (1997) Incorporation of 1-deoxy-d-xylulose into isoprene and phytol by higher plants and algae. FEBS Lett 414:129–134

Sharayei P, Azarpazhooh E, Zomorodi S, Einafshar S, Ramaswamy HS (2021) Optimization of ultrasonic-assisted extraction of astaxanthin from green tiger (Penaeus semisulcatus) shrimp shell. Ultrason Sonochem 76:105666

Sharma R, Ghoshal G (2020) Optimization of carotenoids production by *Rhodotorula mucilaginosa* (MTCC-1403) using agro-industrial waste in bioreactor: a statistical approach. Biotechnol Rep 25:e00407. https://doi.org/10.1016/j.btre.2019.e00407

Soldan ACF, Arvelos S, Watanabe EO, Hori CE (2021) Supercritical fluid extraction of oleoresin from Capsicum annuum industrial waste. J Clean Prod 297:126593

Su A, Chi S, Li Y, Tan S, Qiang S, Chen Z, Meng Y (2018) Metabolic redesign of Rhodobacter sphaeroides for lycopene production. J Agric Food Chem 66(23):5879–5885. https://doi.org/10.1021/acs.jafc.8b00855

Sundararajan PJ, Ramasamy SP (2024) Current perspectives on industrial application of microbial carotenoid as an alternative to synthetic pigments. Sustain Chem Pharm 37:101353. https://doi.org/10.1016/j.scp.2023.101353

Takaichi S (2020) Carotenoids in algae: distributions, biosyntheses and functions. Mar Drugs 9:1101–1118

Teller JT (1987) The pink colour of lakes, with an example from Australia. J Arid Environ 12(2):101–103

Wang GS, Grammel H, Abou-Aisha K, Sägesser R, Ghosh R (2012) High-level production of the industrial product Lycopene by the photosynthetic Bacterium rhodospirillum rubrum. Appl Environ Microbiol 78(20):7205–7215. https://doi.org/10.1128/AEM.00545-12

Wang RQ, Chen G, Chen SN, Zhu HL, Xiong WN, Xu M, Jian SP (2021) Metabolic changes of *Neurospora crassa* in the presence of oleic acid for promoting lycopene production. J Biosci Bioeng 132(2):148–153. https://doi.org/10.1016/j.jbiosc.2021.04.003

Wen C, Xu X, Zhou D, Yu Q, Wang T, Zhou Y (2022) The effects of canthaxanthin microencapsulation on yolk color and canthaxanthin deposition in egg yolk of laying hens. Poult Sci 101(6):101889. https://doi.org/10.1016/J.PSJ.2022.101889

Werner N, Ramirez-Sarmiento CA, Agosin E (2019) Protein engineering of carotenoid cleavage dioxygenases to optimize β-ionone biosynthesis in yeast cell factories. Food Chem 299(2019):125089. https://doi.org/10.1016/j.foodchem.2019.125089

Wood EE, Ross ME, Jubeau S, Montalescot V, MacKechnie K, Marchington RE et al (2023) From green to orange: the change in biochemical composition of phototrophic-mixotrophic Chromochloris zofingiensis in pilot-scale photobioreactors. Algal Res 75:103238

Wu Z, Duangmanee P, Zhao P, Juntawong N, Ma C (2016) The effects of light, temperature, and nutrition on growth and pigment accumulation of three *Dunaliella salina* strains isolated from saline soil. Jundishapur J Microbiol 9(1):e26732

Wu T, Ye L, Zhao D, Li S, Li Q, Zhang B, Bi C, Zhang X (2017) Membrane engineering – a novel strategy to enhance the production and accumulation of β-carotene in *Escherichia coli*. Metab Eng 43(2017):85–91. https://doi.org/10.1016/j.ymben.2017.07.001

Xie Y, Xiong X, Chen S (2021) Challenges and potential in increasing lutein content in microalgae. Microorganisms 9(5):1068

Xu S, Gao S, An Y (2023) Research progress of engineering microbial cell factories for pigment production. Biotechnol Adv 65:108150. https://doi.org/10.1016/j.biotechadv.2023.108150

Yokoyama A, Izumida H, Miki W (1994) Production of astaxanthin and 4-ketozeaxanthin by the marine bacterium, agrobacterium aurantiacum. Biosci Biotechnol Biochem 58(10):1842–1844. https://doi.org/10.1271/bbb.58.1842

Yoon S-H et al (2007) Engineering the lycopene synthetic pathway in E. coli by comparison of the carotenoid genes of *Pantoea agglomerans* and *Pantoea ananatis*. Appl Microbiol Biotechnol 74(1):131–139. https://doi.org/10.1007/s00253-006-0623-z

Young AJ, Lowe GL (2018) Carotenoids—antioxidant properties. Antioxidants 7(2). https://doi.org/10.3390/antiox7020028

Zepka LQ, Do Nascimento TC, Jacob-Lopes E (eds) (2021) Bioactive compounds: biosynthesis, characterization and applications. BoD–Books on Demand

Zhu L, Wu X, Li O, Qian C, Gao H (2012) Cloning and characterization of genes involved in nostoxanthin biosynthesis of sphingomonas elodea ATCC 31461. PLoS One 7(4). https://doi.org/10.1371/journal.pone.0035099

Chapter 10
Composition and Health Benefits of Carotenoids in Foods and Feeds

Hamdy A. Shaaban, Haitham H. Abdel-Gawad, and Mahassen M. Sleem

1 Introduction

Carotenoids are widely occurring compounds in nature, primarily synthesized by photosynthetic organisms like cyanobacteria, algae, and plants, as well as certain fungi and bacteria. While most animals cannot produce carotenoids, they can incorporate and modify them through dietary intake. Notably, specific invertebrate animals, including hemipteran and dipteran insects, mites, and gall midges, exhibit the ability to synthesize carotenoids de novo (Rodrıguez-Concepcion et al. 2018).

Since the 1980s, there has been a growing interest in carotenoids as potential health-promoting compounds. While the nutritional significance of provitamin A carotenoids is acknowledged, proving their direct impact on health is challenging due to dietary complexity and the intricacies of the human body. Nevertheless, evidence from diverse studies, including epidemiological, chemical, animal, and cell culture studies, suggests that the regular consumption of carotenoids in normal diets could yield health benefits.

Optimal carotenoid intake is associated with a reduced risk of certain cancers (cervical, ovarian, colorectal, prostate, breast), cardiovascular disease, bone, skin, or eye disorders. Recent research also indicates potential significance in mental health, metabolic health, during pregnancy, early life, and even cosmetic benefits (Melendez-Martınez 2019; Melendez-Martınez et al. 2018). Although the commonly attributed beneficial actions of carotenoids involve antioxidant mechanisms,

H. A. Shaaban (✉)
Department of Chemistry of Flavor and Aroma, National Research Centre, Dokki, Giza, Egypt

H. H. Abdel-Gawad
Egyptian Drug Authority, EDA, Giza, Egypt

M. M. Sleem
Food Technology Research Institute, Agriculture Research Centre, Giza, Egypt

R. Campos Chisté et al. (eds.), *Carotenoids*,
https://doi.org/10.1007/978-3-031-75322-0_10

it's important to note the existence of other potential mechanisms, including pro-oxidant actions, enhancement of intercellular communication, modulation of signaling pathways, absorption of visible light, or modulation of membrane properties, acting in conjunction. Moreover, there is accumulating evidence that oxidative cleavage derivatives of carotenoids, apart from retinoids, can be biologically active in humans, contributing to the health benefits associated with carotenoids (Melendez-Martınez 2019; Rodrıguez-Concepcion et al. 2018).

Considering the significance of dietary carotenoids in nutrition and health, along with evolutionary aspects, it's noteworthy that humans and their ancestors have traditionally consumed green leaves containing substantial amounts of β-carotene and lutein. While β-carotene can be converted into vitamin A compounds, lutein accumulates notably in tissues like the macula lutea and the brain. Interestingly, lutein and other carotenoids are part of the newborns' diet as they are present in mothers' milk, contributing to the yellowish color of colostrum, the initial breastfed baby food (Johnson 2014; Sommerburg et al. 2000).

The primary aim of this Chapter is to consolidate recent information on the bioactive properties of carotenoids and underscore their potential health benefits in foods and feeds.

2 Definition, Chemical Structure and Classification of Carotenoids

Carotenoids, natural pigments widely dispersed in nature, contribute to the vibrant colors seen in fruits, vegetables, and flowers. The term "carotenoids" is derived from carrots (*Daucus carota*), an exceptional plant that accumulates high levels of these pigments in its root (Stange and Rodriguez-Concepcion 2015). These compounds serve various purposes, acting as food colorants, feed supplements, and finding applications in cosmetics and pharmaceuticals.

Typically composed of eight condensed 5-carbon-containing isoprene units arranged in a head-to-tail pattern, carotenoids form a polyene chain structure with 9–11 conjugated double bonds, sometimes terminating in rings (Fig. 10.1). This structural arrangement dictates their characteristic colors, ranging from yellow to red (Mattea et al. 2009; Vershinin 1999). The number and configuration of these double bonds result in stereoisomers, denoted as *E*- and *Z*-isomers, where *E* means *trans* configuration and *Z* indicates *cis* configuration (Vincent et al. 2017). Synthesized by photosynthetic organisms, as well as certain non-photosynthetic bacteria and fungi (Rodriguez-Amaya 2015), carotenoids play a crucial role in capturing light energy for photosynthesis, absorbing wavelengths between 400 and 550 nm (violet to green light) (Vershinin 1999; Zakynthinos and Varzakas 2016), resulting in their vivid yellow, orange, or red hues.

Over 750 naturally occurring carotenoids have been identified, synthesized de novo by plants, algae, fungi, and bacteria, although only around 50 exhibit

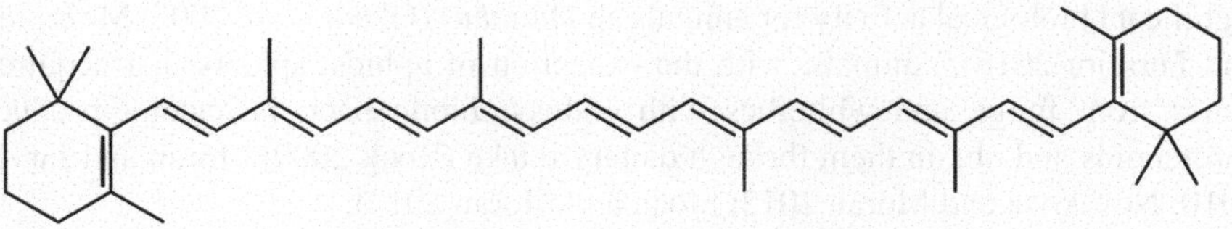

β-carotene

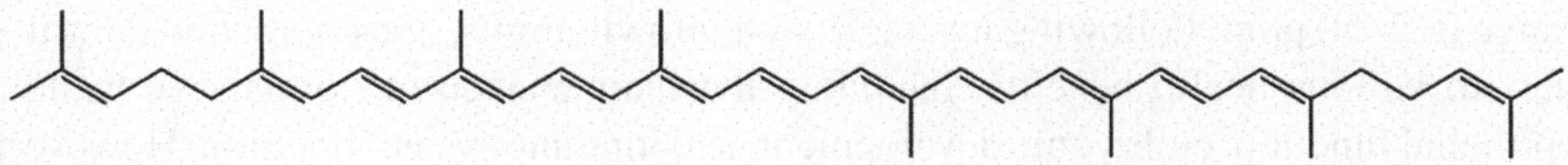

Lycopene

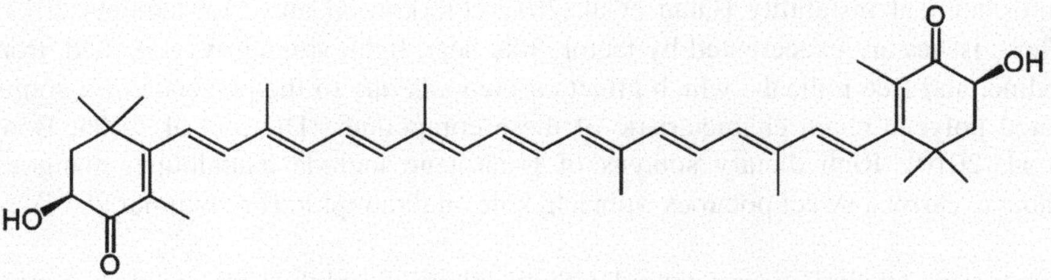

Astaxanthin

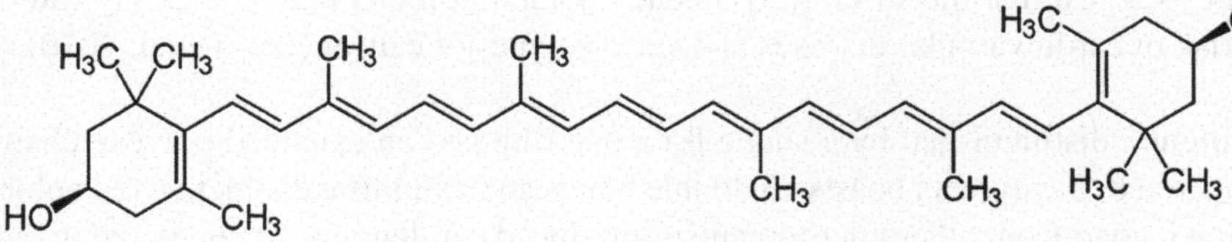

Zeaxanthin

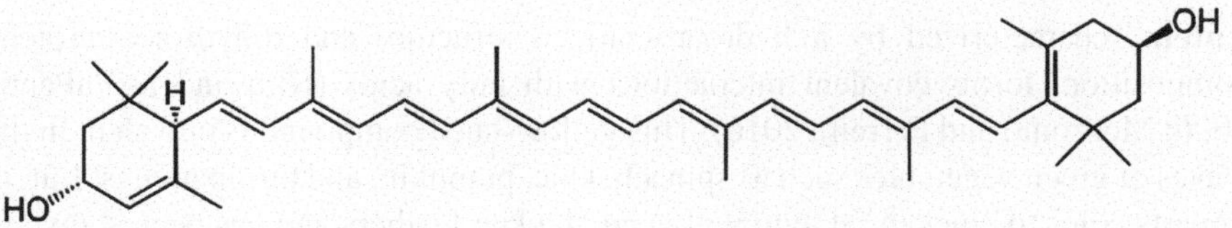

Lutein

Fig. 10.1 Chemical structure of some common carotenoids

significant biological activity for animals and humans (Okada et al. 2008; Mezzomo and Ferreira 2016). Animals, with the exception of aphids, spiders that acquired genes from fungi, and whiteflies with endosymbiotic bacteria, cannot produce carotenoids and obtain them through dietary intake (Rock 2009; Moran and Jarvik 2010; Nováková and Moran 2012; Sloan and Moran 2012).

β-carotene, utilized as both a food colorant and supplement with antioxidant capabilities and provitamin A activity, is typically employed within a concentration range of 2–50 ppm. Following its conversion into vitamin A, the compound contributes to various health benefits, including the maintenance of normal eye health, epithelial function, embryonic development, and immune system function. However, the broader application of carotenoids in food formulations is presently restricted due to their inherent challenges such as poor water solubility, high melting point, and chemical instability (Qian et al. 2012; Piorkowski and McClements 2014). These issues are exacerbated by factors like heat, light, singlet oxygen, acid, iron, iodine, and free radicals, which affect carotenoids due to the presence of a conjugated polyene chain characteristic of these compounds (Dutta et al. 2005; Boon et al. 2010). Rich dietary sources of β-carotene include cantaloupe, mangoes, papaya, carrots, sweet potatoes, spinach, kale, and pumpkin (Premkumar 2014).

Lycopene, an aliphatic carotenoid with oil solubility, exhibits enhanced absorption by the digestive system in the presence of oils (Story et al. 2010). Abundant in tomatoes, watermelons, pink grapefruit, and papaya, lycopene stands out as the most effective singlet oxygen scavenger in vitro (Sies and Stahl 1995). Epidemiological studies suggest that diets rich in lycopene-containing foods may potentially lower the risk of cardiovascular diseases and various types of cancer (Boon et al. 2010).

Torulene, distinguished by a single β-ionone ring and an extended polyene chain compared to β-carotene, boasts 13 double bonds and exhibits a distinctive red color, setting it apart from other microorganism-produced carotenoids. Encountered in red yeasts like Rhodotorula glutinis and Sporobolomyces ruberrimus, its structural attributes suggest potential antitumoral and immune-enhancing properties, rendering it suitable for applications as a food, feed, and cosmetic additive (Zoz et al. 2015).

Lutein, characterized by a hydroaromatic α structure and dihydroxy-carotene composition, forms covalent interactions with fatty acids (Kim and DellaPenna 2006; Mezzomo and Ferreira 2016). This yellow-hued compound is prevalent in the leaves of green vegetables such as spinach, kale, pumpkin, and turnip greens. Lutein imparts color to chicken fat, egg yolk, and chicken feathers and has been shown to enhance the efficacy of vaccination against the infectious bronchitis virus in laying hens (McGraw et al. 2003). Notably, lutein, along with zeaxanthin, plays a crucial role in eye health as the sole carotenoids located in the macula lutea of the retina, potentially offering benefits in age-related macular degeneration (Premkumar 2014).

Astaxanthin, a dark-red organic pigment, stands out as one of the most valuable compounds found in microalgae. The Haematococcu spluvialis strain is recognized as the richest source of astaxanthin (Mezzomo and Ferreira 2016; Zaheer 2017). This pigment imparts the characteristic pinkish-red color to salmonids (salmons and trout), shrimps, lobsters, and crabs, playing a vital role in their immune system and fertility. Astaxanthin is a nutritional powerhouse, serving as one of the most potent antioxidants in nature, effectively scavenging free radicals within the human body (Koller et al. 2014). Much like lutein, astaxanthin has demonstrated positive contributions to various human health issues, including UV-light protection, anti-inflammatory properties, support for eye health (assisting with diabetic retinopathy, macular degeneration, eye strain, fatigue, and fine detail vision), and immune-boosting effects (enhancing the ability of protective white blood cells to surround and eliminate infecting organisms, particularly fungi). Additionally, astaxanthin has shown promise in preventing different types of cancer, such as colon and breast cancer (Amaya et al. 2014a, b).

Canthaxanthin, an orange-red oxycarotenoid present in crustaceans, fungi, and algae, is commonly employed for color enhancement in feeding. Notably, canthaxanthin exhibits antioxidant properties and promotes junctional communication between cells, either directly or by forming 4-oxo-retinoic acid, which also possesses the ability to activate the retinoic acid receptor (Amaya et al. 2014a, b).

3 Sources of Carotenoids

Carotenoid pigments are prevalent in the plant kingdom and are also present in bacteria, yeast, molds, and fungi. While the human and animal body cannot naturally synthesize carotenoids, they are found in human and animal tissues as they are absorbed and deposited through food intake (Singh et al. 2015).

3.1 Plants

Carotenoids play crucial roles in essential processes for normal growth and development in plants, contributing to the vibrant array of yellow to red hues in leaves and flowers, as well as imparting various aromas. In many flowers, carotenoids are prominently displayed, accumulating in chromoplasts in the absence of chlorophyll. Common carotenoids found in terrestrial plants, such as lutein, zeaxanthin, capsanthin, violaxanthin, and neoxanthin, belong to the xanthophylls cycle and contribute to photosynthetic protection. The marigold flower serves as a natural example, showcasing the presence of lutein in its vibrant yellow hue (Amaya et al. 2014a, b).

The content and types of carotenoids in plants are influenced by various factors, both before and after harvesting, including genotype, ripening time, cultivation methods, climatic conditions, and processing. Additionally, different parts of the same plant may exhibit variations in the types and amounts of carotenoids present. For example, fruit peels generally contain a higher concentration of carotenoids compared to the pulp (Kumar Saini et al. 2015).

3.2 Algae

On the flip side, microalgae serve as a valuable source of various beneficial products, including carotenoid pigments, polyunsaturated fatty acids, vitamins, lipids, and proteins. Algae exist as simple unicellular organisms (microalgae) either independently or grouped in colonies, and more complex pluricellular tissues (macroalgae). They can be categorized into different phyla based on their predominant photosynthetic pigment, encompassing green algae (Chlorophyta), brown algae (Phaeophyta), and red algae (Rhodophyta) (Amaya et al. 2014a, b). Microalgae cultivation offers specific advantages over traditional plant-based sources, such as a quicker cultivation, processing, and harvesting cycle, and the capability to be cultivated on waste materials.

The major commercially valuable carotenoids—astaxanthin, β-carotene, lutein, lycopene, and canthaxanthin—are found in microalgae, highlighting their significance. Despite these advantages, the large-scale and cost-effective production of carotenoids from algae presents current challenges, particularly in terms of production and downstream extraction and purification. An integrated bioprocessing approach involving microalgae should address both the upstream production of microalgae and the downstream harvesting and extraction of carotenoids. The presence of robust cell walls in many algal species poses challenges, hindering the complete recovery of bioactive compounds and potentially slowing down the overall bioprocess significantly (Gong and Bassi 2016).

3.3 Bacteria

Numerous bacteria possess the capability to synthesize carotenoids, and these pigments are typically stored in chromatophores, closely associated with the cytoplasmic membrane. Among these bacteria, Halobacterium stands out as it exclusively contains carotenoids, specifically in the form of β-carotene, resulting in a vibrant red coloration of its cell suspension (Amaya et al. 2014a, b). *Paracoccuscarotinifaciens*, obtained from soil, is a Gram-negative, aerobic bacterium characterized by its orange pigmentation. It exhibits a rod-shaped morphology, motility facilitated by peritrichous flagella, and the production of astaxanthin. Unlike some other bacteria, *Paracoccuscarotinifaciens* does not synthesize bacteriochlorophyll. The organism primarily generates carotenoids, particularly astaxanthin, and utilizes the ubiquinone system Q-10. Sequence analysis of the 16S rRNA of *Paracoccuscarotinifaciens* categorizes it within the alpha-3 subclass of Proteobacteria, forming a cluster with other species within the *Paracoccus* genus.

3.4 Fungi

In contrast to various organisms, certain fungal groups demonstrate the ability to synthesize and accumulate notably high concentrations of carotenoids intracellularly, a characteristic shared with specific microalgae. A noteworthy example is the mushroom *Cantharellus cinnabarinus*, which stands out for its natural occurrence of the orange-red oxycarotenoid canthaxanthin (Amaya et al. 2014a, b). Unlike many other organism categories such as bacteria, plants, and animals, numerous fungal phyla, including representatives of *Zygomycetes*, *Ascomycetes*, *Basidiomycetes*, and the asexual *Deuteromycetes*, exhibit the capacity to produce substantial quantities of carotenoids (Echavarri-Erasum and Johnson 2002).

4 Carotenoids and Human Health

The α- and β-carotene are primarily recognized as precursors of vitamin A. Provitamin A activity refers to the capacity of natural carotenoids to produce retinol (Vitamin A) through the action of the dioxygenase enzyme (Lintig and Vogt 2000). Any pigment containing at least one unaltered β-ionone ring in its structure can be classified as a provitamin A carotenoid (Send and Sundholm 2007). Commonly colored fruits and vegetables, such as carrots, sweet potatoes, apricots, pumpkins, and beans, are abundant sources of these carotenoids. β-Carotene exhibits a higher conversion efficacy to vitamin A (retinol) than β-Cryptoxanthin and α-carotene (Tang 2014). Apart from their role as vitamin A precursors, these pigments effectively inhibit the occurrence of chronic diseases such as breast and skin cancer. While β-carotene has been assessed for its potential to reduce the risk of lung cancer in certain studies, other in vitro studies suggest that α-carotene is more effective than β-carotene in suppressing the multiplication of cancer cells (Gallicchio et al. 2008; Murakoshi et al. 1992). Additionally, these pigments have been reported to activate cell communication and the immune system (Singh et al. 2015).

Lycopene is a predominant carotenoid extensively found in fruits and vegetables, with tomatoes and their products being key sources (Arain et al. 2018). Numerous in vivo, ex vivo, and in vitro studies have been conducted to explore the biological effects of lycopene against oxidative stress. Lycopene can also reduce the risk of various cancers, including ovarian, breast, prostate, cervical, and liver cancer (Singh et al. 2015). Conversely, Giovannucci et al. (1995) found that higher concentrations of lycopene effectively reduce the chances of cardiovascular diseases. Camara et al. (2013) demonstrated that lycopene prevents oxidative damage to the skin caused by UV-light exposure.

β-Cryptoxanthin is also considered a provitamin A carotenoid and is extensively found in citrus fruits, peaches, papayas, and mangoes. Due to the dominance of other provitamin A carotenoids, health benefits of these carotenoids have been scarcely investigated. β-cryptoxanthin acts as a therapeutic agent, particularly against bone-related diseases. Several studies have shown that β-cryptoxanthin supplementation enhances the activity of alkaline phosphatase and calcium content in metaphyseal tissue and cortical bone, reducing the chances of osteoporosis (Yamaguchi 2008). Additionally, being recognized as a precursor of provitamin A, β-cryptoxanthin also upregulates cell communication and improves immunity (Burri 2015; San Millán et al. 2015). In general, various biological effects of carotenoids have been well documented in the published literature, including anti-inflammatory, antioxidant, immunomodulatory, anticancer, and inhibition of mutagenesis (Rao and Rao 2007). A brief description of the biological effects and mechanisms of action of carotenoid pigments used for the treatment of several diseases is shown in Fig. 10.2a, b.

4.1 Antioxidant Activities

Antioxidants are biologically active compounds that can inhibit or mitigate oxidative stress caused by reactive oxygen species (ROS) and unstable compounds produced by the body in response to environmental stresses. They also serve as effective agents against various diseases (Abd El-Baky and El-Baroty 2013; Arain et al. 2018). Two main types of antioxidants are available for feed supplementation: synthetic and natural. Natural antioxidants are widely distributed in photosynthetic organisms and their products, such as herbs, fruits, and vegetables. Active compounds in this context include carotenoids, polyphenols, betaine, and certain vitamins. In livestock nutrition, antioxidant properties are valued for mitigating heat stress and promoting animal health. Yuan et al. (2011) reported that the ketocarotenoid astaxanthin derived from microalgae mitigates oxidative stress through various mechanisms, including the downregulation of inflammatory cytokine production by decreasing the activation of nuclear factor-kB, inhibition of the renin–angiotensin system, suppression of transforming growth factor b1 production, and antimicrobial effects. Carotenoids and other phytochemicals are suggested to protect the body against various ROS-mediated conditions, such as neurological, cancer, eye-related, and photosensitive disorders (Fiedor and Burda 2014).

Among all carotenoids, such as α-tocopherol, β-carotene, β-cryptoxanthin, zeaxanthin, and lutein, lycopene is the most potent pigment for protecting against cellular damage caused by ROS (Arain et al. 2018; Giovannucci 2002). The primary role of lycopene is to safeguard DNA from oxidative stress, scavenging oxygen and suppressing mutations that can lead to chronic diseases (Goralczyk and Siler 2003). Studies have shown that lycopene is the most effective free radical quencher among 600 natural carotenoids due to its conjugated double bonds and elongated structure

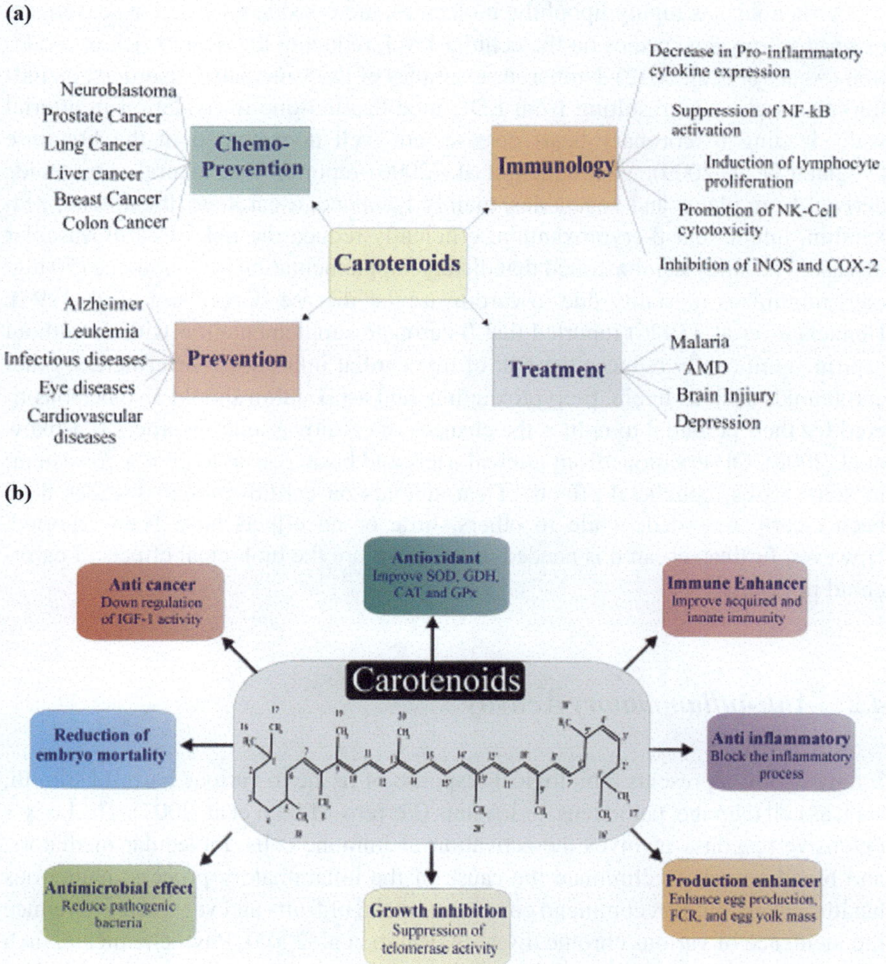

Fig. 10.2 Biological activities of the carotenoids used in the treatment of various diseases, (**a**) showed general effects, and (**b**) showed molecular mechanism

(Van Breemen and Pajkovic 2008). Research by Palozza et al. (2012) indicated that lycopene improves immune response, regulates cell growth, influences gene transcription, and modulates phase I and II enzyme detoxification. Additionally, lycopene enhances the production of antioxidant enzymes such as superoxide dismutase (SOD), glutathione transferase (GST), and quinine reductase through the activation of ARE pathways (Sahin et al. 2013). Another study reported that lycopene reduces heat stress in commercial poultry by increasing the production of phase II cytoprotective enzymes (Sahin et al. 2011). Supplementary lycopene primarily acts as an antioxidant agent by activating the Nrf_2/ARE transcription pathway to alleviate oxidative stress and promote the health of poultry birds (Sahin et al. 2011).

Carotenoids, as highly lipophilic molecules, are considered efficient scavengers of reactive oxygen species on the cellular level, reducing the overall risk of oxidation (Agarwal et al. 2012). Numerous examples of ROS-mediated disorders, including atherosclerosis, resulting from LDL modification due to oxidation in arterial walls leading to coronary heart disease, are well documented in the literature (Vogiatzi et al. 2009). Voutilainen et al. (2006) reported that natural carotenoids derived from plants and vegetables, mainly lycopene, α-carotene, β-carotene, zeaxanthin, lutein, and β-cryptoxanthin, efficiently reduce the risk of cardiovascular disease. The study demonstrated that dietary supplementation of β-carotene (50 mg/day) minimizes mortality due to cardiovascular disease (Greenberg et al. 1996). Hennekens et al. (1996) reported that β-carotene supplementation with or without aspirin significantly reduces the risk of myocardial infarction. Nevertheless, other carotenoids such as lutein, β-cryptoxanthin, and astaxanthin should also be considered for their potential to reduce the chances of cardiovascular disorders (Pashkow et al. 2008). Observations from clinical trials and basic research are not consistent; in some cases, beneficial effects of carotenoids on cardiovascular diseases have been clearly recorded, while in others, little or no effects have been reported. However, further research is needed to fully explore the biological effects of carotenoid pigments.

4.2 Anti-inflammatory Activity

Inflammation represents a biological response of tissue to various harmful stimuli, such as cell damage, pathogens, or irritants (Ferrero-Miliani et al. 2007). The body's defensive response involves the activation of immune cells, molecular mediators, and blood vessels to eliminate the cause of the inflammatory process. Numerous health organizations recommend consistent intake of fruits and vegetables to reduce the incidence of various chronic diseases (Russo et al. 2005). Phytochemicals, such as lipophilic carotenoids found abundantly in fruits and plants, are suggested to possess excellent anti-inflammatory and antioxidant properties. Many of these properties have been demonstrated to inhibit or block inflammatory processes and oxidative stress (Kaulmann and Bohn 2014). Rubin et al. (2012) reported that carotenoid supplementation (lycopene, lutein, and β-carotene) in the infant diet decreases the inflammatory response and increases plasma concentrations of carotenoids.

5 Carotenoids in Feed

Compared to information about the carotenoid content in foods, data about carotenoids in feed are limited. High-carotenoid feeds, essential for animal product pigmentation, have been extensively evaluated. Carotenoid supplementation in diets is believed to enhance the oxidative stability of animal products, providing advantages for human health through the carry-over of carotenoids in the food chain (Golzar Adabi et al. 2010; Ma and Lin 2010).

Cereals and their products are among the most commonly used feedstuffs in animal nutrition. Despite their high prevalence in animal diets, grains contribute minimally to carotenoid intake due to their low carotenoid content (Zhai et al. 2016). Maize typically has the highest total carotenoid content, followed by barley and wheat (1.2, 1.5, and 11.1 mg/g dry weight, respectively; Panfili et al. 2004). Lutein, zeaxanthin, β-cryptoxanthin, and β-carotene contents vary among commercial maize hybrids (Kljak and Grbesa 2015). Maize grain has recently been recognized as a natural source of carotenoids in poultry diets, leading to biofortification with increased β-cryptoxanthin and β-carotene contents (Liu et al. 2012). Processing of cereal grains decreases carotenoid content (Blandino et al. 2017); maize feed flour and germ contain less total xanthophylls than whole grain. However, removal of starch during processing results in the concentration of carotenoids in the resulting products: maize gluten meal and distillers dried grains with soluble (DDGS) contain significantly higher carotenoid levels (Moros et al. 2002; Robertson et al. 2005; Shin et al. 2016).

Oilseeds and their products do not significantly contribute to carotenoids in animal diets, despite being important sources of fat, protein, and fiber. Rapeseed and its products are exceptions, containing higher lutein contents (Franke et al. 2010). Legume seeds, used in both human and animal nutrition, contain lutein, zeaxanthin, β-carotene, and violaxanthin (Ashokkumar et al. 2015; Marles et al. 2013). Lupin seeds are a rich source of carotenoids (Wang et al. 2008).

By-products of the food industry, such as DDGS, gluten meal, citrus pulp, and carrot pulp, are good sources of carotenoids. Citrus pulp, for example, has been reported to contain up to 30.6% of β-cryptoxanthin and 12% of lutein (Agocs et al. 2007), while carrot pulp and dried carrot meal are used in cattle and poultry nutrition, respectively (Chen and Tang 1998).

In contrast to dry concentrates, feeds and forages used in ruminant nutrition are rich sources of carotenoids. Cattle in grass-based production systems generally have carcass fat that is more yellow than that of their concentrate-fed counterparts due to carotenoids from lush green forages (Daley et al. 2010). Carotenoid contents in forages and roughages are influenced by plant species, stage of growth, harvest and postharvest methods, and season. Major carotenoids in forages and roughages are lutein, zeaxanthin, and β-carotene, while neoxanthin, violaxanthin, and antheraxanthin are found in lower concentrations (Noziere et al. 2006). Forages grown in mixtures can affect carotenoid content: a mixture of birds' foot trefoil and timothy contained more β-carotene than a mixture of red clover and timothy or meadow fescue (Lindqvist et al. 2012).

Preserving or processing fresh forage, such as making silage, haylage, or hay, can lead to the destruction of as much as 80% of carotenoid content (Chauveau-Duriot et al. 2005). Grass fresh material consisting of timothy, meadow fescue, and couch grass resulted in varying carotenoid contents in silage and haylage, depending on the duration of wilting (Müller et al. 2007). Maize silage, whether whole crop or grain, is considered a poorer source of carotenoids than grass forage (Noziere et al. 2006). Wilting often reduces carotenoid content, especially in sunny weather (Ballet et al. 2000).

Artificial drying of forages, mainly alfalfa, enables their use in non-ruminant nutrition. These dried products are rich sources of carotenoids, with alfalfa protein concentrate containing substantial amounts of lutein, β-carotene, zeaxanthin, violaxanthin, and antheraxanthin (Calderon et al. 2007). Due to their high carotenoid content, alfalfa-dried products are often used as natural sources for pigmentation in poultry.

Although seaweed was historically used in livestock nutrition, there is renewed interest in algae as sources of protein, whether macroalgae or microalgae. Algae products, such as dried biomass and meal, are used as sources of pigments for fish and poultry products. The carotenoid profile in algae feed products is highly species-dependent; for example, biotechnological production focuses on astaxanthin from *Haematococcus*, β-carotene from *Dunaliella*, and lutein from *Scenedesmus* (Zaťková et al. 2011).

Fishmeal, primarily used as a protein source in animal nutrition, has low carotenoid content, usually consisting mainly of astaxanthin (Garcia-Romero et al. 2014; Kalinowski et al. 2005). Marine crab and echinoderm meals contain higher levels of total carotenoids (Garcia-Romero et al. 2014).

In the nutrition of specific animals, such as poultry and fish, diets are supplemented with synthetic or natural carotenoids to achieve desirable egg yolk and flesh color. The European Union Register of Feed Additives (2018) lists eight carotenoids that can be added to concentrate feeds. While a wide range of synthetic pigments is available on the market, consumer concerns about synthetic additives have increased interest in natural alternatives. Dried alfalfa and algae products are natural alternatives, and other sources, such as petals, plant extracts, or by-products, could also be used as pigment additives, such as marigold extract (6178 mg total carotenoids/g dry weight; Karadas et al. 2006).

6 Concluding Remarks

Carotenoids are natural pigments that are of high interest to food scientists, nutritionists, and food industries due to their positive impact on human health, and they are currently being studied to understand their effect on human chronic degenerative diseases. Carotenoids are responsible for the attractive color of most fruit and vegetables, having diverse biological functions and activities. Antioxidant properties of carotenoids are linked with their ability to scavenge pro-oxidant species, including free radicals and singlet oxygen.

Carotenoids color our foods and provide pleasure to our eating experience. They contribute to our vitamin A requirement and the decrease of oxidative stress. They may even contribute to seeing those pigmented foods more clearly, even if it is no longer in the bush but rather in the supermarket. Dark green vegetables, particularly leafy greens, are nutrient-dense foods that may have a range of health benefits when included in a varied, balanced diet. Apart from the content of these bioactive compounds in food, their bioavailability is very important for beneficial functions in the body, which is strongly influenced by the molecular structure, as well as the interaction with other nutrients (i.e., proteins, lipids, and fibers) and dietary phytochemicals. Prevalently, carotenoid consumption and absorption help to fight oxidative stress and reduce the risk of several chronic diseases, including cardiovascular and neurological disorders, type-2 diabetes, and different types of cancer.

Plants have evolved a vast number of carotenoids as essential compounds for their development and survival. Animals and humans, as plant consumers, have also evolved to take advantage of the unique properties of carotenoids. In fact, there is an economic potential to foster the use of plant pigments, such as carotenoids, as natural antioxidants to enhance the quality and extend the shelf life of food products. Because of an increased environmental conscience and an increasing market for natural feed/food additives in developed countries and the chance for developing countries to use sound knowledge of effective plant-derived feed additives, available without high costs, the topic of plant pigments in nutrition should be given high priority.

References

Abd El-Baky HH, El-Baroty GS (2013) The potential use of microalgal carotenoids as dietary supplements and natural preservative ingredients. J Aquat Food Product Technol 22(4):392–406. https://doi.org/10.1080/10498850.2011.654381

Agarwal M, Parameswari RP, Vasanthi HR, Das DK (2012) Dynamic action of carotenoids in cardioprotection and maintenance of cardiac health. Molecules 17(4):4755–4769. https://doi.org/10.3390/molecules17044755

Agocs A, Nagy V, Szabó Z, Mark L, Ohmacht R, Deli J (2007) Comparative study on the carotenoid composition of the peel and the pulp of different citrus species. Innovative Food Sci Emerg Technol 8(3):390–394

Amaya E, Becquet P, Carné S, Peris S, Miralles P (2014a) Carotenoids in animal nutrition. FEFANA (European Union Association of Specialty Feed Ingredients and their Mixtures)

Amaya E, Philippe B, Sergi C, Silvia P, Pilar M (2014b) Carotenoids in animal analysis, occurrence, bioavailability and biological activities. Food Res Int 76:735–750

Arain MA, Mei Z, Hassan F, Saeed M, Alagawany M, Shar A, Rajput I (2018) Lycopene: a natural antioxidant for prevention of heat-induced oxidative stress in poultry. Worlds Poult Sci J 74(1):89–100. https://doi.org/10.1017/S0043933917001040

Ashokkumar K, Diapari M, Jha AB, Tar'an B, Arganosa G, Warkentin TD (2015) Genetic diversity of nutritionally important carotenoids in 94 pea and 121 chickpea accessions. J Food Compos Anal 43:49–60. https://doi.org/10.1016/j.jfca.2015.04.014

Ballet N, Robert JC, Williams PEV (2000) Vitamins in forages. In: Givens DI, Owen E, Omed HM, Axford RFE (eds) Forage evaluation in ruminant nutrition. CABI, Wallingford, pp 399–431. Biotechnology Advances 34:1396–1412

Blandino M, Alfieri M, Giordano D, Vanara F, Redaelli R (2017) Distribution of bioactive compounds in maize fractions obtained in two different types of large scale milling processes. J Cereal Sci 77:251–258. https://doi.org/10.1016/j.jcs.2017.08.006

Boon CS, McClements DJ, Weiss J, Decker EA (2010) Factors influencing the chemical stability of carotenoids in foods. Crit Rev Food Sci Nutr 50:515–532

Burri BJ (2015) Beta-cryptoxanthin as a source of vitamin A. J Sci Food Agric 95(9):1786–1794. https://doi.org/10.1002/jsfa.6942

Calderon F, Chauveau-Duriot B, Martin B, Graulet B, Doreau M, Noziere P (2007) Variations in carotenoids, vitamins A and E, and color in cow's plasma and milk during late pregnancy and the first three months of lactation. J Dairy Sci 90(5):2335–2346. https://doi.org/10.3168/jds.2006-630

Camara M, de Cortes Sánchez-Mata M, Fernández-Ruiz V, Cámara RM, Manzoor S, Caceres JO (2013) Lycopene: a review of chemical and biological activity related to beneficial health effects. In: Studies in natural products chemistry, vol 40. Elsevier, pp 383–426. https://doi.org/10.1016/B978-0-444-59603-1.00011-4

Chauveau-Duriot B, Thomas D, Portelli J, Doreau M (2005) Carotenoids content in forages: variation during conservation. Renc Rech Ruminants 12:117

Chen BH, Tang YC (1998) Processing and stability of carotenoid powder from carrot pulp waste. J Agric Food Chem 46(6):2312–2318. https://doi.org/10.1021/jf9800817

Daley CA, Abbott A, Doyle PS, Nader GA, Larson S (2010) A review of fatty acid profiles and antioxidant content in grass-fed and grain-fed beef. Nutr J 9(1):10. https://doi.org/10.1186/1475-2891-9-10

Dutta D, Chaudhuri UR, Chakraborty R (2005) Structure, health benefits, antioxidant property and processing and storage of carotenoids. Afr J Biotechnol 4(13):1510–1520

Echavarri-Erasum C, Johnson EA (2002) Fungal carotenoids. Appl Mycol Biotechnol. https://doi.org/10.1038/jp.2011.87jama.1996.03530330043027

European Union Register of Feed Additives (2018) European Union register of feed additives persuant to regulation (EC) No 1831/2003. Annex I: list of additives, Eddition 1/2018 (260). Directorate-General for Health and Food Safety, European Commission, Brussels

Ferrero-Miliani L, Nielsen O, Andersen P, Girardin S (2007) Chronic inflammation: importance of NOD2 and NALP3 in interleukin- 1β generation. Clin Exp Immunol 147(2):227–235

Fiedor J, Burda K (2014) Potential role of carotenoids as antioxidants in human health and disease. Nutrients 6(2):466–488. https://doi.org/10.3390/nu6020466

Franke S, Fröhlich K, Werner S, Böhm V, Schöne F (2010) Analysis of carotenoids and vitamin E in selected oilseeds, press cakes and oils. Eur J Lipid Sci Technol 112(10):1122–1129. https://doi.org/10.1002/ejlt.200900251

Gallicchio L, Boyd K, Matanoski G, Tao XG, Chen L, Lam TK, Alberg AJ (2008) Carotenoids and the risk of developing lung cancer: a systematic review. Am J Clin Nutr 88(2):372–383. https://doi.org/10.1093/ajcn/88.2.372

Garcia-Romero J, Gines R, Izquierdo MS, Haroun R, Badilla R, Robaina L (2014) Effect of dietary substitution of fish meal for marine crab and echinoderm meals on growth performance, ammonia excretion, skin colour, and flesh quality and oxidation of red porgy (Pagrus pagrus). Aquaculture 422–423:239–248. https://doi.org/10.1016/j.aquaculture.2013.11.024

Giovannucci E (2002) A review of epidemiologic studies of tomatoes, lycopene, and prostate cancer. Exp Biol Med (Maywood) 227(10):852–859. https://doi.org/10.1177/153537020222701003

Giovannucci E, Ascherio A, Rimm EB, Stampfer MJ, Colditz GA, Willett WC (1995) Intake of carotenoids and retino in relation to risk of prostate cancer. JNCI J Natl Cancer Inst 87(23):1767–1776. https://doi.org/10.1093/jnci/87.23.1767

Golzar Adabi SH, Kamali MA, Davoudi J, Cooper RG, Hajbabaei A (2010) Quantification of lutein in egg following feeding hens with a lutein supplement and quantification of lutein in human plasma after consumption of lutein enriched eggs. Arch Geflügelk 74(3):158–163

Gong M, Bassi A (2016) Carotenoids from microalgae: a review of recent developments. Biotechnol Adv 34(8):1396–1412. https://doi.org/10.1016/j.biotechadv.2016.10.005

Goralczyk R, Siler U (2003) The role of lycopene in health and disease. Phytochem Health Dis 3:285–309

Greenberg ER, Baron JA, Karagas MR, Stukel TA, Nierenberg DW, Stevens MM, Haile RW (1996) Mortality associated with low plasma concentration of beta carotene and the effect of oral supplementation. JAMA 275(9):699–703. https://doi.org/10.1001/jama.1996.03530330043027

Hennekens CH, Buring JE, Manson JAE, Stampfer M, Rosner B, Cook NR et al (1996) Lack of effect of long-term supplementation with beta carotene on the incidence of malignant neoplasms and cardiovascular disease. N Engl J Med 334(18):1145–1149. https://doi.org/10.1056/NEJM199605023341801

Johnson EJ (2014) Role of lutein and zeaxanthin in visual and cognitive function throughout the lifespan. Nutr Rev 72(9):605–612. https://doi.org/10.1111/nure.12133

Kalinowski CT, Robaina LE, Fernandez-Palacios H, Schuchardt D, Izquierdo MS (2005) Effect of different carotenoid sources and their dietary levels on red porgy (Pagrus pagrus) growth and skin colour. Aquaculture 244(1–4):223–231. https://doi.org/10.1016/j.aquaculture

Karadas F, Grammenidis E, Surai PF, Acamovic T, Sparks NHC (2006) Effects of carotenoids from lucerne, marigold and tomato on egg yolk pigmentation and carotenoid composition. Br Poult Sci 47(5):561–566. https://doi.org/10.1080/00071660600962976

Kaulmann A, Bohn T (2014) Carotenoids, inflammation, and oxidative stress implications of cellular signaling pathways and relation to chronic disease prevention. Nutr Res 34(11):907–929. https://doi.org/10.1016/j.nutres.2014.07.010

Kim J, DellaPenna D (2006) Defining the primary route for lutein synthesis in plants: the role of Arabidopsis carotenoid beta-ring hydroxylase CYP97A3. Proc Natl Acad Sci USA 103:3474–3479

Kljak K, Grbesa D (2015) Carotenoid content and antioxidant activity of hexane extracts from selected Croatian corn hybrids. Food Chem 167:402–408. https://doi.org/10.1016/j.foodchem.2014.07.002

Koller M, Muhr A, Braunegg G (2014) Microalgae as versatile cellular factories for valued products. Algal Res 6:52–63

Kumar Saini R, Nile SH, Park SW (2015) Carotenoids from fruits and vegetables: chemistry, analysis, occurrence, bioavailability and biological activities. Food Res Int 76(Pt 3):735–750

Lindqvist H, Nadeau E, Jensen SK (2012) Alpha-tocopherol and b-carotene in legume–grass mixtures as influenced by wilting, ensiling and type of silage additive. Grass Forage Sci 67(1):119–128. https://doi.org/10.1111/j.1365-2494.2011.00827.x

Lintig JV, Vogt K (2000) Filling the gap in vitamin A research. Molecular identification of an enzyme cleaving beta-carotene to retinal. J Biol Chem 275(16):11915–11920. https://doi.org/10.1074/jbc.275.16.11915

Liu YQ, Davis CR, Schmaelzle ST, Rocheford T, Cook ME, Tanumihardjo SA (2012) b-Cryptoxanthin biofortified maize (Zea mays) increases b-cryptoxanthin concentration and enhances the color of chicken egg yolk. Poult Sci 91(2):432–438. https://doi.org/10.3382/ps.2011-01719

Ma L, Lin X-M (2010) Effects of lutein and zeaxanthin on aspects of eye health. J Sci Food Agric 90(1):2–12. https://doi.org/10.1002/jsfa.3785

Marles MA, Warkentin TD, Bett KE (2013) Genotypic abundance of carotenoids and polyphenolics in the hull of field pea (Pisum sativum L.). J Sci Food Agric 93(3):463–470. https://doi.org/10.1002/jsfa.5782

Mattea F, Martin A, Cocero MJ (2009) Carotenoid processing with supercritical fluids. J Food Eng 93:255–265

McGraw KJ, Beebee MD, Hill GE, Parker RS (2003) Lutein-based plumage coloration in songbirds is a consequence of selective pigment incorporation into feathers. Comp Biochem Physiol Part B 135:689–696

Melendez-Martınez AJ (2019) An overview of carotenoids, apocarotenoids and vitamin A in agro-food, nutrition, health and disease. Mol Nutr Food Resarch 180:1045

Melendez-Martınez AJ, Mapelli-Brahm P, Stinco CM (2018) The colourless carotenoids phytoene and phytofluene: from dietary sources to their usefulness for the functional foods and nutricosmetics industries. J Food Compos Anal 67:91–103. https://doi.org/10.1016/j.jfca.2018.01.002

Mezzomo N, Ferreira S (2016) Carotenoids functionality, sources, and processing by supercritical technology: a review. J Chem:Article ID 3164312, 1–16. Hindawi Publishing Corporation

Moran NA, Jarvik T (2010) Lateral transfer of genes from fungi underlies carotenoid production in aphids. Science 328(5978):624–627

Moros EE, Darnoko D, Cheryan M, Perkins EG, Jerrell J (2002) Analysis of xanthophylls in corn by HPLC. J Agric Food Chem 50(21):5787–5790. https://doi.org/10.1021/jf020109l

Müller CE, Möller J, Jensen SK, Udén P (2007) Tocopherol and carotenoid levels in baled silage and haylage in relation to horse requirements. Anim Feed Sci Technol 137(1–2):182–197. https://doi.org/10.1016/j.anifeedsci.2006.10.007

Murakoshi M, Nishino H, Satomi Y, Takayasu J, Hasegawa T, Tokuda H, Kitano H (1992) Potent preventive action of α-carotene against carcinogenesis: spontaneous liver carcinogenesis and promoting stage of lung and skin carcinogenesis in mice are suppressed more effectively by alpha-carotene than by beta-carotene. Cancer Res 52(23):6583–6587

Nováková E, Moran NA (2012) Diversification of genes for carotenoid biosynthesis in aphids following an ancient transfer from a fungus. Mol Biol Evol 29(1):313–323

Noziere P, Graulet B, Lucas A, Martin B, Grolier P, Doreau M (2006) Carotenoids for ruminants: from forages to dairy products. Anim Feed Sci Technol 131(3–4):418–450. https://doi.org/10.1016/j.anifeedsci.2006.06.018

Okada T, Nakai M, Maeda H, Hosokawa M, Sashima T, Miyashita K (2008) Suppressive effect of neoxanthin on the differentiation of 3t3-11 adipose cells. J Oleo Sci 57(6):345–351

Palozza P, Catalano A, Simone RE, Mele MC, Cittadini A (2012) Effect of lycopene and tomato products on cholesterol metabolism. Ann Nutr Metab 61(2):126–134. https://doi.org/10.1159/000342077

Panfili G, Fratianni A, Irano M (2004) Improved normal-phase high-performance liquid chromatography procedure for the determination of carotenoids in cereals. Journal of Agricultural and Food Chemistry 52(21):6373–7.

Pashkow FJ, Watumull DG, Campbell CL (2008) Astaxanthin: a novel potential treatment for oxidative stress and inflammation in cardiovascular disease. Am J Cardiol 101(10A):58D–68D. https://doi.org/10.1016/j.amjcard.2008.02.010

Piorkowski DT, McClements DJ (2014) Beverage emulsions: recent developments in formulation, production, and applications. Food Hydrocoll 42(Part 1):5–41

Premkumar L (2014) Fascinating facts about phytonutrients in spices and healthy food: scientifically proven facts. Xlibris Publisher

Qian C, Decker ED, Xiao H, McClements DJ (2012) Physical and chemical stability of beta caroteneenriched nanoemulsions: influence of pH, ionic strength, temperature and emulsifier type. Food Chem 132:1221–1229

Rao AV, Rao LG (2007) Carotenoids and human health. Pharmacol Res 55(3):207–216. https://doi.org/10.1016/j.phrs.2007.01.012

Robertson KD, Kalbfleisch JL, Pan W, Charbeneau RA (2005) Effect of corn distiller's dried grains with solubles at various levels on performance of laying hens and egg yolk color. Int J Poult Sci 4(2):44–51. https://doi.org/10.3923/ijps.2005.44.51

Rock CL (2009) Carotenoids and cancer. In: Carotenoids vol. 5: Nutrition and health. Birkhauser Verlag Basel, pp 269–286

Rodriguez-Amaya DB (2015) Carotenes and xanthophylls as antioxidants. In: Handbook of antioxidants for food preservation, Woodhead Publishing series in food science, technology and nutrition, pp 17–50

Rodrıguez-Concepcion M, Avalos J, Bonet ML, Boronat A, Gomez-Gomez L, Hornero-Mendez D, Limon MC, Melendez-Martinez AJ, Olmedilla-Alonso B, Palou A et al (2018) A global perspective on carotenoids: metabolism, biotechnology, and benefits for nutrition and health. Prog Lipid Res 70:62–93. https://doi.org/10.1016/j.plipres.2018.04.004

Rubin LP, Chan GM, Barrett-Reis BM, Fulton AB, Hansen R, Ashmeade T, Hartmann E (2012) Effect of carotenoid supplementation on plasma carotenoids, inflammation and visual development in preterm infants. J Perinatol 32(6):418–424

Russo M, Tedesco I, Iacomino G, Palumbo R, Galano G, Russo G (2005) Dietary phytochemicals in chemoprevention of cancer. Curr Med Chem Immunol Endocr Metab Agents 5(1):61–72. https://doi.org/10.2174/1568013053005454

Sahin K, Orhan C, Akdemir F, Tuzcu M, Ali S, Sahin N (2011) Tomato powder supplementation activates Nrf-2 via ERK/Akt signaling pathway and attenuates heat stress-related responses in quails. Anim Feed Sci Technol 165(3–4):230–237
Sahin K, Orhan C, Tuzcu M, Sahin N (2013) The effects of lycopene on the meat lycopene levels, antioxidant enzymes and Nrf2 pathway in broiler chickens. In: Paper presented at the 2nd international poultry meat congress
San Millán C, Soldevilla B, Martín P, Gil-Calderón B, Compte M, Pérez-Sacristán B, Granado-Lorencio F (2015) β-Cryptoxanthin synergistically enhances the antitumoral activity of oxaliplatin through ΔNP73 negative regulation in colon cancer. Clin Cancer 21(19):4398–4409
Send R, Sundholm D (2007) The role of the beta-ionone ring in the photochemical reaction of rhodopsin. J Phys Chem A 111(1):27–33. https://doi.org/10.1021/jp065510f
Shin HS, Kim JW, Kim JH, Lee DG, Lee S, Kil DY (2016) Effect of feeding duration of diets containing corn distillers dried grains with solubles on productive performance, egg quality, and lutein and zeaxanthin concentrations of egg yolk in laying hens. Poult Sci 95(10):2366–2371. https://doi.org/10.3382/ps/pew127
Sies H, Stahl W (1995) Vitamins E and C, betacarotene, and other carotenoids as antioxidants. Am J Clin Nutr 62(6 Suppl):1315S–1321S
Singh A, Ahmad S, Ahmad A (2015) Green extraction methods and environmental applications of carotenoids-a review. RSC Adv 5(77):62358–62393. https://doi.org/10.1039/C5RA10243J
Sloan DB, Moran NA (2012) Endosymbiotic bacteria as a source of carotenoids in whiteflies. Biol Lett 8(6):986–989
Sommerburg O, Meissner K, Nelle M, Lenhartz H, Leichsenring M (2000) Carotenoid supply in breast-fed and formula-fed neonates. Eur J Pediatr 159(1–2):86–90. https://doi.org/10.1007/PL00013811
Stange KC, Rodriguez-Concepcion M (2015) Carotenoids in carrot. In: Pigments in fruits and vegetables. Springer, New York
Story EN, Kopec RE, Schwartz SJ, Harris GK (2010) An update on the health effects of tomato lycopene. Annu Rev Food Sci Technol 1:1–16
Tang G (2014) Vitamin A value of plant food provitamin A evaluated by the stable isotope technologies. Int J Vitamin Nutr Res 84(Suppl 1):25–29. https://doi.org/10.1024/0300-9831/a000183
Van Breemen RB, Pajkovic N (2008) Multitargeted therapy of cancer by lycopene. Cancer Lett 269(2):339–351. https://doi.org/10.1016/j.canlet.2008.05.016
Vershinin A (1999) Biological functions of carotenoids – diversity and evolution. Biofactors 10(2–3):99–104
Vincent U, Serano F, Von Holst C (2017) Development and validation of a multi-analyte method for the regulatory control of carotenoids used as feed additives in fish and poultry feed. Food Addit Contam Part A Chem Anal Control Expo Risk Assess 34(8):1285–1297
Vogiatzi G, Tousoulis D, Stefanadis C (2009) The role of oxidative stress in atherosclerosis. Hell J Cardiol 50(5):402–409
Voutilainen S, Nurmi T, Mursu J, Rissanen TH (2006) Carotenoids and cardiovascular health. Am J Clin Nutr 83(6):1265–1271. https://doi.org/10.1093/ajcn/83.6.1265
Wang S, Errington S, Yap HH (2008) Studies on carotenoids from lupin seeds. In: Lupins for health and wealth. Proceedings of the 12th international Lupin conference
Yamaguchi M (2008) β-Cryptoxanthin and bone metabolism: the preventive role in osteoporosis. J Health Sci 54(4):356–369. https://doi.org/10.1248/jhs.54.356
Yuan JP, Peng J, Yin K, Wang JH (2011) Potential health-promoting effects of astaxanthin: a high-value carotenoid mostly from microalgae. Molecular nutrition & food research 55(1):150–65
Zaheer K (2017) Hen egg carotenoids (lutein and zeaxanthin) and nutritional impacts on human health: a review. Cyta J Food 15(3:474–487
Zakynthinos Z, Varzakas T (2016) Carotenoids: from plants to food industry. Curr Res Nutr Food Sci 4(Special Issue 1):38–51
Zaťková I, Sergejevová M, Urban J, Vachta R, Stys D, Masojídek J (2011) Carotenoid-enriched microalgal biomass as feed supplement for freshwater ornamentals: Albinic form of Wels catfish (Silurus glanis). Aquac Nutr 17(3):278–286. https://doi.org/10.1111/j.1365-2095.2009.00751.x

Zhai SN, Xia XC, He ZH (2016) Carotenoids in staple cereals: metabolism, regulation, and genetic manipulation. Front Plant Sci 7:1197. https://doi.org/10.3389/fpls.2016.01197

Zoz L, Carvalho JC, Socco VT, Casagrande TC, Cardoso L (2015) Torularhodin and torulene: bioproduction, properties and prospective applications in food and cosmetics – a review. Braz Arch Biol Technol 58(2):278–288

Chapter 11
Therapeutic Effects of Carotenoids: Trends and Advances

Ingryd Rodrigues Martins, Fernanda Wariss Figueiredo Bezerra, Jonilson de Melo e Silva, Johnatt Allan Rocha de Oliveira, and Luiza Helena da Silva Martins

1 Introduction

Carotenoids are natural liposoluble pigments in plants, fungi, bacteria, and algae. These bioactive compounds are classified as carotene hydrocarbons, structured in molecules containing only carbon and hydrogen and xanthophylls containing oxygen atoms, making them less hydrophobic. The intensity of color is directly related to the amount of this pigment, which varies from yellow to red, with β-carotene being the most exploited hydrocarbon (Ahmed et al. 2020).

Carotenoids demonstrate diverse physiological properties, including antioxidant activity and potential health benefits (Wal et al. 2023). Research from epidemiological and clinical studies suggests that diets rich in carotenoids, primarily from fruits and vegetables, may contribute to preventing various human diseases (Gebregziabher et al. 2023). Lycopene, α-carotene, β-carotene, zeaxanthin, and β-cryptoxanthin are the carotenoids most abundant in human plasma post-consumption (Martini et al.

I. R. Martins
Instituto Federal de Educação, Ciência e Tecnologia do Pará (IFPA), Núcleo de Estudos em Engenharia, Ciência e Tecnologia de Alimentos (NEECTA), Castanhal, Pará, Brazil

F. W. F. Bezerra
Universidade Federal do Pará (UFPA), Faculdade de Engenharia de Alimentos (FEA), Belém, Pará, Brazil

J. de Melo e Silva · J. A. R. de Oliveira
Universidade Federal do Pará (UFPA), Faculdade de Nutrição (FANUT), Belém, Pará, Brazil
e-mail: johnatt@ufpa.br

L. H. da Silva Martins (✉)
Universidade Federal Rural da Amazônia (UFRA), Instituto de Saúde e Produção Animal (ISPA), Belém, Pará, Brazil
e-mail: luiza.martins@ufra.edu.br

R. Campos Chisté et al. (eds.), *Carotenoids*,
https://doi.org/10.1007/978-3-031-75322-0_11

2022). Specific carotenoids such as α-carotene, β-carotene, and β-cryptoxanthin serve as dietary sources of pro-vitamin A, while lutein and zeaxanthin aid in reducing macular degeneration (Ahmed et al. 2020). Additionally, as bioactive antioxidants, carotenoids are associated with lower incidences of chronic diseases, cardiovascular diseases, neurodegenerative conditions, and various cancers (Saini et al. 2022).

Martini et al.'s research (2022) is the first investigation to show current trends in registered clinical trials using carotenoids, either through the consumption of foods rich in this bioactive compound, with most trials using tomatoes, followed by eggs, corn, and green vegetables, or as a dietary supplement, with the most explored being lutein, zeaxanthin, lycopene, and β-carotene. This occurrence has generated interest from food and nutraceutical industries to invest in innovations and technologies for the extraction and stability of carotenoids, aiming to ensure availability and maximize their benefits in food and disease prevention.

Investigations focusing on the therapeutic benefits of using bioactive compounds began with clinical trials on cell oxidative stress. As a result, more research is being directed toward carotenoids, which can potentially prevent or reduce the adversities of different chronic diseases (Crupi et al. 2023). Therefore, this Chapter provides an overview of the therapeutic capacity of carotenoids based on recent beneficial evidence for health.

2 Food Bioactive Compounds

These compounds can be defined as components from foods that can have a beneficial effect on human cells or tissues. We cannot confuse them with nutrients; these are different since such compounds are not essential to the functioning of the body, but despite this, they have beneficial and therapeutic actions on human health due to their vast chemical composition that presents antioxidant, anti-inflammatory, antitumor, among others (Astley and Finglas 2016). Among these compounds are carotenoids, polyphenols, and fatty acids (Astley and Finglas 2016).

There is still no consensus on the action of these compounds to benefit human health; the mechanisms still need to be wholly consolidated. However, it is known that some of these compounds can act as antioxidants, others can have an immunoglobular action, and others prevent the cell against oxidative stresses. More consolidated evidence still needs to be provided for a recommendation regarding these substances' intake, efficiency, and safety, especially when considering their supplements. However, it is accepted that consuming these substances in the diet, with a balance, brings significant benefits for health (Astley and Finglas 2016).

The way bioaccessibility and bioavailability differ significantly about each class of bioactive compounds, and the action of bioavailability on human health is not yet well elucidated, as, for example, not always eating a large amount of fruit will guarantee that we will have a higher concentration of active metabolites in our tissues (Galanakis 2018).

Among the wide range of existing foods, those made from plants, such as fruits, vegetables, grains, etc., are undoubtedly the most significant sources. It can also be found in animal matrices, such as some crustaceans with the carotenoid astaxanthin in their chemical composition (Galanakis 2018).

The understanding of secondary metabolites in plant foods and their bioactive compounds can be illuminated by studying microRNAs (miRNAs). MiRNAs, single-stranded RNAs ranging from 21 to 25 nucleotides in length, are primarily recognized for their role as post-transcriptional silencers. They pair with mRNAs to regulate their translation and stability, impacting plant growth, development, and genetic regulation (Owusu et al. 2021).

Research has investigated miRNAs to elucidate the mechanisms underlying the formation of secondary metabolites in plants, which comprise a diverse array of biologically active compounds with potential implications for human health. These miRNAs perpetually modulate the functions of secondary metabolites in plants, including their protective roles.

According to Owusu et al. (2021), to better distinguish the plant metabolites, it's critical to examine how primary and secondary metabolites work together. Primary metabolites vary from secondary metabolites because they are essential to plant growth at every stage. Secondary metabolic pathways operate at the post-transcriptional level, interacting with microRNAs. The explanation of this is more apparent in Fig. 11.1.

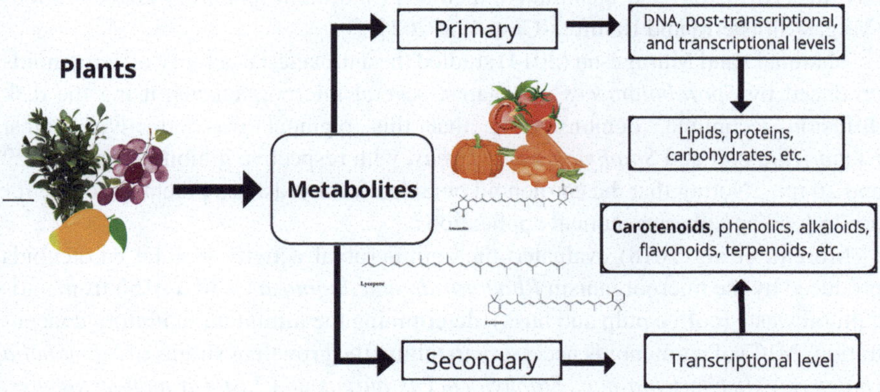

Fig. 11.1 A systemic system of how plants' primary and secondary metabolites occur at the systemic level, and we can observe that secondary metabolites, such as carotenoids, play role at the transcriptional level

3 Carotenoids: What are They? an Overview

According to Maoka (2020), the pigments that give plants their red, orange, and yellow hues are known as carotenoids. These pigments are found in all living things, including animals, photosynthetic bacteria, and plants. Tetraterpenes are a class of compounds composed of eight isoprene units, presenting a carbon chain of 40 carbons with double and conjugated bonds, consisting of a polyene chain with nine bonds. These compounds have two categories: xanthophylls and carotenes (Table 11.1).

4 Food as Main Sources of Carotenoids

Many foods are sources of carotenoids, as shown in Table 11.1. Each molecule has an action on the body and produces therapeutic effects that will be discussed later. Carotenoids are plentiful in the human diet, primarily sourced from natural fruits and vegetables. However, they are also found in seafood, eggs, dairy products, and other animal-derived foods (Gebregziabher et al. 2023).

5 Antimicrobial Activity of Carotenoids

Carotenoids can also act as antimicrobial agents, with studies reporting their action against different pathogens, which has been investigated by both qualitative methods (disk diffusion) and quantitative methods (minimum inhibitory concentration) (Vargas-Sinisterra and Ramírez-Castrillón 2021).

Manimala and Murugesan (2014) studied the antibacterial activity of carotenoids produced by *Sporobolomyces sp.* against several microorganisms, using the disk diffusion technique, demonstrating that this pigment was effective against *Escherichia coli* and *Staphylococcus aureus*, with respective inhibition halos of 29 and 26 mm. Noting that the carotenoid obtained from yeast is a potential source for use in food and pharmaceutical applications.

Moreira et al. (2018) evaluated the antimicrobial activity of total carotenoids produced by the microorganism *Rhodotorula mucilaginosa* CCMA 0156 from agricultural waste (coffee pulp and husk), determining the minimum inhibitory concentration (MIC) of carotenoids necessary to inhibit the growth of strains of *Salmonella cholerasus*, *Escherichia coli*, *Staphylococcus aureus* and *Listeria monocytogenes*. Total carotenoids were effective against the tested microorganisms, although to different degrees. For all organisms tested, antimicrobial activity was observed even at low concentrations of carotenoids, achieving complete inhibition of the growth of the four microorganisms, with MIC ranging from 612 to 1225 μg/mL.

Table 11.1 Therapeutic effects of selected carotenoids

Class	Carotenoid	Molecular structure	Natural source	Therapeutic effect	Reference
Carotenes	α-Carotene	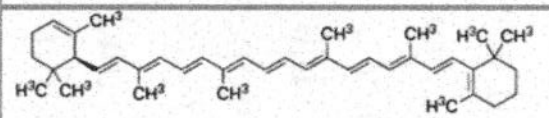	Orange, tomatoes, tangerine, napa cabbage, bananas, avocados, and sweet potatoes	Blood pressure, lung cancer, neuroblastoma, and breast cancer	Lai et al. (2022) and Saini et al. (2022)
	β-Carotene	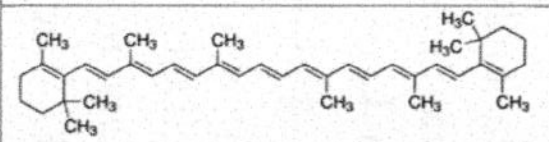	Kale, broccoli, watercress, spinach, carrots, papaya, pumpkin, and mango	Fibromyalgia, depression, anxiety, stress, head and neck cancer	Farhadnejad et al. (2020) and Elkholy et al. (2023)
	γ-Carotene	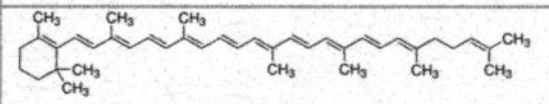	Fruits and vegetables, and salmonid fish	Colon cancer; gastrointestinal diseases	Milani et al. (2017) and Hegde et al. (2022)
	Lycopene	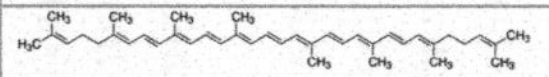	Tomatoes, guava, watermelon, papaya, and grapefruit	Obesity, diabetes, gastrointestinal cancer	Al-Ishaq et al. (2020) and Zhu et al. (2020)
Xanthophylls	β-Cryptoxanthin	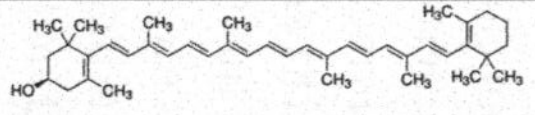	Papayas, oranges, tangerines, peaches, and yellow and orange corn	Colon câncer and cervical carcinoma	Cilla et al. (2015) and Gansukh et al. (2019)
	Astaxanthin	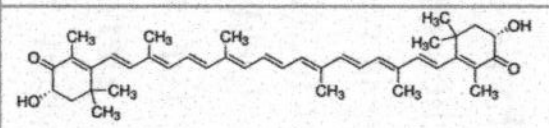	Krill, shrimp, crayfish, salmon, trout, algae, and yeast	Skin diseases and diabetes	Fathalipour et al. (2020) and Baswan et al. (2021)
	Lutein	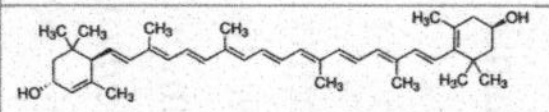	Green leafy vegetables, egg yolks, corn, and pumpkin	Fibromyalgia and macular degeneration	Sawa et al. (2020) and Elkholy et al. (2023)
	Zeaxanthin	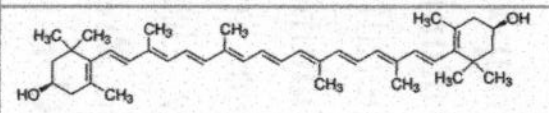	Same for lutein	Several neurological, skin, eye disorders, and gastric cancer	Sheng et al. (2020) and Bouyahya et al. (2021)
	Fucoxanthin	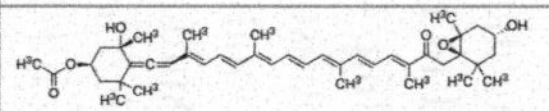	Brown seaweeds and Bacillariophyta	Lung cancer, leucemia and hepatic cancer	Garg et al. (2019) and Ming et al. (2021)
	Peridinin	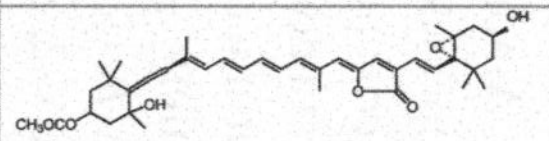	Variety of marine and freshwater dinoflagellates	Multiple myeloma, colorectal cancer	Supasri et al. (2021) and Broijl et al. (2022)

Haddad et al. (2017) found that the carotenoid pigment produced by the bacterium *Rhodococcus rhodochrous* has antimicrobial activity, which was analyzed using the agar well diffusion method, being obtained against the microorganisms *Staphylococcus aureus* and *Candida albicans*, for which the diameter of the inhibition zone was 21 and 35 mm, respectively—indicating that this compound with antimicrobial activity can be a promising biological source for applications in the pharmaceutical, biotechnology, food, cosmetics, and sanitation industries.

Soni et al. (2023) investigated an orange pigment extracted from marine sources and found that *Bacillus infantis* (16S rRNA gene) exhibited antimicrobial properties against gram-positive and gram-negative strains. The inhibition zones against *Staphylococcus aureus, Serratia marcescens, and Shigella dysenteriae* were reported as 12 ± 0.28 mm, 17 ± 0.25 mm, and 12 ± 0.32 mm, with a value of 350 µg/mL of MIC. The researchers suggested potential pharmaceutical applications of the orange pigment in combating microbial infections.

Other studies have investigated the action of specific carotenoids, such as the work of (Karpiński and Adamczak 2019), who evaluated the antimicrobial potential of fucoxanthin, a carotenoid produced by brown algae and diatoms, confirming its action against 13 aerobic bacteria. This study also observed that this carotenoid had a more substantial impact on gram-positive bacteria, reaching MICs of 62.5–250 µg/mL, while on gram-negative bacteria the MICs ranged from 125 to 500 µg/mL.

The study by Guerfali et al. (2022) showed that by-products from the food industry, such as the production of biodieselvegetable oils, can also have promising effects on health, such as therapeutic and antimicrobial benefits, as well as adding value to environmentally sustainable processes. The extracted carotenoids, torularhodin, and torulene, isolated from the oleaginous yeast strain R. babjevae Y-SL7, demonstrated action against the gram-positive bacterium *Staphylococcus aureus* at a MIC of 78 µg/mL.

Among the group of xanthophylls from carotenoids, astaxanthin still needs to be improved in the number of studies on microbial action mechanisms compared to fucoxanthin. Generally, 2–24 mg is recommended (Karpiński et al. 2022).

In contrast, Shanmugapriya et al. (2018) and Karpiński et al. (2021) have shown antimicrobial doses above the tolerable limit for consumption, as European Food Safety Authority determined, for microorganisms such as *Bacillus subtilis*, *Escherichia coli*, *Pseudomonas aeruginosa*, *Streptococcus mutans*, and SARS-CoV-2, with MIC ranging between 500 and 2000 µg/mL. These results indicate impracticality for use in clinical treatments of infections.

Extracts obtained from seeds and peels of ten tomato varieties were evaluated for their antimicrobial activity against seven species of bacteria. The Tiny Tim tomato variety presented the best performances but still had high MIC values against *Staphylococcus aureus*, *Bacillus subtilis*, *Listeria monocytogenes*, and *Klebsiella pneumonia*, reaching respective values of 625, 1.250, 2.500, and 2.500 mg/mL (Szabo et al. 2019).

Given this, it was observed that in some studies carotenoids showed moderate antimicrobial activity. In contrast, others prove their high effectiveness against a wide variety of microorganisms, emphasizing the need for more in-depth research to investigate the mechanisms of action responsible for the antibacterial activity of carotenoids.

6 Effects of Carotenoids in Some Diseases

Carotenoids in the human diet bring many health benefits; they can reduce the risks of some diseases, such as eye diseases and some types of cancer; among the existing carotenoids, we take a unique look at β-carotene, lycopene, lutein, and zeaxanthin, in addition to their antioxidant effects. Of the carotenoids explained in this topic, β-carotene, converted into vitamin A, also has other benefits. Lutein and zeaxanthin act against eye diseases as they help absorb harmful light when entering the eyes (Johnson 2002).

The physicochemical characteristics of carotenoids, determined by their molecular structure, are intimately related to their biological role. To fulfill their roles, carotenoids must fit into cellular and subcellular structures in the proper location and orientation. This requires consideration of general molecular geometry, which includes size, shape, and the presence of functional groups. Moreover, the conjugated double-bond system of carotenoids is the chromophore group responsible for these molecules' photochemical characteristics and electrochemical reactivity, which serve as the foundation for their recognition and antioxidant activity (Britton 1995). There are around 850 carotenoids with chemical composition, but only 50 occur in the human food chain (Table 11.1) (Gebregziabher et al. 2023).

Carotenoids requires at least seven conjugated double bonds to produce color; the hue intensifies and might go from a light yellow to a deep red color as the conjugated system rises (Rodriguez-Amaya 2001). Carotenoids that contain nine or more conjugated doublets improving their ability to absorb singlet oxygen, exhibit the most potent antioxidant action (Rodriguez-Amaya 2001).

Various steps of the oxidative process, which correspond to the subsequent stages of the chain's initiation, propagation, and completion, can be affected by antioxidants from carotenoids. As a result, these compounds' method of action has the potential to stop or interrupt this chain reaction. Lipid peroxidation is a classic example of a chain reaction; a radical releases or sequesters an electron, forming a second radical that does the same action on another molecule. This oxidation process continues until a free radical is formed, neutralized by an antioxidant, or degraded to an inert product (Pisoschi and Pop 2015).

The biological benefits of high carotenoids have been linked to a decrease in a few degenerative conditions in recent decades, as demonstrated by epidemiological research. These effects have been attributed to the antioxidant qualities of carotenoids (Cooperstone and Schwartz 2016).

We must emphasize that there are also carotenoids that do not present color, which are phytoene and phytofluene, abundantly found in a number of fruits, but also in blood plasma and human tissues (Gebregziabher et al. 2023).

Studies carried out in vitro and vivo, and the same clinical trials have demonstrated that carotenoids can successfully promote human health due to the nature of their chemical composition previously mentioned (Gebregziabher et al. 2023).

Regarding health issues, there is a connection between eating habits and tumor formation, depending on the type of diet the person follows (if it is not particularly healthy), and these dietary elements have a much greater causal relationship with cancer than the disease itself. Naturally, occurring phytochemicals found in food have demonstrated beneficial and chemo preventive action in relation to smoking, pollution, and occupational risks. Among these compounds are carotenoid pigments, which have been linked to a reduction in the appearance of tumors in diets that include them in their food composition (Tanaka et al. 2012).

Although research on human tissues has shown that just six of the 40 carotenoids typically consumed through diet are absorbed, this indicates that different carotenoids are absorbed differently by the intestines. However, research on the content of mammalian breast milk found that it contained about 34 different types of carotenoids (Kotake-Nara and Nagao 2011).

This intestinal epithelium's ability to absorb selectively carotenoids through enhanced diffusion—a mechanism still being thoroughly investigated—is responsible for the pigments' excretion from the intestinal lumen (Kotake-Nara and Nagao 2011).

Because they have a strong antioxidant effect and eliminate free radicals and other oxidants involved in chronic diseases, carotenoids like β-carotene may help prevent certain types of cancer, like lung cancer. They primarily act on reactive oxygen species (ROS), which include superoxide anion radical (O_2^-), hydroxyl radical (HO), hydrogen peroxide (H_2O_2), and singlet oxygen ($^1O_2$) (Fiedor and Burda 2014).

According to Milani et al. (2017), carotenoids like β-cryptoxanthin, γ-carotene, and β-carotene have reduced breast cancer. Treatment of human breast cell lines and MCF-7 cells also increased apoptosis and an inhibition of cell proliferation and cell cycle arrest in various phases.

Due to its ability to accumulate in the brain, lutein has been linked to cognitive functions in recent studies. Cannavale et al. (2019) research shows that lutein plasma levels contribute to improved hippocampus function in obese people. Zuniga and Moran (2018) found a correlation between inflammatory markers and total plasma carotenoids and cognitive performance in cancer survivors. There was an increased prevalence of self-reported cognitive impairment in these patients with low plasma levels of carotenoids. The carotenoid buildup in neural tissues, such as the brain and eyes, may cause these effects (Amengual 2019).

When we evaluate the therapeutic effects of carotenoids, we must emphasize that we have scientific evidence that makes associations between diet and chronic diseases. Such evidence helped to formulate dietary guidelines that strive to prevent long-term health conditions such as diabetes, osteoporosis, cardiovascular diseases, cancer, among others.

Therefore, it is recommended that there be an increase in the consumption of plant-derived foods, including vegetables, fruits and grains, since plant sources have a wide range of carotenoid contents and other phytochemicals that induce a beneficial effect on health, acting on metabolism, hormonal induction, and as immunoglobular agents (Rao and Rao 2007).

7 Technologies for Obtaining Carotenoids

Technologies for obtaining carotenoids have played a fundamental role in the food and health sectors. Carotenoids are natural pigments in various fruits and vegetables, known for their health benefits, such as antioxidant action and combating chronic diseases (Fiedor and Burda 2014; Vasarri and Degl'Innocenti 2022).

Efficiently obtaining carotenoids is challenging, as these compounds are sensitive to heat, light, and oxygen, which can cause them to lose their nutritional and therapeutic properties during processing and storage. However, several technologies have been developed to guarantee the preservation and extraction of these compounds to maximize their benefits (de Oliveira et al. 2022; Carneiro et al. 2023).

One of the most common technologies for obtaining carotenoids is solvent extraction. This method uses a suitable solvent to extract carotenoids from plant material. It is essential to use safe solvents for health and the environment and consider factors such as time, temperature, and pH during extraction (Saini and Keum 2018; Wen et al. 2018).

The extraction process begins with the appropriate choice of the plant matrix, as different species present variations in the composition and concentration of carotenoids. The most employed solvents include hexane, petroleum ether, acetone, and ethanol due to their affinity with fat-soluble pigments (Saini and Keum 2018; Shankar et al. 2022; Generalić Mekinić et al. 2023). The success of the extraction depends on carefully selecting the solvent, considering the polarity of the desired carotenoids.

Another promising technology is ultrasound-assisted extraction. In this method, plant material is subjected to high-frequency sound waves, which promote the breakdown of cells and the release of carotenoids. This technology is advantageous because it is a fast and efficient process, reducing extraction time and increasing the yield of carotenoids (Umair et al. 2021; Sharma and Bhat 2021).

Furthermore, microencapsulation technology has been widely used to protect and stabilize carotenoids. In this process, carotenoids are encapsulated in microscopic particles, such as liposomes or microspheres, which protect the compounds against external factors and increase their stability during storage and processing (Moser et al. 2020; de Freitas Santos et al. 2021).

The application of fermentation techniques to obtain carotenoids has also gained prominence. In this process, microorganisms are used to produce pigments naturally. This technique is advantageous for generating high-purity and high-quality carotenoids and is a sustainable method with low environmental impact (Nagaraj et al. 2022; Paul et al. 2023).

Another innovative technology is plant genetic engineering. This technique can genetically modify plants to produce high amounts of carotenoids. This approach can improve carotenoid production more efficiently and economically (Kim et al. 2012; Velmurugan and Muthukaliannan 2022).

In addition to technologies for obtaining carotenoids, it is worth highlighting the use of technologies for analyzing and quantifying these compounds. Advanced carotenoid analysis and quantification technologies revolutionize nutritional and agricultural research. These tools, such as liquid chromatography and mass spectrometry, provide an accurate and rapid assessment of food pigments. This technological approach improves the quality of analysis and drives the discovery of new compounds and applications, strengthening studies related to health, nutrition, and food development (Aguilar-espinosa et al. 2023).

In conclusion, technologies for obtaining carotenoids have played a fundamental role in advancing the food and health industry. These technologies allow efficient extraction of carotenoids, preserving their nutritional and therapeutic properties through solvent extraction, ultrasound-assisted extraction, microencapsulation, fermentation, and plant genetic engineering. They also allow large-scale production, ensuring the availability of these beneficial compounds for human consumption. With these advances, it is possible to maximize the benefits of carotenoids, promoting a healthier diet and helping to prevent chronic diseases.

8 Concluding Remarks

Carotenoids are promising agents for human health as they offer a variety of therapeutic benefits, including the prevention of various chronic diseases. However, further clarification on the mechanisms of action and bioavailability for positive health effects is still needed, along with their clinical significance in studies, which are based only on substantial scientific evidence but corroborate with their protective role for the human body.

Concurrently, technological innovations are crucial to ensure efficiency and quality in obtaining carotenoids, ranging from extraction to plant genetic engineering, contributing to availability for consumption and, consequently, to advances in health.

Thus, by maximizing the benefits of carotenoids through innovative technologies and further research, it is possible to promote healthier eating habits, contributing to more effective disease prevention, with an impact on the well-being and quality of life of the global population.

References

Aguilar-Espinosa M et al (2023) Advancing carotenoid quantification: a new method for semi-quantitative assessment of β-carotene and lycopene content in food extracts. J Chromatogr B Analyt Technol Biomed 1231:123929. https://doi.org/10.1016/j.jchromb.2023.123929

Ahmed A et al (2020) Potential adverse effect of ultrasound on the electroretinogram of the retina and the protective role of different types of antioxidants. Eur J Biophys 8(2):20–25. https://doi.org/10.11648/j.ejb.20200802.12

Al-Ishaq RK et al (2020) Phytochemicals and gastrointestinal cancer: cellular mechanisms and effects to change cancer progression. Biomol Ther 10(1):105. https://doi.org/10.3390/biom10010105

Amengual J (2019) Bioactive properties of carotenoids in human health. Nutrients 11(10):2388. https://doi.org/10.3390/nu11102388

Astley S, Finglas P (2016) Nutrition and health. In: Reference module in food science. Elsevier. https://doi.org/10.1016/B978-0-08-100596-5.03425-9

Baswan SM et al (2021) Role of ingestible carotenoids in skin protection: a review of clinical evidence. Photodermatol Photoimmunol Photomed 37(6):490–504. https://doi.org/10.1111/phpp.12690

Bouyahya A et al (2021) Sources, health benefits, and biological properties of zeaxanthin. Trends Food Sci Technol 118:519–538. https://doi.org/10.1016/j.tifs.2021.10.017

Britton G (1995) Structure and properties of carotenoids about function. FASEB J 9(15):1551–1558. https://doi.org/10.1096/fasebj.9.15.8529834

Broijl A et al (2022) VS38c and CD38-multiepitope antibodies provide highly comparable minimal residual disease data in patients with multiple myeloma. Am J Clin Pathol 157(4):494–497. https://doi.org/10.1093/ajcp/aqab163

Cannavale CN et al (2019) Serum lutein is related to relational memory performance. Nutrients 11(4):768. https://doi.org/10.3390/nu11040768

Carneiro CR et al (2023) Potential challenges of the extraction of carotenoids and fatty acids from pequi (Caryocar brasiliense) oil. Food Secur 12(9):1907. https://doi.org/10.3390/foods12091907

Cilla A et al (2015) Anti-proliferative effect of main dietary phytosterols and β-cryptoxanthin alone or combined in human colon cancer Caco-2 cells through cytosolic Ca+ 2–and oxidative stress-induced apoptosis. J Funct Foods 12:282–293. https://doi.org/10.1016/j.jff.2014.12.001

Cooperstone JL, Schwartz SJ (2016) Recent insights into health benefits of carotenoids. In: Reinhold C, Ralf MS (eds) Handbook on natural pigments in food and beverages, Series in food science, technology and nutrition. Woodhead Publishing, pp 473–497. https://doi.org/10.1016/B978-0-08-100371-8.00020-8

Crupi P et al (2023) Overview of the potential beneficial effects of carotenoids on consumer health and well-being. Antioxidants (Basel) 12(5):1069. https://doi.org/10.3390/antiox12051069

de Freitas Santos PD et al (2021) Microencapsulation of carotenoid-rich materials: a review. Food Res Int 147:110571. https://doi.org/10.1016/j.foodres.2021.110571

de Oliveira JAR et al (2022) High concentrate flavonoids extract from citrus pomace using enzymatic and deep eutectic solvents extraction. Food Secur 11(20):3205. https://doi.org/10.3390/foods11203205

Elkholy NS et al (2023) Assessment of the therapeutic potential of lutein and beta-carotene nanodispersions in a rat model of fibromyalgia. Sci Rep 13(1):19712. https://doi.org/10.1038/s41598-023-46980-6

Farhadnejad H et al (2020) Antioxidant vitamin intakes and risk of depression, anxiety and stress among female adolescents. Clin Nutr ESPEN 40:257–262. https://doi.org/10.1016/j.clnesp.2020.09.010

Fathalipour M et al (2020) The therapeutic role of carotenoids in diabetic retinopathy: a systematic review. Diabetes Metab Syndr Obes 13:2347–2358. https://doi.org/10.2147/DMSO.S255783

Fiedor J, Burda K (2014) Potential role of carotenoids as antioxidants in human health and disease. Nutrients 6(2):466–488. https://doi.org/10.3390/nu6020466

Galanakis CM (2018) Polyphenols: properties, recovery, and applications. Woodhead Publishing

Gansukh E et al (2019) Chemopreventive effect of β-cryptoxanthin on human cervical carcinoma (HeLa) cells is modulated through oxidative stress-induced apoptosis. Antioxidants 9(1):28. https://doi.org/10.3390/antiox9010028

Garg S et al (2019) Marine carotenoid fucoxanthin possesses anti-metastasis activity: molecular evidence. Mar Drugs 17(6):338. https://doi.org/10.3390/md17060338

Gebregziabher BS et al (2023) Carotenoids: dietary sources, health functions, biofortification, marketing trend and affecting factors-a review. J Agri Food Res 14:100834. https://doi.org/10.1016/j.jafr.2023.100834

Generalić Mekinić I et al (2023) Algal carotenoids: chemistry, sources, and application. Food Secur 12(14):2768. https://doi.org/10.3390/foods12142768

Guerfali M et al (2022) Concomitant production of multifunctional metabolites on biodiesel-derived crude glycerol by the oleaginous yeast *Rhodotorula babjevae* Y-SL7. Biomass Convers Biorefin 1–14:10237–10250. https://doi.org/10.1007/s13399-022-03028-5

Haddad M et al (2017) Antimicrobial and antioxidant activity of carotenoid pigment produced by native Rhodococcus spp. isolated from soil. Int J Mol Clin Microbiol 7(1):809–815. https://journals.iau.ir/article_537781.html?lang=en

Hegde PS et al (2022) Impact of carotenoids on gut microbiome: implications in human health and disease. J Appl Nat Sci 14(3):1085–1099. https://doi.org/10.31018/jans.v14i3.3582

Johnson EJ (2002) The role of carotenoids in human health. Nutr Clin Care 5(2):56–65. https://doi.org/10.1046/j.1523-5408.2002.00004.x

Karpiński TM, Adamczak A (2019) Fucoxanthin-an antibacterial carotenoid. Antioxidants 8(8):239. https://doi.org/10.3390/antiox8080239

Karpiński TM et al (2021) Studies of selected xanthophylls as potential candidates against SARS-CoV-2 targeting main protease (Mpro) and papain-like protease (PLpro). Herba Pol 67(2):1–8. https://doi.org/10.2478/hepo-2021-0009

Karpiński TM et al (2022) What do we know about antimicrobial activity of Astaxanthin and Fucoxanthin? Mar Drugs 20(1):36. https://doi.org/10.3390/md20010036

Kim M-J et al (2012) Genetic modification of the soybean to enhance the β-carotene content through seed-specific expression. PLoS One 7(10):e48287. https://doi.org/10.1371/journal.pone.0048287

Kotake-Nara E, Nagao A (2011) Absorption and metabolism of xanthophylls. Mar Drugs 9(6):1024–1037. https://doi.org/10.3390/md9061024

Lai JS et al (2022) Perinatal plasma carotenoid and vitamin E concentrations with maternal blood pressure during and after pregnancy. Nutr Metab Cardiovasc Dis 32(12):2811–2821. https://doi.org/10.1016/j.numecd.2022.07.019

Manimala MRA, Murugesan R (2014) In vitro antioxidant and antimicrobial activity of carotenoid pigment extracted from Sporobolomyces sp. isolated from natural source. J Appl Nat Sci 6(2):649–653. https://doi.org/10.31018/jans.v6i2.511

Maoka T (2020) Carotenoids as natural functional pigments. J Nat Med 74(1):1–16. https://doi.org/10.1007/s11418-019-01364-x

Martini D et al (2022) What is the current direction of the research on carotenoids and human health? An overview of registered clinical trials. Nutrients 14(6):1191. https://doi.org/10.3390/nu14061191

Milani A et al (2017) Carotenoids: biochemistry, pharmacology and treatment. Br J Pharmacol 174(11):1290–1324. https://doi.org/10.1111/bph.13625

Ming JX et al (2021) Fucoxanthin extracted from Laminaria Japonica inhibits metastasis and enhances the sensitivity of lung cancer to Gefitinib. J Ethnopharmacol 265:113302. https://doi.org/10.1016/j.jep.2020.113302

Moreira MD et al (2018) Solid coffee waste as alternative to produce carotenoids with antioxidant and antimicrobial activities. Waste Manag 82:93–99. https://doi.org/10.1016/j.wasman.2018.10.017

Moser P et al (2020) Functional properties of chickpea protein-pectin interfacial complex in buriti oil emulsions and spray dried microcapsules. Food Hydrocoll 107:105929. https://doi.org/10.1016/j.foodhyd.2020.105929

Nagaraj YN et al (2022) Identification, quantification and kinetic study of carotenoids and lipids in Rhodotorula toruloides CBS 14 cultivated on wheat straw hydrolysate. Fermentation 8(7):300. https://doi.org/10.3390/fermentation8070300

Owusu A et al (2021) MicroRNAs roles in plants secondary metabolism. Plant Signal Behav 16(7):1915590. https://doi.org/10.1080/15592324.2021.1915590

Paul D et al (2023) Yeast carotenoids: cost-effective fermentation strategies for health care applications. Fermentation 9(2):147. https://doi.org/10.3390/fermentation9020147

Pisoschi AM, Pop A (2015) The role of antioxidants in the chemistry of oxidative stress: a review. Eur J Med Chem 97:55–74. https://doi.org/10.1016/j.ejmech.2015.04.040

Rao AV, Rao LG (2007) Carotenoids and human health. Pharmacol Res 55(3):207–216. https://doi.org/10.1016/j.phrs.2007.01.012

Rodriguez-Amaya DB (2001) A guide to carotenoid analysis in foods, vol 71. ILSI press, Washington

Saini RK, Keum SY (2018) Significance of genetic, environmental, and pre- and postharvest factors affecting carotenoid contents in crops: a review. J Agric Food Chem 66:5310–5324. https://doi.org/10.1021/acs.jafc.8b01613

Saini RK et al (2022) Carotenoids: dietary sources, extraction, encapsulation, bioavailability, and health benefits-a review of recent advancements. Antioxidants (Basel) 11(4):795. https://doi.org/10.3390/antiox11040795

Sawa M et al (2020) Effects of lutein supplementation in Japanese patients with unilateral age-related macular degeneration: the Sakai lutein Study. Sci Rep 10(1):5958. https://doi.org/10.1038/s41598-020-62483-0

Shankar U et al (2022) Comparison of different solvent extraction compositions for the analysis of phytoplankton pigments and lipids using a rapid and sensitive LC-MS method. Res Sq. https://doi.org/10.21203/rs.3.rs-1706520/v1

Shanmugapriya K et al (2018) Astaxanthin-alpha tocopherol nanoemulsion formulation by emulsification methods: investigation on anticancer, wound healing, and antibacterial effects. Colloids Surf B Biointerfaces 172:170–179. https://doi.org/10.1016/j.colsurfb.2018.08.042

Sharma M, Bhat R (2021) Extraction of carotenoids from pumpkin peel and pulp: comparison between innovative green extraction technologies (ultrasonic and microwave-assisted extractions using corn oil). Food Secur 10(4):787. https://doi.org/10.3390/foods10040787

Sheng YN et al (2020) Zeaxanthin induces apoptosis via ROS-regulated MAPK and AKT signaling pathway in human gastric cancer cells. Onco Targets Ther 13:10995–11006. https://doi.org/10.2147/OTT.S272514

Soni N et al (2023) Carotenoid from marine *Bacillus infantis*: production, extraction, partial characterization, and its biological activity. Arch Microbiol 205:161. https://doi.org/10.1007/s00203-023-03505-z

Supasri KM et al (2021) Characterisation and bioactivity analysis of peridinin-chlorophyll a-Protein (PCP) isolated from symbiodinium tridacnidorum CS-73. J Mar Sci Eng 9(12):1387. https://doi.org/10.3390/jmse9121387

Szabo K et al (2019) Antimicrobial and antioxidant properties of tomato processing byproducts and their correlation with the biochemical composition. LWT 116:108558. https://doi.org/10.1016/j.lwt.2019.108558

Tanaka T et al (2012) Cancer chemoprevention by caroteno. Molecules 17(3):3202–3242. https://doi.org/10.3390/molecules17033202

Umair M et al (2021) Ultrasound-assisted extraction of carotenoids from carrot pomace and their optimization through response surface methodology. Molecules 26(22):6763. https://doi.org/10.3390/molecules26226763

Vargas-Sinisterra AF, Ramírez-Castrillón M (2021) Yeast carotenoids: production and activity as antimicrobial biomolecule. Arch Microbiol 203(3):873–888. https://doi.org/10.1007/s00203-020-02111-7

Vasarri M, Degl'Innocenti D (2022) News and updates from 2022 on antioxidant and anti-inflammatory properties of marine products. Mar Drugs 21(1):26. https://doi.org/10.3390/md21010026

Velmurugan A, Muthukaliannan GK (2022) Genetic manipulation for carotenoid production in microalgae an overview. Curr Res Biotechnol 4:221–228. https://doi.org/10.1016/j.crbiot.2022.03.005

Wal P et al (2023) A detailed analysis of the carotenoids and their derivatives, including their multiple health advantages. Open J Med Chem 17(1):e187410452309140. https://doi.org/10.2174/18741045-v17-e230916-2023-8

Wen C et al (2018) Advances in ultrasound assisted extraction of bioactive compounds from cash crops–a review. Ultrason Sonochem 48:538–549. https://doi.org/10.1016/j.ultsonch.2018.07.018

Zhu R et al (2020) Lycopene in protection against obesity and diabetes: a mechanistic review. Pharmacol Res 159:104966. https://doi.org/10.1016/j.phrs.2020.104966

Zuniga KE, Moran NE (2018) Low serum carotenoids are associated with self-reported cognitive dysfunction and inflammatory markers in breast cancer survivors. Nutrients 10(8):1111. https://doi.org/10.3390/nu10081111

Chapter 12
Neuroprotective Potential of Carotenoids

N. Meena, Srutee Rout, Suraj N. Mali, and Mozaniel Oliveira

1 Introduction

Carotenoids are a group of naturally occurring pigments found in various plants and microorganisms. After Wackenroder discovered carotene in carrots in 1831, this group was given the name carotenoid (Namitha and Negi 2010). There are two main classes of carotenoids: nonpolar carotenoids including carotenes (lycopene, α-carotene, and β-carotene), and polar carotenoids—xanthophylls (β-cryptoxanthin, zeaxanthin, astaxanthin, and lutein). Further, carotenoids are divided into non-provitamin A (astaxanthin, lutein, lycopene, and zeaxanthin) and provitamin A (α-carotene, β-cryptoxanthin, and β-carotene). According to reports, 700 carotenoids are produced from various source organisms (plants, algae and fungi), nonetheless only 50 (7.1%) of them are contained in the human diet (Tan and Norhaizan 2019). They are essential for numerous physiological functions of living organisms. They impart colour to fruits (oranges, tomatoes, peach, peppers, apricot etc.) and flowers (chrysanthemum, rapeseed etc.), favors pollination, are responsible for beaks and feathers of few birds, skin or muscle of some fish and the exoskeleton of shellfish (Zia-Ul-Haq et al. 2021). However, the birds and animals derive these pigments from food as they cannot synthesize carotenoids in vivo.

N. Meena · S. Rout
Agricultural and Food Engineering Department, Indian Institute of Technology, Kharagpur, India

S. N. Mali (✉)
School of Pharmacy, D.Y. Patil University (Deemed to be University), Sector 7, Nerul, Navi Mumbai, India

M. Oliveira (✉)
Programa de Pós-Graduação em Ciências Biológicas - Botânica Tropical Universidade Federal Rural da Amazônia/Museu Paraense Emílio Goeldi, Belém, Brazil

R. Campos Chisté et al. (eds.), *Carotenoids*,
https://doi.org/10.1007/978-3-031-75322-0_12

Carotenoids possess numerous therapeutic properties *viz.* anti-inflammatory, immunomodulatory, antidiabetic, antibacterial, anticancer, and neuroprotective (Chisté et al. 2014; Háda et al. 2010; Papandreou et al. 2006; Shaish et al. 2008). Alzheimer's disease, Parkinson's disease, multiple sclerosis, central nervous system injuries are the most devastating neuronal disorders (Cho et al. 2018). Carotenoids can protect tissues from oxidative stress and significantly alleviates neurodegenerative diseases. This chapter evaluates the therapeutic effects of carotenoids on neuroprotective disease and their ability in the prevention and treatment of these neuronal disorders.

2 Chemistry of Carotenoids

2.1 *Structure*

Carotenoids are terpenoids (made up of poly-isoprenoid skeleton) that have an extense conjugated polyene (double bonds) with a bilateral symmetry (Fig. 12.1). Out of the *cis*- and *trans*-carotenoid isomers, the *cis*-isomers have lower melting point as they are less likely to form crystal. Furthermore, the *cis*-carotenoids have a lower absorption coefficient and can absorb light at shorter wavelength (Britton et al. 2008). It is hypothesized that all the carotenoids are derived from lycopene $C_{40}H_{56}$, having long central chain made up of 11 conjugated double bonds. Several other carotenoids are formed from β-carotene as a precursor; hence it is a familiar carotenoid found in both plant and animal kingdom.

The structure of carotenoids is composed of eight (C_5) isoprenoid units and is made up of a central main chain and end ring system. The isoprenoid units are connected head to tail except at the centre where it is connected tail to tail. The alternate double bonds of polyene which excites the delocalized electrons, act as chromophore by absorbing visible light, which is accountable for the characteristic colour of carotenoids. The chromophore can be modified by modifying the chain structure (Preedy 2012). For example, β-carotene exhibits bright orange colour with nine double bonds, ζ-carotene with seven double bonds has pale yellow colour and canthaxanthin with 13 double bonds has a deep red colour (Meléndez-Martínez et al.

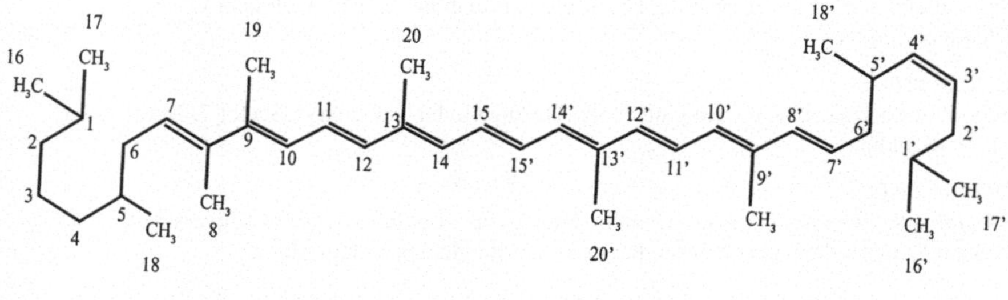

Fig. 12.1 The carotenoid skeleton numbered according to IUPAC

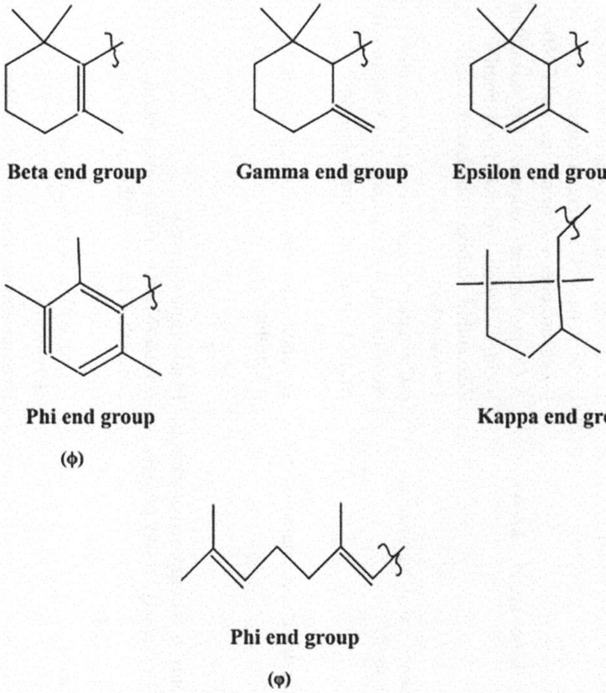

Fig. 12.2 Functional end groups of carotenoids. (Source: Manochkumar et al. 2021)

2007). The longer the chromophore more is the resonance through electronic transitions, which results in lowering the molecular orbital energy, thus stabilizing the system. The end rings may be saturated or unsaturated and commonly substituted by epoxy or hydroxy or keto group (Fig. 12.2). It has been reported that oxidation reactions are more prevalent than reduction reactions (Preedy 2012).

2.2 Classification

Carotenoids are classified based on their chemical structure, function, cyclization, and modification in structure as shown in Table 12.1.

Table 12.1 Classification of carotenoids

Classified based on	Name	Description	Example
Chemical structure	(i) Acyclic carotenes	Carotenes with both sides having open ends	Lycopene (red), ζ-Carotene, Phytoene (colourless), Phytofluene and Neurosporene
	(ii) Cyclic carotenes	Containing bicyclic (two cyclic) or monocyclic (one) structure	B-Carotene (orange), α-Carotene (orange), δ-Carotene, γ- Carotene (orange), β-Zeacarotene, α-Zeacarotene, and E160a
	(iii) Carotenols	Contains at least one hydroxyl group, hence are also known as hydroxy carotenoids	A-Cryptoxanthin (yellow), β-Cryptoxanthin (orange), Lutein (yellow), Lycoxanthin, Zeaxanthin, Rubixanthin, Lycofill and Zeinoxanthin
	(iv) Epoxycarotenoids	Contains at least one epoxy group, hence are also known as epoxide, and is a subgroup of xanthophylls	β-Carotene-5,6-epoxide, Antheraxanthin, Neoxanthin, Auroxanthin, and Violaxanthin
	(v) Species-specific carotenoids		Bixin, Capsanthin, Capsorubin, and Crocetin
	(i) Hydrocarbon Carotenoids or Carotenes	It is either cyclic or linear polyunsaturated chain, composed of only hydrogen and carbon. Its chemical formula is C_{40} H_{56}. It is soluble in hydrocarbon solvents	Lycopene, β-cryptoxanthin, β-Carotene, α-Carotene.
	(ii) Xanthophylls	They are also known as oxycarotenoids. It is composed of carbon, hydrogen, and oxygen forming hydroxy, epoxy, or oxy groups. Its chemical formula is C_{40} H_{56} $O_{1\text{-}4}$. It is soluble in hydrophilic solvents (ethanol)	Plant origin—Lutein Animal/Marine origin—astaxanthin, capsanthin, cryptoxanthin, violaxanthin, and zeaxanthin
Structural modification	(i) Allenic carotenoids	It consists of continuous double bonds	Crocoxanthin, neoxanthin, and diatoxanthin
	(ii) Acetylenic carotenoids	It consists of triple bonds	Dehydroapocarotenoids, fucoxanthin, peridinin
	(iii) Apocarotenoid	It contains less than 40 carbons. Apocarotenoids are generated by shortening the carotenoid skeleton at one end, whereas removal of fragments at both the ends generates dia-apocarotenoids (Zia-Ul-Haq et al. 2021)	Bixin, and Crocetin

Function	(i) Primary carotenoids	These carotenoids contain components useful for photosynthesis, hence also called as photosynthetic pigments	Lutein, β-Carotene, antheraxanthin, violaxanthin, neo-xanthin, zeaxanthin
	(ii) Secondary carotenoids	These are not necessary for plant survival. However, essential for other functions	Lycopene, bixin, astaxanthin and capsanthin (responsible for cell-protection)
Partition character—classification based on the ability to separate two immiscible solvents (90% aqueous methanol and light petrol)	(i) Hydrophasic carotenoids	It contains two or more hydroxyl groups. Distributed in lower (90% aqueous methanol) phase	
	(ii) Epiphasic carotenoids	It contains an oxo or ether group Distributed in upper (light petrol) phase	Xanthophyll esters, Hydrocarbons
	(iii) Biphasic carotenoids	It contains one carboxyl or one hydroxyl or two oxo groups. Distributed between the two phases	
Based on Provitamin A	(i) Provitamin A carotenoids	The carotenoids which can be converted to retinol by the body	α-Carotene, β-cryptoxanthin, and β-carotene
	(ii) Non-provitamin A carotenoids	These cannot be converted to retinol by the body	Lycopene

Source: Zia-Ul-Haq et al. (2021)

2.3 *Physico-Chemical Properties*

2.3.1 Size and Shape

The size and shape of the carotenoids are accountable for their formation into cellular or subcellular structures with proper alignment and precise location for its regular functioning. There is a significant difference in biological functions and properties of Z- and E isomers due to their difference in size and shape. The Z-isomers are bent and have a lesser tendency to crystallize, thus exhibits increased absorption, solubility, and transportation, and lower melting points than corresponding E isomers, which are rigid and linear structures (Hurst 2008).

2.3.2 Melting Points

The melting points of carotenoids varies between 130 and 220 °C and the small crystals are generally better soluble in benzene and chlorinated organic solvents (Zia-Ul-Haq et al. 2021).

2.3.3 Z-Isomerization

The physicochemical properties *viz.*, colour, solubility, melting point, stability, and crystallinity are influenced by Z-isomerization. The visibility and colour of carotenoids are decreased as the absorption maxima is shifted towards shorter wavelength. Z-isomerization also decreases crystallinity and stability. Honda et al. (2019) reported that z-isomers are less stable than E-isomers.

2.3.4 Absorption of Light and Colour

The conjugated alternate π bonds regulates the absorption of light by carotenoids. The excitation energy required for the π electrons is low, ambit to absorption of blue and violet light (400–500 nm), thereby exhibiting yellow to red colours. Carotenoids exhibits a different absorption maximum at three wavelengths due to their conjugated double bond (CDB) structure producing a tri-peak spectrum (from 430 to 480 nm); consequently exhibiting vivid colours (Britton et al. 2004). It has been reported that 7 CDB is a prerequisite for a carotenoid to produce colour, thus phytofluene and phytoene with $\leq$5 CDB are colourless (Meléndez-Martínez et al. 2018). The colour of carotenoids changes from light yellow (7 CDB, ζ-carotene), orange (11 CDB, β-carotene) to red (13 CDB, astaxanthin) as the maximum absorption wavelength increase with CDB (Britton et al. 2004).

2.4 *Chemical Properties*

2.4.1 Oxidation

Carotenoids are subjected to oxidation during processing, storage, or other metabolic activities. The various oxidizing agents used e.g., ozone, oxygen, chromic acid, and potassium permanganate, influences the products of oxidation. The use of potassium permanganate and chromic acid results in gradual and stepwise degradation (Gross 2012).

$$\textbf{Carotenoids} \xrightarrow[\text{Potassium Permanganate}]{\text{Chromic acid or}} \textbf{Apocarotenols} + \textbf{Ketones}$$

$$\beta - \textbf{Carotene} \xrightarrow{\text{MnO}_2} \underset{(10\%)}{\textbf{Retinal}} + \underset{(2\%)}{\beta - \textbf{apo} - \textbf{10'} - \textbf{carotenol}} + \underset{(\text{traces})}{\beta - \textbf{apo} - \textbf{8'} - \textbf{carotenol}}$$

$$\beta - \textbf{Carotene} \xrightarrow{\text{H}_2\text{O}_2} \textbf{Retinal}$$

Typically, carotenoids undergo autooxidation in the presence of air, even in their crystallized form or in solution. Additionally, moisture, light, and high temperature exaggerate the condition (Gross 2012).

$$\beta - \text{carotene} \xrightarrow{\text{autooxidation}} \text{Mono and diaepoxide of carotene}$$

2.4.2 Halogenation

Halogens react with carotenes to produce halogenated products, for instance:

$$C_{40}H_{56} \xrightarrow{Cl_2} C_{40}H_{56}Cl_{22}$$

2.4.3 Cyclization

In this reaction, the linear chain of carotenoids undergoes cyclization during biosynthesis, converting them into a cyclic structure at one or both ends. A few enzymes that have been reported as responsible for cyclization are cyclases (in lycopene), neoxanthin synthase (in tomato), and capsanthin-capsorubin synthase (in pepper) (Cunningham and Gantt 2001). There is a noticeable colour difference between cyclized and straight-chain carotenoids, with the latter being more vibrant. For example, although lycopene and β-carotene have 11 carbon double bonds, the former one absorbs at 470–500 nm and possesses bright red colour, while the latter appears orange-yellow colour and absorbs at 450 nm (due to cyclization) (Zia-Ul-Haq et al. 2021).

2.5 *Miscellaneous Role of Carotenoids*

In addition to their vital role in the physiological functioning of living organisms, carotenoids also play an essential role in energy transduction and light harvesting pathways of photosynthesis in plants, bacteria, fungi, and algae (Rivera-Madrid et al. 2020). Besides, it also protects the mitochondria and chlorophyll from photo-oxidative damage. Moreover, carotenoids serve a key function in photoreception by converting surplus energy into heat and regulating the excitation energy to environmental conditions (Domonkos et al. 2013). Figure 12.3 outlines the key health benefits of carotenoids.

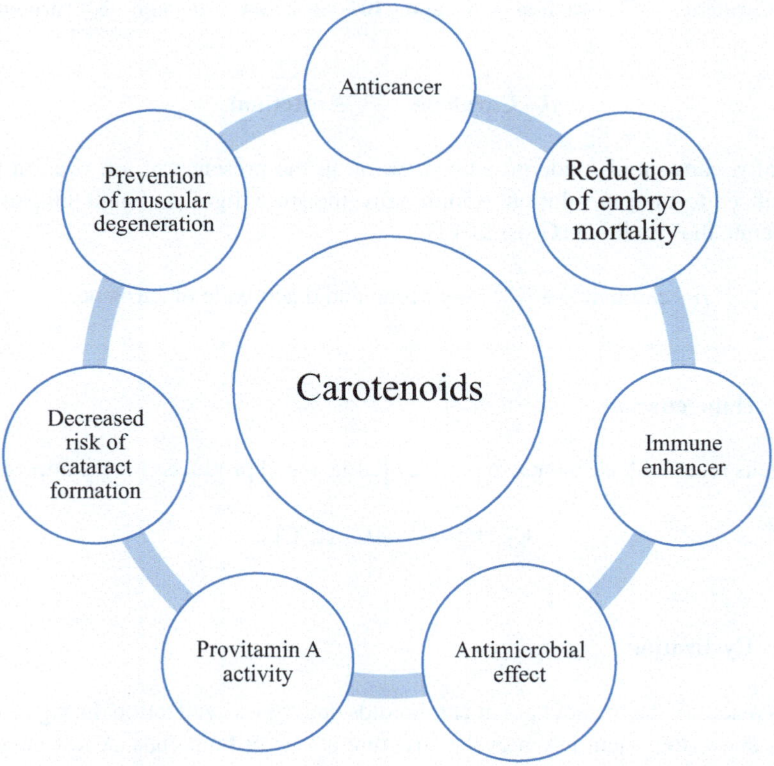

Fig. 12.3 Primary health benefits of carotenoids

3 Neuroprotective Potential of Carotenoids

3.1 *Antiapoptotic Effects*

Carotenoids have been shown to have a range of health benefits, including their antiapoptotic effect. Apoptosis is a natural process of programmed cell death that occurs in the body as part of normal growth and development, as well as in response to various types of cellular damage. It is regulated by a variety of genes and proteins. However, excessive, or uncontrolled apoptosis can lead to tissue damage and contribute to the development of various diseases, including cancer. It can be triggered by various stimuli, such as interleukin-1α (IL-1α), gamma radiation, ultraviolet irradiation, and tumor necrosis factor-α (TNF-α) receptors. The apoptotic pathway is generally classified as intrinsic and extrinsic. In the intrinsic pathway, apoptosis mitochondria play a crucial role, whereas death receptors (DR) play a vital role in extrinsic pathway apoptosis and is activated due to the binding of death ligands to DRs. DR3-DR6 and TNF-α are omnipresent and trigger TNF receptor-associated death domain (TRADD) and Fas-associated death domain (FADD) (Locksley et al. 2001). Death-inducing signaling complex (DISC) induces apoptosis and the subsequent release of cytochrome c from mitochondria leads to the intrinsic pathway of apoptosis (Walczak and Krammer 2000; Wallach et al. 1999).

Astaxanthin (AST) is a ubiquitous reddish-orange fat-soluble pigment (Galasso et al. 2018). It is produced by microorganisms like marine *Agrobacterium aurantiacum*, yeast, and microalgae when they are stressed (Grimmig et al. 2017). AST exhibits both hydrophilic and lipophilic characteristics, making it easy to transport to various organs and tissues. It comprises two hydroxyl groups with two hydrophilic rings featuring a hydrophilic part, and a lipophilic portion comprised of a polyunsaturated carbon chain (Hussain et al. 2016). Studies reported that AST reduced the risk of neurological diseases. As shown by Kim et al. (2009), AST reduced caspase activation and cytochrome c levels, thereby inhibiting apoptosis. Another study reported that AST had the potential to mitigate mitochondrial membranes by preventing neuronal apoptosis (Lu et al. 2019). AST significantly reduces Reactive oxygen species (ROS) production, consequently, blocks caspase-3 and caspase-9 activations decreases cytochrome c release, and improves mitochondrial dysfunction (Gomez-Lazaro et al. 2008; Ikeda et al. 2008).

Another carotenoid that makes it through the blood-brain barrier (BBB) is lutein. In addition to its antidiabetic property, lutein's anti-inflammatory, antioxidant, and antiapoptotic effects have been reported to be effective in the treatment of various diseases, including neurodegenerative disorders, retinal ischemia injury, and diabetes. Lutein protected retinal neurons from being damaged by stroke and n-methyl-D-aspartate (NMDA). Also, it prevented further brain ischemia and reperfusion injury and helped repair the BBB (Zeni et al. 2019). It has been reported that Lutein increased the striatal dopamine level and restored the loss of neurons in mice. Furthermore, lutein enhanced the anti-apoptotic marker by suppressing pro-apoptotic markers (caspases-3, 8 and 9, Bax) and prevented apoptosis (Nataraj et al. 2016).

Lycopene is a potential carotenoid having therapeutic effects against neurodegenerative diseases and can pass BBB. It has been established that lycopene has neuroprotective effects, hypocholesterolemic, antiproliferative, and anticancer activities. Lycopene-rich foods protected neurons from damage caused by free radicals and apoptosis in a mouse model of Alzheimer's disease. Studies reported that lycopene inhibited cytochrome c release, DNA fragmentation, and caspase-3 activation in neurons. It reduced mitochondrial ROS and apoptosis in neuronal cells (Lim et al. 2017).

3.2 Antineuroinflammatory Effects

In reaction to hazardous stimuli, the body undergoes a complex cascade known as inflammation, which serves a defensive purpose by activating the immune system. Nevertheless, inflammation is a crucial part of many biological processes. Neuroinflammation is a type of inflammation that occurs in the central nervous system (CNS) and is associated with several neurological disorders (Artemiadis 2018). Microglia and astrocytes produce a variety of proinflammatory chemokines (CCL2, CCL3, CCL5, and CXCL1), secondary mediators (NO and prostaglandins), cytokines (IL-1, IL-6, and TNF-), and ROS that orchestrate neuroinflammatory responses (Becher et al. 2017; Nuzziello and Liguori 2019).

Carotenoids inhibit inflammatory mediators. The therapeutic effects of carotenoids include the prevention of cardiovascular disease, cancer, age-related macular degeneration, and neurological disorders caused by inflammation (Cho et al. 2018). Carotenoids are inversely related to inflammatory indicators such as IL-1α (major initiator of immune response), TNF-α, IL-6, VCAM-1, and MCP-1 (Brough et al. 2011; Canas et al. 2012). Studies reported that carotenoids significantly reduced the activity of proinflammatory cytokines (IL-1β and TNF-α), and intercellular adhesion molecules (ICAM) (Cho et al. 2018).

The anti-inflammatory properties of AST have also been demonstrated to exhibit neuroprotection effects (Wu et al. 2015). AST prevents the release of downstream signaling mediators *viz.*, TNF-α, ILs, and ICAM1. Further, it inhibits proinflammatory cytokine such as macrophage migration inhibitory factor (MIF) (Fakhri et al. 2019b). As another carotenoid, lycopene exhibits anti-inflammation effects by reducing inflammation–induced cognitive deficiency and depression-like behaviors in murine models. Lycopene ameliorated brain water content, neurological impairments, BBB rupture, and neuronal death. In addition, lycopene reduces the levels of ICAM-1, IL-1β, and TNF-α (Wu et al. 2015).

Evidence demonstrated that β-carotene can reduce inflammation by inhibiting the production of inflammatory cytokines. It also inhibited NF-κB thereby inhibiting proinflammatory genes (Bai et al. 2005). Moreover, it has been discovered to inhibit nitric oxide (NO) and prostaglandin E2 (PGE2) production, as well as cyclooxygenase-2 (COX-2), IL-1β, and TNF-α (Gilroy et al. 2004). As another carotenoid, lutein can prevent the production of TNF-α, COX, and NO. It is also shown to inhibit NF-κB pathway, the generation of proinflammatory cytokines, lipid peroxidation, and enhances antioxidant enzyme activity.

3.3 Antioxidant Effects

An imbalance between production and scavenging reactive oxygen species (ROS) is what leads to oxidative stress (Durackova 2009). Proinflammatory chemokines and cytokines are generated when ROS activates kinases such as protein kinase C (PKC), JNK, and p38 MAPK (Chapple 1997). Neurodegenerative disorders are also triggered by stress-induced damage, such as lipid peroxidation, glycated products, and oxidized proteins (Hussain et al. 2016).

Studies revealed that AST is a powerful antioxidant having strong singlet quenching activity among other carotenoids (β-carotene, lutein, and canthaxanthin) and protects against cardiovascular, diabetes, inflammation, gastrointestinal, and cancer (Bidaran et al. 2018; Yamashita 2013). AST regulates inducible nitric oxide synthase (iNOS) and heme-oxygenase 1 (HO-1), therefore, prevents the production of cytokines (Iwata et al. 2018). Additionally, fucoxanthin enhances antioxidant enzyme activity and reduces the levels of oxidative stress (Pangestuti et al. 2013). Evidence demonstrated that fucoxanthin could decline the levels of oxidative stress in fibroblast cells, microglia cells, and hepatic cells (Yu et al. 2017). As another carotenoid, β-carotene protects cell membranes exhibiting antioxidant activity. Studies reported that β-carotene improved neural and cognitive deficits. In the traumatic brain injury model, β-carotene modulated the Nrf2-/Keap1-mediated antioxidant system and hence reduced oxidative stress (Chen et al. 2019).

Fucoxanthin reduced oxidative stress-induced cell damage by inhibiting the ERK signaling pathway (key mediator of oxidative stress) (Kim et al. 2010). Besides, lycopene increases the levels of antioxidant enzymes SOD, CAT, and GPx by exhibiting free radical scavenging activity (Wertz et al. 2004). Lycopene-containing foods improved SOD function by repairing damaged enzymes, raising glutathione (GSH) levels, and inhibiting GSH oxidation (Alshatwi et al. 2010; Kaur et al. 2011).

4 Role of Carotenoids on Neurodegenerative Disorder

4.1 Alzheimer's Disease

Alzheimer's disease is one of a progressive neurodegenerative disorder, that is characterized by the accumulation of neurofibrillary tangles and amyloid plaques in certain parts of the brain. It occurs due to the aggregation of phosphophorylated microtubule-associated protein tau (MPAT) and misfolded Aβ-peptides (Cho et al. 2018; Zia-Ul-Haq et al. 2021). Studies found that individuals with higher levels of lutein and zeaxanthin in their blood had better cognitive function and a reduced risk of developing Alzheimer's disease (Kim et al. 2016; Mewborn et al. 2019). Another study found that a higher intake of carotenoids, particularly lutein, zeaxanthin, and beta-carotene, was associated with a reduced rate of cognitive decline in older adults (Morris et al. 2015; Nolan et al. 2015). It was also revealed that their antioxidant properties may help reduce oxidative stress and inflammation in the brain and prevent neuroblastoma cell death (Santa-Cecília et al. 2019). Studies demonstrated that the antioxidant effects of lutein and zeaxanthin also benefits the cognitive processes like visual-motor coordination and executive decision-making (Hammond et al. 2017).

The antioxidant effects of β-carotene have also been demonstrated. Studies revealed that β-carotene preserved telomeres and telomere length by reducing oxidative stress and proinflammatory cytokines in mononuclear cells (Boccardi et al. 2020). Fucoxanthin and lycopene have also been found to exhibit neuroprotective effects. Besides modulating acetylcholine levels, fucoxanthin reduced Aβ neurotoxicity and oxidative stress (Lin et al. 2017). It was demonstrated that lycopene significantly suppressed mitochondrial reactive oxygen species (ROS), amyloid Aβ, and apoptotic cell death in human neural cells (Hwang et al. 2017).

4.2 *Parkinson's Disease*

Parkinson's disease (PD) is a neurodegenerative disorder that affects the dopamine-producing neurons in the brain. It affects the motor neurons by suppressing the effect of dopamine in the corpus striatum. The symptoms of PD include muscle rigidity, posture instability, bradykinesia, tremor, and neurological dysfunction (lack of balance and coordination) (Santa-Cecília et al. 2019). Oxidative stress, mitochondrial dysfunction, protein misfolding, and neuroinflammation have been shown to play a vital role in neuronal death. Studies showed that the antioxidant effect of carotenoids inhibits inflammation (aggregation of synphilin-1 and synuclein) and is beneficial in reducing oxidative stress by scavenging ROS (Sutachan et al. 2012). Liu et al. revealed that the antioxidant potential of astaxanthin prevents the generation of ROS and mitochondrial dysfunction (Liu et al. 2009). Furthermore, fucoxanthin suppressed extracellular signal-regulated kinase (ERK), an important mediator of oxidative stress, and activates the phosphatidyl inositol-3-kinase pathway, thereby aids in treating PD (Huang et al. 2019).

4.3 *Multiple Sclerosis*

Multiple sclerosis (MS) is an autoimmune neuronal disease of the central nervous system. It is characterized by neuroinflammation, axon demyelination (destruction of the protective myelin sheath or oligodendrocytes bounding the axon leading to poor insulation of the neurons), and brain atrophy (Santa-Cecília et al. 2019). In MS, the Th17 helper cells are amplified, and on the other hand, the regulatory T cells are substantially declined (Abdolahi et al. 2015). The synthesis of pro-inflammatory mediators like IL-6 and IL-17 contributes to the development of the disease. Bidaran et al. revealed that astaxanthin is a potential candidate for reducing IL-6, IFN-γ, IL-7 and elevating the levels of IL-10. Furthermore, it also prevents lipid peroxidation due to its increased antioxidant activity than β-carotene (Bidaran et al. 2018).

Studies have shown that increasing the levels of retinoic acid aids in neutralizing these immune-protective and proinflammatory cells resulting in increased regeneration in neurons. Evidence has reported that astrocytes increase retinoic acid production by regulating the enzyme-retinaldehyde dehydrogenase 2. The upregulation of retinoic acid synthesis by astrocytes is an endogenous response to neuroinflammation that serves to maintain the integrity of the blood-brain barrier (Mirshafiey and Mohsenzadegan 2009). Additionally, it was also shown that retinoic acid suppresses IL-17 and RORγt genes.

4.4 Depression

Depression is a complex disorder of the central nervous system. It is believed to be related to an imbalance of neurotransmitters in the brain, inflammation, and oxidative stress. Various experimental models depicted astaxanthin as an antidepressant. Zhou et al. observed that astaxanthin declined the levels of IL-6, IL-1α, glial fibrillary acidic protein (GFAP), cleaved caspase-3, and COX-2 in the brain of mice, consequently inhibiting the neuroinflammation in depression (Zhou et al. 2017). Jiang et al. demonstrated that astaxanthin reduced the overexpression of IL-6, TNF-α, COX-2, and neuronal NOS in mice (Jiang et al. 2018). Moreover, the Akt/ GSK-3β/ CREB signaling pathway of astaxanthin was also involved in its antidepressant effects (Fakhri et al. 2019b). Besides, β-carotenoid curbs oxidative stress, and hence exhibits effective anti-depressant like activity. Treatment with lycopene has been shown to ameliorate depressive symptoms by decreasing plasma levels of IL-6, TNF-α, and LPS (Zhang et al. 2016c). As another carotenoid, fucoxanthin exhibited antidepressant properties that are linked to its ability to dampen NF-κB activation (Jiang et al. 2019).

4.5 Central Nervous System (Brain/Spinal Cord) Injuries

The central nervous system is responsible for coordinating and controlling many of the body's functions, including movement, sensation, and cognition. The secondary phase of spinal cord injury (SCI) is characterized by the activation of destructive pathways such as oxidative stress, inflammation, apoptosis, and others, resulting in edema and neuronal cell damage (Galgano et al. 2015). Following SCI, astaxanthin diminished sensory-motor impairment by elevating pAkt/Akt and decreasing TNF-α, NMDAR 2B (NR2B), p-p38MAPK (Fakhri et al. 2018b), and pERK/ERK (Fakhri et al. 2019a). Astaxanthin also downregulated NAD(P) H-related enzymes: NQO-1, HO-1, and GST-α1. It also enhanced Akt and Bad phosphorylation, reducing apoptosis and caspase-3. Further, it reduces mitochondrial dysfunction, caspase, and the Akt/GSK-3β signaling pathway, making it neuroprotective (Fakhri et al. 2018a, 2019b).

Astaxanthin reduced early brain injury by dramatically reducing brain edema, and caspase-3 levels (Zhang et al. 2017). By increasing GSH and SOD and decreasing malondialdehyde (MDA) astaxanthin also demonstrated substantial antioxidant activity. Moreover, it significantly slowed apoptosis, improved neurological impairment, and decreased infarct volume (Zhang et al. 2018). Zhang et al. revealed that AST reduced damage by suppressing TLR4 and downstream signaling pathways (Zhang et al. 2016a, b, c). β-carotene, another carotenoid, inhibited NF-κB activity and reduced reactive oxygen species and nitric oxide. Severe SCI rats treated with β-carotene showed significant increases in spinal cord activity. Moreover, β-carotene dramatically decreased TNF-α, IL-1β, and IL-18 production in spinal cord-injured rats (Zhou et al. 2018). Besides, crocin inhibited TNF-α induced cell death by regulating Bcl-2 mRNA expression (Terraf et al. 2017).

Fucoxanthin downregulates apoptosis and oxidative stress using Nrf2-ARE pathways, proving its neuroprotective properties (Hua et al. 2019). Furthermore, lycopene therapy has been shown to inhibit apoptosis (Zhang et al. 2016a, b, c) and significantly reduced TNF-α and NF-B activation as well (Zhang et al. 2016c). Scientific investigations demonstrated that by increasing HO-1 and decreasing COX-2, lycopene reduced neuroinflammation, SCI, and neuronal cell death (Hu et al. 2017). Moreover, it was also found that lycopene suppressed the pain-promoting genes in rats with neuropathic pain, making it a promising potential treatment for the condition (Mohammadzadeh Honarvar et al. 2017; Reza Dorosty-Motlagh et al. 2016; Bonomini et al. 2015).

4.6 Aging/Autism

The brain becomes especially susceptible to neurodegenerative disorders like aging (Melo Pereira et al. 2019). According to research, memory loss has been linked to oxidative stress in older people (Eren et al. 2019). Antioxidant studies on astaxanthin have produced encouraging results, with the compound being able to prevent oxidative damage to biological membranes (Petyaev 2016). Evidence found that astaxanthin containing foods improve human skin when examined visually by a dermatologist (Yamashita 2013). Besides, the key aging-related mechanism, the mTOR/AMPK pathway, is also inhibited by lycopene (Rodríguez-Luna et al. 2018). As another carotenoid, fucoxanthin reduced aging by boosting Nrf2/GSH-dependent antioxidant responses. Researchers have found that fucoxanthin acts as a sunblock against UVA rays, making it a photoprotective molecule (Chang et al. 2019). Moreover, carotenoids inhibit prostaglandin production and melanogenesis.

The neurological illness autism produces significant behavioral deficits. Certain inflammatory cytokines and autism are linked, and carotenoids can reduce autism (Krajcovicova-Kudlackova et al. 2009; Zarei et al. 2015). Evidence demonstrated that AST increases CAT activity, and reduces lipid peroxidation and NO (Fakhri et al. 2019b).

5 Conclusion

Carotenoids are natural lipid-soluble pigments, which are widespread. Its therapeutic properties aid to fight against neurodegenerative diseases. In this chapter, we discussed the diverse nature of carotenoids and their therapeutic potential in attenuating several pathogenic pathways of neurodegeneration. Nonetheless, only a few carotenoids have been explored for their neuroprotective potential. The antioxidant and nutraceutical properties of carotenoids are considered vital to combat Alzheimer's disease. Besides, they can pass through the blood-brain barrier. Further, carotenoids and their derivatives are being investigated for their neuroprotective properties.

References

Abdolahi M, Yavari P, Honarvar NM, Bitarafan S, Mahmoudi M, Saboor-Yaraghi AA (2015) Molecular mechanisms of the action of vitamin A in Th17/Treg Axis in multiple sclerosis. J Mol Neurosci 57:605–613. https://doi.org/10.1007/s12031-015-0643-1

Alshatwi AA, Al Obaaid MA, Al Sedairy SA, Al-Assaf AH, Zhang JJ, Lei KY (2010) Tomato powder is more protective than lycopene supplement against lipid peroxidation in rats. Nutr Res 30:66–73. https://doi.org/10.1016/j.nutres.2009.12.002

Artemiadis AK (2018) Neuropathic pain in acute and SubacuteNeuropathies: a systematic review. Pain Physician 1:111–120. https://doi.org/10.36076/ppj.2018.2.111

Bai S-K, Lee S-J, Na H-J, Ha K-S, Han J-A, Lee H, Kwon Y-G, Chung C-K, Kim Y-M (2005) β-Carotene inhibits inflammatory gene expression in lipopolysaccharide-stimulated macrophages by suppressing redox-based NF-κB activation. Exp Mol Med 37:323–334. https://doi.org/10.1038/emm.2005.42

Becher B, Spath S, Goverman J (2017) Cytokine networks in neuroinflammation. Nat Rev Immunol 17:49–59. https://doi.org/10.1038/nri.2016.123

Bidaran S, Ahmadi AR, Yaghmaei P, Sanati MH, Ebrahim-Habibi A (2018) Astaxanthin effectiveness in preventing multiple sclerosis in animal model. Bratisl Lek Listy 119:160–166. https://doi.org/10.4149/bll_2018_031

Boccardi V, Arosio B, Cari L, Bastiani P, Scamosci M, Casati M, Ferri E, Bertagnoli L, Ciccone S, Rossi PD, Nocentini G, Mecocci P (2020) Beta-carotene, telomerase activity and Alzheimer's disease in old age subjects. Eur J Nutr 59:119–126. https://doi.org/10.1007/s00394-019-01892-y

Bonomini F, Rodella LF, Rezzani R (2015) Metabolic syndrome, aging and involvement of oxidative stress. Aging Dis 6:109–120. https://doi.org/10.14336/AD.2014.0305

Britton G, Liaaen-Jensen S, Pfander H (2004) Carotenoids: Handbook. Springer Science & Business Media
Britton G, Liaaen-Jensen S, Pfander H (eds) (2008) Carotenoids. Birkhäuser, Basel. https://doi.org/10.1007/978-3-7643-7499-0
Brough D, Tyrrell PJ, Allan SM (2011) Regulation of interleukin-1 in acute brain injury. Trends Pharmacol Sci 32:617–622. https://doi.org/10.1016/j.tips.2011.06.002
Canas JA, Damaso L, Altomare A, Killen K, Hossain J, Balagopal P, (Babu) (2012) Insulin resistance and adiposity in relation to serum β-carotene levels. J Pediatr 161:58–64.e2. https://doi.org/10.1016/j.jpeds.2012.01.030
Chang P-M, Li K-L, Lin Y-C (2019) Fucoidan–Fucoxanthin ameliorated cardiac function via IRS1/GRB2/ SOS1, GSK3β/CREB pathways and metabolic pathways in senescent mice. Mar Drugs 17:69. https://doi.org/10.3390/md17010069
Chapple ILC (1997) Reactive oxygen species and antioxidants in inflammatory diseases. J Clin Periodontol 24:287–296. https://doi.org/10.1111/j.1600-051X.1997.tb00760.x
Chen P, Li L, Gao Y, Xie Z, Zhang Y, Pan Z, Tu Y, Wang H, Han Q, Hu X, Xin X (2019) β-carotene provides neuroprotection after experimental traumatic brain injury via the Nrf2-ARE pathway. J Integr Neurosci 18:153–161. https://doi.org/10.31083/j.jin.2019.02.120
Chisté RC, Freitas M, Mercadante AZ, Fernandes E (2014) Carotenoids inhibit lipid peroxidation and hemoglobin oxidation, but not the depletion of glutathione induced by ROS in human erythrocytes. Life Sci 99:52–60. https://doi.org/10.1016/j.lfs.2014.01.059
Cho KS, Shin M, Kim S, Lee SB (2018) Recent advances in studies on the therapeutic potential of dietary carotenoids in neurodegenerative diseases. Oxidative Med Cell Longev 2018:1–13. https://doi.org/10.1155/2018/4120458
Cunningham FX, Gantt E (2001) One ring or two? Determination of ring number in carotenoids by lycopene ε-cyclases. Proc Natl Acad Sci 98:2905–2910. https://doi.org/10.1073/pnas.051618398
Domonkos I, Kis M, Gombos Z, Ughy B (2013) Carotenoids, versatile components of oxygenic photosynthesis. Prog Lipid Res 52:539–561. https://doi.org/10.1016/j.plipres.2013.07.001
Durackova Z (2009) Some current insights into oxidative stress. Physiol Res Acad Sci Bohemoslov 59:459–469. https://doi.org/10.33549/physiolres.931844
Eren B, Tuncay Tanrıverdi S, Aydın Köse F, Özer Ö (2019) Antioxidant properties evaluation of topical astaxanthin formulations as anti-aging products. J Cosmet Dermatol 18:242–250. https://doi.org/10.1111/jocd.12665
Fakhri S, Abbaszadeh F, Dargahi L, Jorjani M (2018a) Astaxanthin: a mechanistic review on its biological activities and health benefits. Pharmacol Res 136:1–20. https://doi.org/10.1016/j.phrs.2018.08.012
Fakhri S, Dargahi L, Abbaszadeh F, Jorjani M (2018b) Astaxanthin attenuates neuroinflammation contributed to the neuropathic pain and motor dysfunction following compression spinal cord injury. Brain Res Bull 143:217–224. https://doi.org/10.1016/j.brainresbull.2018.09.011
Fakhri S, Dargahi L, Abbaszadeh F, Jorjani M (2019a) Effects of astaxanthin on sensory-motor function in a compression model of spinal cord injury: involvement of ERK and AKT signalling pathway. Eur J Pain 23:750–764. https://doi.org/10.1002/ejp.1342
Fakhri S, Yosifova Aneva I, Farzaei MH, Sobarzo-Sánchez E (2019b) The neuroprotective effects of Astaxanthin: therapeutic targets and clinical perspective. Molecules 24:2640. https://doi.org/10.3390/molecules24142640
Galasso C, Orefice I, Pellone P, Cirino P, Miele R, Ianora A, Brunet C, Sansone C (2018) On the neuroprotective role of Astaxanthin: new perspectives? Mar Drugs 16:247. https://doi.org/10.3390/md16080247
Galgano M, Chin LS, Galgano MA, Chin LS (2015) Central cord syndrome in a young patient with early diffuse idiopathic skeletal hyperostosis and ossification of the posterior longitudinal ligament after minor trauma: a case report and review. Cureus 7:e284. https://doi.org/10.7759/cureus.284

Gilroy DW, Lawrence T, Perretti M, Rossi AG (2004) Inflammatory resolution: new opportunities for drug discovery. Nat Rev Drug Discov 3:401–416. https://doi.org/10.1038/nrd1383

Gomez-Lazaro M, Galindo MF, Concannon CG, Segura MF, Fernandez-Gomez FJ, Llecha N, Comella JX, Prehn JHM, Jordan J (2008) 6-Hydroxydopamine activates the mitochondrial apoptosis pathway through p38 MAPK-mediated, p53-independent activation of Bax and PUMA. J Neurochem 104:1599–1612. https://doi.org/10.1111/j.1471-4159.2007.05115.x

Grimmig B, Kim S-H, Nash K, Bickford PC, Douglas Shytle R (2017) Neuroprotective mechanisms of astaxanthin: a potential therapeutic role in preserving cognitive function in age and neurodegeneration. GeroScience 39:19–32. https://doi.org/10.1007/s11357-017-9958-x

Gross J (2012) Pigments in vegetables: chlorophylls and carotenoids. Springer Science & Business Media

Háda M, Nagy V, Gulyás-Fekete G, Deli J, Agócs A (2010) Towards carotenoid dendrimers: carotenoid diesters and triesters with aromatic cores. Helv Chim Acta 93:1149–1155. https://doi.org/10.1002/hlca.200900338

Hammond BR, Miller LS, Bello MO, Lindbergh CA, Mewborn C, Renzi-Hammond LM (2017) Effects of Lutein/Zeaxanthin supplementation on the cognitive function of community dwelling older adults: a randomized, double-masked, placebo-controlled trial. Front Aging Neurosci 9:9

Honda M, Kageyama H, Hibino T, Zhang Y, Diono W, Kanda H, Yamaguchi R, Takemura R, Fukaya T, Goto M (2019) Improved carotenoid processing with sustainable solvents utilizing Z-isomerization-induced alteration in physicochemical properties: a review and future directions. Molecules 24:2149. https://doi.org/10.3390/molecules24112149

Hu W, Wang H, Liu Z, Liu Y, Wang R, Luo X, Huang Y (2017) Neuroprotective effects of lycopene in spinal cord injury in rats via antioxidative and anti-apoptotic pathway. Neurosci Lett 642:107–112. https://doi.org/10.1016/j.neulet.2017.02.004

Hua Y, Xu N, Ma T, Liu Y, Xu H, Lu Y (2019) Anti-inflammatory effect of Lycopene on experimental spinal cord ischemia injury via Cyclooxygenase-2 suppression. Neuroimmunomodulation 26:84–92. https://doi.org/10.1159/000495466

Huang C, Zhang Z, Cui W (2019) Marine-derived natural compounds for the treatment of Parkinson's disease. Mar Drugs 17:221. https://doi.org/10.3390/md17040221

Hurst WJ (ed) (2008) Methods of analysis for functional foods and nutraceuticals, 2nd edn. CRC Press, Boca Raton. https://doi.org/10.1201/9781420007152

Hussain T, Tan B, Yin Y, Blachier F, Tossou MCB, Rahu N (2016) Oxidative stress and inflammation: what polyphenols can do for us? Oxidative Med Cell Longev 2016:e7432797. https://doi.org/10.1155/2016/7432797

Hwang S, Lim JW, Kim H (2017) Inhibitory effect of Lycopene on Amyloid-β-induced apoptosis in neuronal cells. Nutrients 9:883. https://doi.org/10.3390/nu9080883

Ikeda Y, Tsuji S, Satoh A, Ishikura M, Shirasawa T, Shimizu T (2008) Protective effects of astaxanthin on 6-hydroxydopamine-induced apoptosis in human neuroblastoma SH-SY5Y cells. J Neurochem 107:1730–1740. https://doi.org/10.1111/j.1471-4159.2008.05743.x

Iwata S, Imai T, Shimazawa M, Ishibashi T, Hayashi M, Hara H, Nakamura S (2018) Protective effects of the astaxanthin derivative, adonixanthin, on brain hemorrhagic injury. Brain Res 1698:130–138. https://doi.org/10.1016/j.brainres.2018.08.009

Jiang X, Yan Q, Liu F, Jing C, Ding L, Zhang L, Pang C (2018) Chronic trans-astaxanthin treatment exerts antihyperalgesic effect and corrects co-morbid depressive like behaviors in mice with chronic pain. Neurosci Lett 662:36–43. https://doi.org/10.1016/j.neulet.2017.09.064

Jiang X, Wang G, Lin Q, Tang Z, Yan Q, Yu X (2019) Fucoxanthin prevents lipopolysaccharide-induced depressive-like behavior in mice via AMPK- NF-κB pathway. Metab Brain Dis 34:431–442. https://doi.org/10.1007/s11011-018-0368-2

Kaur H, Chauhan S, Sandhir R (2011) Protective effect of Lycopene on oxidative stress and cognitive decline in rotenone induced model of Parkinson's disease. Neurochem Res 36:1435–1443. https://doi.org/10.1007/s11064-011-0469-3

Kim J-H, Choi W-B, Lee J-H, Jeon S-J, Choi Y-H, Kim B-W, Chang H-I, Nam S-W (2009) Astaxanthin inhibits H_2O_2-mediated apoptotic cell death in mouse neural progenitor cells via modulation of P38 and MEK signaling pathways. J Microbiol Biotechnol 19:1355–1363. https://doi.org/10.4014/jmb.0906.06003

Kim K-N, Heo S-J, Yoon W-J, Kang S-M, Ahn G, Yi T-H, Jeon Y-J (2010) Fucoxanthin inhibits the inflammatory response by suppressing the activation of NF-κB and MAPKs in lipopolysaccharide-induced RAW 264.7 macrophages. Eur J Pharmacol 649:369–375. https://doi.org/10.1016/j.ejphar.2010.09.032

Kim JH, Hwang J, Shim E, Chung E-J, Jang SH, Koh S-B (2016) Association of serum carotenoid, retinol, and tocopherol concentrations with the progression of Parkinson's disease. Nutr Res Pract 11:114–120. https://doi.org/10.4162/nrp.2017.11.2.114

Krajcovicova-Kudlackova M, Valachovicova M, Mislanova C, Hudecova Z, Sustrova M, Ostatnikova D (2009) Plasma concentrations of selected antioxidants in autistic children and adolescents. Bratisl Lek Listy 110:247–250

Lim S, Hwang S, Yu JH, Lim JW, Kim H (2017) Lycopene inhibits regulator of calcineurin 1-mediated apoptosis by reducing oxidative stress and down-regulating Nucling in neuronal cells. Mol Nutr Food Res 61:1600530. https://doi.org/10.1002/mnfr.201600530

Lin J, Yu J, Zhao J, Zhang K, Zheng J, Wang J, Huang C, Zhang J, Yan X, Gerwick WH, Wang Q, Cui W, He S (2017) Fucoxanthin, a marine carotenoid, attenuates β-amyloid oligomer-induced neurotoxicity possibly via regulating the PI3K/Akt and the ERK pathways in SH-SY5Y cells. Oxidative Med Cell Longev 2017:e6792543. https://doi.org/10.1155/2017/6792543

Liu X, Shibata T, Hisaka S, Osawa T (2009) Astaxanthin inhibits reactive oxygen species-mediated cellular toxicity in dopaminergic SH-SY5Y cells via mitochondria-targeted protective mechanism. Brain Res 1254:18–27. https://doi.org/10.1016/j.brainres.2008.11.076

Locksley RM, Killeen N, Lenardo MJ (2001) The TNF and TNF receptor Superfamilies: integrating mammalian biology. Cell 104:487–501. https://doi.org/10.1016/S0092-8674(01)00237-9

Lu Y, Wang X, Feng J, Xie T, Si P, Wang W (2019) Neuroprotective effect of astaxanthin on newborn rats exposed to prenatal maternal seizures. Brain Res Bull 148:63–69. https://doi.org/10.1016/j.brainresbull.2019.03.009

Manochkumar J, Doss CGP, El-Seedi HR, Efferth T, Ramamoorthy S (2021) The neuroprotective potential of carotenoids in vitro and in vivo. Phytomedicine 91:153676. https://doi.org/10.1016/j.phymed.2021.153676

Meléndez-Martínez AJ, Britton G, Vicario IM, Heredia FJ (2007) Relationship between the colour and the chemical structure of carotenoid pigments. Food Chem 101:1145–1150. https://doi.org/10.1016/j.foodchem.2006.03.015

Meléndez-Martínez AJ, Mapelli-Brahm P, Stinco CM (2018) Colourless carotenoids phytoene and phytofluene: from dietary sources to their usefulness for the functional foods and nutricosmetics industries. Subtrop Plant Sci 67:91–103

Melo Pereira S, Ribeiro R, Logarinho E (2019) Approaches towards longevity: reprogramming, Senolysis, and improved mitotic competence as anti-aging therapies. Int J Mol Sci 20:938. https://doi.org/10.3390/ijms20040938

Mewborn CM, Lindbergh CA, Hammond BR, Renzi-Hammond LM, Miller LS (2019) The effects of Lutein and Zeaxanthin supplementation on brain morphology in older adults: a randomized, controlled trial. J Aging Res 2019:1–11. https://doi.org/10.1155/2019/3709402

Mirshafiey A, Mohsenzadegan M (2009) Antioxidant therapy in multiple sclerosis. Immunopharmacol Immunotoxicol 31:13–29. https://doi.org/10.1080/08923970802331943

Mohammadzadeh Honarvar N, Saedisomeolia A, Abdolahi M, Shayeganrad A, Taheri Sangsari G, Hassanzadeh Rad B, Muench G (2017) Molecular anti-inflammatory mechanisms of Retinoids and carotenoids in Alzheimer's disease: a review of current evidence. J Mol Neurosci 61:289–304. https://doi.org/10.1007/s12031-016-0857-x

Morris MC, Tangney CC, Wang Y, Sacks FM, Bennett DA, Aggarwal NT (2015) MIND diet associated with reduced incidence of Alzheimer's disease. Alzheimers Dement 11:1007–1014. https://doi.org/10.1016/j.jalz.2014.11.009

Namitha KK, Negi PS (2010) Chemistry and biotechnology of carotenoids. Crit Rev Food Sci Nutr 50:728–760. https://doi.org/10.1080/10408398.2010.499811

Nataraj J, Manivasagam T, Thenmozhi AJ, Essa MM (2016) Lutein protects dopaminergic neurons against MPTP-induced apoptotic death and motor dysfunction by ameliorating mitochondrial disruption and oxidative stress. Nutr Neurosci 19:237–246. https://doi.org/10.1179/1476830515Y.0000000010

Nolan JM, Loskutova E, Howard A, Mulcahy R, Moran R, Stack J, Bolger M, Coen RF, Dennison J, Akuffo KO, Owens N, Power R, Thurnham D, Beatty S (2015) The impact of supplemental macular carotenoids in Alzheimer's disease: a randomized clinical trial. J Alzheimers Dis 44:1157–1169. https://doi.org/10.3233/JAD-142265

Nuzziello N, Liguori M (2019) The MicroRNA centrism in the orchestration of neuroinflammation in neurodegenerative diseases. Cells 8:1193. https://doi.org/10.3390/cells8101193

Pangestuti R, Vo T-S, Ngo D-H, Kim S-K (2013) Fucoxanthin ameliorates inflammation and oxidative reponses in microglia. J Agric Food Chem 61:3876–3883. https://doi.org/10.1021/jf400015k

Papandreou MA, Kanakis CD, Polissiou MG, Efthimiopoulos S, Cordopatis P, Margarity M, Lamari FN (2006) Inhibitory activity on amyloid-β aggregation and antioxidant properties of *Crocus sativus* stigmas extract and its Crocin constituents. J Agric Food Chem 54:8762–8768. https://doi.org/10.1021/jf061932a

Petyaev IM (2016) Lycopene deficiency in ageing and cardiovascular disease. Oxidative Med Cell Longev 2016:e3218605. https://doi.org/10.1155/2016/3218605

Preedy VR (2012) Vitamin A and carotenoids: chemistry, analysis, function and effects. Royal Society of Chemistry

Reza Dorosty-Motlagh A, Mohammadzadeh Honarvar N, Sedighiyan M, Abdolahi M (2016) The molecular mechanisms of vitamin a deficiency in multiple sclerosis. J Mol Neurosci 60:82–90. https://doi.org/10.1007/s12031-016-0781-0

Rivera-Madrid R, Carballo-Uicab VM, Cárdenas-Conejo Y, Aguilar-Espinosa M, Siva R (2020) 1 - overview of carotenoids and beneficial effects on human health. In: Galanakis CM (ed) Carotenoids: properties, processing and applications. Academic, pp 1–40. https://doi.org/10.1016/B978-0-12-817067-0.00001-4

Rodríguez-Luna A, Ávila-Román J, González-Rodríguez ML, Cózar MJ, Rabasco AM, Motilva V, Talero E (2018) Fucoxanthin-containing cream prevents epidermal hyperplasia and UVB-induced skin erythema in mice. Mar Drugs 16:378. https://doi.org/10.3390/md16100378

Santa-Cecília FV, Leite CA, Del-Bel E, Raisman-Vozari R (2019) The neuroprotective effect of doxycycline on neurodegenerative diseases. Neurotox Res 35:981–986. https://doi.org/10.1007/s12640-019-00015-z

Shaish A, Harari A, Kamari Y, Soudant E, Harats D, Ben-Amotz A (2008) A carotenoid algal preparation containing phytoene and Phytofluene inhibited LDL oxidation in vitro. Plant Foods Hum Nutr 63:83–86. https://doi.org/10.1007/s11130-008-0075-y

Sutachan JJ, Casas Z, Albarracin SL, Stab BR, Samudio I, Gonzalez J, Morales L, Barreto GE (2012) Cellular and molecular mechanisms of antioxidants in Parkinson's disease. Nutr Neurosci 15:120–126. https://doi.org/10.1179/1476830511Y.0000000033

Tan BL, Norhaizan ME (2019) Carotenoids: how effective are they to prevent age-related diseases? Molecules 24:1801. https://doi.org/10.3390/molecules24091801

Terraf P, Kouhsari SM, Ai J, Babaloo H (2017) Tissue-engineered regeneration of hemisected spinal cord using human endometrial stem cells, poly ε-Caprolactone scaffolds, and Crocin as a neuroprotective agent. Mol Neurobiol 54:5657–5667. https://doi.org/10.1007/s12035-016-0089-7

Walczak H, Krammer PH (2000) The CD95 (APO-1/Fas) and the TRAIL (APO-2L) Apoptosis systems. Exp Cell Res 256:58–66. https://doi.org/10.1006/excr.2000.4840

Wallach D, Varfolomeev EE, Malinin NL, Goltsev YV, Kovalenko AV, Boldin MP (1999) TUMOR NECROSIS FACTOR RECEPTOR AND Fas SIGNALING MECHANISMS. Annu Rev Immunol 17:331–367. https://doi.org/10.1146/annurev.immunol.17.1.331

Wertz K, Siler U, Goralczyk R (2004) Lycopene: modes of action to promote prostate health. Arch. Biochem. Biophys. Highlight Issue Carotenoids 430:127–134. https://doi.org/10.1016/j.abb.2004.04.023

Wu H, Niu H, Shao A, Wu C, Dixon BJ, Zhang J, Yang S, Wang Y (2015) Astaxanthin as a potential neuroprotective agent for neurological diseases. Mar Drugs 13:5750–5766. https://doi.org/10.3390/md13095750

Yamashita E (2013) Astaxanthin as a medical food. Funct Foods Health Dis 3:254–258. https://doi.org/10.31989/ffhd.v3i7.49

Yu J, Lin J-J, Yu R, He S, Wang Q-W, Cui W, Zhang J-R (2017) Fucoxanthin prevents H2O2-induced neuronal apoptosis via concurrently activating the PI3-K/Akt cascade and inhibiting the ERK pathway. Food Nutr Res 61:1304678. https://doi.org/10.1080/16546628.2017.1304678

Zarei S, Carr K, Reiley L, Diaz K, Guerra O, Altamirano P, Pagani W, Lodin D, Orozco G, Chinea A (2015) A comprehensive review of amyotrophic lateral sclerosis. Surg Neurol Int 6:171. https://doi.org/10.4103/2152-7806.169561

Zeni ALB, Camargo A, Dalmagro AP (2019) Lutein prevents corticosterone-induced depressive-like behavior in mice with the involvement of antioxidant and neuroprotective activities. Pharmacol Biochem Behav 179:63–72. https://doi.org/10.1016/j.pbb.2019.02.004

Zhang FF, Morioka N, Kitamura T, Fujii S, Miyauchi K, Nakamura Y, Hisaoka-Nakashima K, Nakata Y (2016a) Lycopene ameliorates neuropathic pain by upregulating spinal astrocytic connexin 43 expression. Life Sci 155:116–122. https://doi.org/10.1016/j.lfs.2016.05.021

Zhang M, Cui Z, Cui H, Cao Y, Wang Y, Zhong C (2016b) Astaxanthin alleviates cerebral edema by modulating NKCC1 and AQP4 expression after traumatic brain injury in mice. BMC Neurosci 17:60. https://doi.org/10.1186/s12868-016-0295-2

Zhang Q, Wang J, Gu Z, Zhang Q, Zheng H (2016c) Effect of lycopene on the blood-spinal cord barrier after spinal cord injury in mice. Biosci Trends 10:288–293. https://doi.org/10.5582/bst.2016.01062

Zhang L, Wang H, Fan Y, Gao Y, Li X, Hu Z, Ding K, Wang Y, Wang X (2017) Fucoxanthin provides neuroprotection in models of traumatic brain injury via the Nrf2-ARE and Nrf2-autophagy pathways. Sci Rep 7:46763. https://doi.org/10.1038/srep46763

Zhang L, Previn R, Lu L, Liao R-F, Jin Y, Wang R-K (2018) Crocin, a natural product attenuates lipopolysaccharide-induced anxiety and depressive-like behaviors through suppressing NF-kB and NLRP3 signaling pathway. Brain Res Bull 142:352–359. https://doi.org/10.1016/j.brainresbull.2018.08.021

Zhou X, Zhang F, Hu X, Chen J, Tang R, Zheng K, Song Y (2017) Depression can be prevented by astaxanthin through inhibition of hippocampal inflammation in diabetic mice. Brain Res 1657:262–268. https://doi.org/10.1016/j.brainres.2016.12.018

Zhou L, Ouyang L, Lin S, Chen S, Liu Y, Zhou W, Wang X (2018) Protective role of β-carotene against oxidative stress and neuroinflammation in a rat model of spinal cord injury. Int Immunopharmacol 61:92–99. https://doi.org/10.1016/j.intimp.2018.05.022

Zia-Ul-Haq M, Dewanjee S, Riaz M (eds) (2021) Carotenoids: structure and function in the human body. Springer International Publishing, Cham. https://doi.org/10.1007/978-3-030-46459-2

Index

R. Campos Chisté et al. (eds.), *Carotenoids*,
https://doi.org/10.1007/978-3-031-75322-0

SPRINGER NATURE

GPSR Compliance

The European Union's (EU) General Product Safety Regulation (GPSR) is a set of rules that requires consumer products to be safe and our obligations to ensure this.

If you have any concerns about our products, you can contact us on ProductSafety@springernature.com

In case Publisher is established outside the EU, the EU authorized representative is:

Springer Nature Customer Service Center GmbH
Europaplatz 3
69115 Heidelberg, Germany

The manufacturer's authorised representative in the EU is Springer
Nature Customer Service Centre GmbH, Europaplatz 3, 69115 Heidelberg,
Germany. If you have any concerns regarding our products, please
contact ProductSafety@springernature.com

Printed and bound by CPI Group (UK) Ltd, Croydon, CR0 4YY
30/09/2026
02225569-0009